The Science of AI for Energy Storage

The Science of AI for Energy Storage presents a comprehensive exploration of the intersection between artificial intelligence (AI) and energy storage technologies. This book provides a detailed analysis of energy storage systems, including mechanical, electrochemical, chemical, electrical, and thermal solutions, while emphasizing the transformative role of AI and machine learning in optimizing these technologies. By integrating foundational principles with cutting-edge advancements, the book delivers a robust framework for understanding the design, application, and sustainability of modern energy storage systems. Written in the author's hallmark conversational yet scientifically rigorous style, the work bridges the gap between theory and practice, offering invaluable insights for energy professionals, engineers, policymakers, and academics.

- Explores the transformative role of artificial intelligence (AI) and machine learning in optimizing energy storage systems, including batteries, supercapacitors, and pumped hydro storage.
- Examines a wide range of storage technologies—mechanical, electrochemical, chemical, electrical, and thermal—ensuring a thorough understanding of current and future solutions.
- Highlights the role of energy storage in achieving sustainability goals and addresses the challenges of integrating AI to meet growing global energy demands efficiently.

The Science of AI for Energy Storage

Frank R. Spellman

CRC Press
Taylor & Francis Group
Boca Raton London New York

CRC Press is an imprint of the
Taylor & Francis Group, an **informa** business

Designed cover image: Shutterstock

First edition published 2026
by CRC Press
2385 NW Executive Center Drive, Suite 320, Boca Raton FL 33431

and by CRC Press
4 Park Square, Milton Park, Abingdon, Oxon, OX14 4RN

CRC Press is an imprint of Taylor & Francis Group, LLC

ISBN: 978-1-041-06920-1 (hbk)
ISBN: 978-1-041-06921-8 (pbk)
ISBN: 978-1-003-63785-1 (ebk)

DOI: 10.1201/9781003637851

Typeset in Times
by KnowledgeWorks Global Ltd.

Contents

PART III Physical Security and Cybersecurity of Energy

Preface

This book, *The Science of AI in Storing Energy*, is the 17th volume in the acclaimed series that includes *The Science of Energy: Principles, Concepts, and Applications*, *The Science of Fluid Mechanics: Concepts and Applications in Water and Wastewater*, *The Science of Water Reuse*, *The Science of Land Subsidence*, *The Science of Green Energy*, *The Science of Carbon Capture and Storage (CCS)*, *The Science of Ocean Pollution*, *The Science of Lithium (Li)*, *The Science of Electric Vehicles (EVs): Concepts and Applications*, *The Science of Rare Earth Elements: Concepts and Applications*, *The Science of Water*, *The Science of Air*, *The Science of Environmental Pollution*, *The Science of Renewable Energy*, *The Science of Waste*, *The Science of Wind Power*, all of which bring this highly successful series fully into the 21st century. I always state in one of my new books that the others and this book continue the series' mantra based on sound rather than feel-good science. It also continues to be presented in the author's trademark conversational style, ensuring communication is inevitable, not a failure …. I like to ensure that the reader sees both the forest and the trees, and maybe because I want to converse with the reader. I aim to be comprehensive in coverage, understandable in what I deliver, and clear when using verbiage. Moreover, this work is thorough but with a bit of personal perspective, making it readable, engaging, concise, comprehensive, and opinionated.

Key features:

- Provides a guided tour of the rapidly evolving world of artificial intelligence (AI)-connected mindsets.
- Provides a comprehensive overview of energy storage systems.
- Points out and stresses the point that AI is the ability to solve problems too intricate for most humans to assimilate and digest.
- Offers a practical approach that includes in-depth information about current trends and developments in energy storage.
- Provides information on integrating AI and ML in the development of autonomous systems for energy storage. Based on its contents and message, this book defines intelligence as the ability to learn. The goal is to learn about how energy is produced and stored with the influence of using AI, but the limitations and challenges of AI are also pointed out.
- In today's fast-changing world, our critical energy infrastructure, energy operational technology, and energy storage operations facilities and workforce continue to face new threats and challenges, as new digital technology, opportunities, and pathways develop over time. Because energy is essential to maintaining our modern lifestyles and personal security, we must strengthen supply chains' resilience, cyber and physical security, and climate resilience, and address the risk associated with energy operations and the workforce. The bottom line is that we need to know the threats and challenges and an integrated risk management approach. Many of these threats and challenges require us to integrate safety and security actions to protect energy infrastructure and the workforce. Thus, the last three chapters of this work address the threats from terrorism, including insider terrorist activities.

Following the successful format of the other editions in this series, this no-holds-barred book addresses the present puzzle of how to capture, store, and protect energy. This capture and storage of energy involves producing energy for later use. The primary device we use to store energy is the battery (aka the accumulator). However, I also discuss in detail ultra- or supercapacitors and pumped hydro storage in this book because they are or are becoming major players in storing energy for later use.

Based on science and operational experience, the information presented is intended to provide an understanding of the current science underpinning natural and human-made energy storage techniques. Moreover, this book covers energy—the technologies that produce, store, transmit, and use the power that provides heat, light, and mobility to modern society.

This book's goal is not to make energy engineers out of anyone but to provide the "basic" principles required to understand energy, energy production, energy usage, energy sustainability, and energy storage. Its purpose is to ensure correlation to modern practice and design, and to illustrate problems regarding commonly used energy parameters.

This book is designed for and is accessible to energy managers, engineers, designers, policy and decision-makers, corporate planners, energy industries, students, and anyone else interested in energy, energy technologies, lesser-known energy technologies, and AI (artificial intelligence) as it is employed in energy production, usage, the goal of reaching sustainability, and optimal energy storage.

Note that the Associated Press, in a business briefing released in November 2024, reported that the United States is now the leader in developing AI technology, surpassing No. 2 China in research and other vital measures of AI innovation. We are entering an AI-powered world, and because of this, I continue to point to the need for us to board the AI Train but not on the caboose—we need to be the engine pulling the rest of the world into the AI-Age. Eventually, growth in the globe's population and the material economy will confront humanity. When this occurs (and it will), we will either adjust and survive or join the ranks of the dinosaurs, dodo birds, passenger pigeons, golden toads, or several other species presently experiencing the Sixth Extinction. To paraphrase Dickens, this will be the worst of times for those clinging to the old technology, yet potentially the best of times for those embracing and driving the digital transformation. Riding or driving the digital locomotive or the caboose.

The combination of AI and its subsections creates massive new opportunities for energy storage and for energy practitioners and companies to improve efficiency, optimize performance, drive innovation, and accelerate growth.

Because AI used in energy production, distribution, and storage can be both a shield and a sword, this book's final chapters deal with security, safety, resiliency, and emergency response procedures for the energy storage sector. Security is a fundamental component in the energy storage sector, and it is critical.

The bottom line: A word about the fear and anxiety many people feel about AI replacing them in their occupational endeavors. AI is simply a tool that can provide a scalable solution to many constantly changing data-rich operations that are difficult or impossible for humans to manage quickly. This book explains the nexus between AI and energy storage. The thing to remember about the human-AI interface is that mindset is everything; thus, the human-AI interface is a "connected" mindset. Use AI for good, not harm. Use it to protect our energy infrastructure and workforce. Use it to effect change toward the distant goal of a sustainable future for all.

Frank R. Spellman
Norfolk, Virginia

About the Author

Frank R. Spellman, PhD, is a retired assistant professor of environmental health at Old Dominion University, Norfolk, Virginia, and the author of more than 165 books covering topics ranging from concentrated animal feeding operations (CAFOs) to all areas of environmental science and occupational health. Many of his texts are readily available online, and several have been adopted for classroom use at major universities throughout the United States, Canada, China, Europe, South Korea, and Russia; two have been translated into Spanish for South American markets. Dr. Spellman has been cited in more than 1,250 publications. He is a professional expert witness for three law groups and an incident/accident investigator for the U.S. Department of Justice and a Northern Virginia law firm. In addition, he consults on homeland security vulnerability assessments for critical infrastructures, including water/wastewater facilities nationwide. He conducts nationwide pre-Occupational Safety and Health Administration (OSHA)/Environmental Protection Agency (EPA) audits. Dr. Spellman receives frequent requests to co-author with well-recognized experts in several scientific fields; for example, he is a contributing author of the prestigious text *The Engineering Handbook* (2nd ed., CRC Press). Dr. Spellman lectures on sewage treatment, water treatment, homeland security, and safety topics throughout the country and teaches water/wastewater operator short courses at Virginia Tech (Blacksburg, Virginia). He holds a BA in public administration, a BS in business management, Summa Cum Laude, an MBA, and an MS and PhD in environmental engineering.

The author stands in front of the remains of a giant, formerly 300+ feet tall Sequoia in Sequoia National Park, California.

Acronyms/Abbreviations/Units/Conversions

°C	degrees centigrade or Celsius
°F	degrees Fahrenheit
µ	micron
µg	microgram
µm	micrometer
AC	alternating current
ACEE	American Council for an Energy-Efficient Economy
AI	artificial intelligence
Amp	ampere
APPA	American Public Power Association
ASCE	American Society of Civil Engineers
BEP	best efficiency point
bhp	brake horsepower
CCS	carbon capture and storage
CEC	California Energy Commission
CEE	Consortium for Energy Efficiency
DC	direct current
DOE	Department of Energy
ECM	energy conservation measure
EIS	environmental impact statement
ENR	*Engineering News-Record*
EPA	Environmental Protection Agency
EPACT	Energy Policy Act
EPC	Energy Performance Contracting
EPRI	Electric Power Research Institute
ESCO	Energy Services Company
EV	electric vehicle
GHG	greenhouse gas
GIS	geographic information system
GPD	gallons per day
GPM	gallons per minute
GW	gigawatt
HVAC	heating, ventilation, and air conditioning
Hz	hertz
I&C	instrumentation and control
IoT	Internet of Things
kW	kilowatt-hour
kWh/year	kilowatt-hours per year
M	mega
ML	machine learning
MPN	most probable number
MW	molecular weight
NEPA	National Environmental Policy Act
NREL	National Renewable Energy Laboratory
O&M	operation and maintenance

Pa	pascal
PG&E	Pacific Gas and Electric
psi	pounds per square inch
psig	pounds per square inch gauge
rpm	revolutions per minute
SCADA	Supervisory Control and Data Acquisition
SCFM	standard cubic feet per minute
TVA	Tennessee Valley Authority
VFD	variable-frequency drive
VSS	volatile suspended solids
W	watt

UNITS/CONVERSIONS

Most calculations made in energy storage operations/equations involve using *units*. While the number tells us how many, the units tell us what we have. Examples of units include inches, feet, square feet, cubic feet, gallons, pounds, watts, kilowatts, liters, pounds per square inch, and miles per hour. *Conversions* are a process of changing the units of a number to make the number usable in a specific instance. Multiplying or dividing by another number to change the units of the number accomplishes conversions. Common conversions are described below.

Note: In many instances, the conversion factor cannot be derived—it must be known. Therefore, we use tables like the one below to determine the common conversions.

Note: Conversion factors change measurements or calculated values from one unit of measure to another. In making the conversion from one unit to another, you must know two things:

1. The exact number that relates the two units
2. Whether to multiply or divide by that number

Common Conversions

Linear Measurements

1 inch = 2.54 cm
1 foot = 30.5 cm
1 meter = 100 cm 3.281 ft = 39.4 in.
1 acre = 43,560 ft2
1 yard = 3 feet

VOLUME

1 gal = 3.78 liters
1 ft3 = 7.48 gal
1 l = 1,000 ml
1 acre foot = 43,560 cubic feet
1 gal = 32 cups
1 pound = 16 oz dry weight

Weight

1 ft3 of water = 62.4 lbs
1 gal = 8.34 lbs
1 lb =453.6 g
1 kg = 1,000 g = 2.2 lbs
1% = 10,000 mg/l

PRESSURE

1 ft of head = 0.433 psi
1 psi = 2.31 ft of head

FLOW

1 cfs = 448 gpm
1 gpm = 1,440 gpd

TEMPERATURE CONVERSIONS

$$°C = 5/9(°F - 32°)$$
$$°F = 5/9(°C + 32°)$$

These conversions are easy to perform. The difficulty arises when we must recall these formulas from memory. The easiest way to recall these crucial formulas is to remember three basic steps for both Fahrenheit and Celsius conversions:

1. Add 40°.
2. Multiply by the appropriate fraction (5/9 or 9/5).
3. Subtract 40°.

The only variable in this method is the choice of 5/9 or 9/5 in the multiplication step. To make the proper choice, you must be familiar with two scales. On the Fahrenheit scale, the freezing point of water is 32°, and 0° on the Celsius scale. The boiling point of water is 212° on the Fahrenheit scale and 100° on the Celsius scale.

Why is it important? Note, for example, that higher numbers are associated with the Fahrenheit scale at the same temperature and lower numbers with the Celsius scale. This vital relationship helps you decide whether to multiply by 5/9 or 9/5. Let us look at a few conversion problems to see how the three-step process works.

Problem #1: Convert 220°F to Celsius. Using the three-step process, we proceed as follows:

Solution:

1. *Step 1*: Add 40°F

$$\begin{array}{r} 220^\circ \\ +\ \ 40^\circ \\ \hline 260^\circ \end{array}$$

2. *Step 2*: 260°F must be multiplied by either 5/9 or 9/5. Since the conversion is to the *Celsius* scale, you will move to a number *smaller* than 260. Through reason and observation, we see that multiplying 260 by 9/5 would be the same as multiplying by 2, which would double 260 rather than make it smaller. On the other hand, multiplying by 5/9 is about the same as multiplying by 1/2, which would cut 260 in half. Since, in this problem, you wish to move to a smaller number, you should multiply by 5/9:

$$(5/9)(260^\circ) = 144.4^\circ\text{C}$$

3. *Step 3*: Now subtract 40°

$$\begin{array}{r} 144.4^\circ\text{C} \\ -\ \ 40.0^\circ\text{C} \\ \hline 104.4^\circ \end{array}$$

Thus, 220°F = 104.4°C.

Problem #2: Convert 22°C to Fahrenheit

1. *Step 1*: Add 40°

$$\begin{array}{r} 22^\circ\text{F} \\ +\ \ 40^\circ \\ \hline 62^\circ \end{array}$$

Because we are converting from Celsius to Fahrenheit, you are moving from a smaller to a larger number and should use 9/5 in the multiplication:

2. *Step 2*:

$$(9/5)(62°) = 112°$$

3. *Step 3*: Subtract 40°

$$\begin{array}{r} 112° \\ -\ 40° \\ \hline 72° \end{array}$$

Thus, 22°C = 72°F.

Knowing how to make these temperature conversion calculations is useful. However, in practical (real-world) operations, you may wish to use a temperature conversion table.

Note: Additional units are used in energy and are listed in the Energy Storage Technician's Toolbox below.

ENERGY STORAGE PRACTITIONER'S TOOLBOX

As with all professionals, the energy storage professional who is adequately trained and has learned via on-the-job experience has a toolbox. The accessories in the toolbox include a practical suite of proven and cost-effective tools, measures, and activities to improve energy efficiency, reduce environmental damage, and, when needed, devise appropriate mitigation strategies. This section discusses different energy sources, types, and computations related to the environment; these are elements contained in the toolbox.

ENERGY UNITS

Watts (W) is the term we usually use when measuring energy. In practical usage, a single watt is too tiny, so to make energy numbers usable, scientists use multiples of thousands of watts, like kilowatt, megawatt, gigawatt, and terawatt. **Kilowatts (kW)** are 1,000 watts, most commonly used for domestic energy. **Megawatts (MW)** are 1,000 kW and are most widely used for power generation and distribution. **Gigawatts (GW)** are 1,000 Megawatts, which are used to describe notional demand. **Terawatts (TW)** are 1,000 gigawatts, which describe country-level consumption.

MKS SYSTEM (SI UNITS)

The *MKS system* (so named because it uses the meter, kilogram, and second as base units) comprises *SI units*. All other units are derived from the base units listed in Table TB. This is an entirely consistent system; there is only one recognized unit for each variable (physical quantity).

TABLE TB
SI Base Units

Quantity	Name	Symbol
Length	Meter	m
Mass	Kilogram	kg
Time	Seconds	s
Electric current	Ampere	A
Temperature	Kelvin	K
Amount of substance	Mole	mol
Luminous intensity	Candela	cd

DID YOU KNOW?

U.S. energy storage practitioners find identifying specific quantities used in the U.S. confusing. For example, if the question was asked, how many gallons are contained in a barrel of U.S. liquid compared to a barrel of petroleum? How many would know the correct answer? Along with answering this question and to clear the air on other quantities, several are listed below:

- Atmosphere standard atmosphere
- Barrel 31.5 gallons (U.S., liquid)
- Barrel 42 gallons (petroleum)
- Btu tradition (thermochemical) value
- Calories thermochemical
- Gallons U.S., liquid
- Horsepower U.S. or mechanical
- Ounces avoirdupois
- Pints U.S., liquid
- Pounds avoirdupois
- Quarts U.S., liquid
- Tons short tons of 2,000 pounds

GREENHOUSE GAS EMISSION NUMBERS TO EQUIVALENT UNITS AND CARBON SEQUESTRATION

DID YOU KNOW?

An abampere, in electricity, is a centimeter-gram-second unit of electromagnetic current, equivalent to 10 amperes. A statampere is the electric unit of current equal to the current produced by an electromotive force of one statvolt acting through a resistance of one statohm. An abampere is multiplied by 2.99793×10^{10} to obtain a statampere.

This section describes the calculations used to convert greenhouse gas emission numbers—an area of increasing concern for energy and environmental practitioners—into different equivalent units.

Electricity Reductions (Kilowatt-Hours)

USEPA's Greenhouse Gas Equivalencies calculator uses the Emissions & Generation Resource Integrated Database (eGRID) U.S. annual non-baseload CO_2 output emission rate to convert reductions of kilowatt-hours into avoided units of carbon dioxide emissions. Most users of the Equivalencies Calculator who seek equivalencies for electricity-related emissions want to know equivalencies for emissions reductions from energy efficiency or renewable energy programs. These programs are not assumed to affect baseload emissions (the emissions from power plants that run constantly) but rather non-baseload generation (power plants brought online as necessary to meet demand). Therefore, the Equivalencies Calculator uses a non-baseload emission rate (USEPA 2012a).

Emission Factor

$$7.0555 \times 10^{-4} \text{ metric tons } CO_2 \text{ / kWh}$$

Notes:

- This calculation does not include any greenhouse gases other than CO_2.
- This calculation does not include line losses.

Passenger Vehicles Per Year

Passenger vehicles are two-axle, four-tire vehicles, including passenger cars, vans, pickup trucks, and sport/utility vehicles. In 2010, the weighted average combined rule of cars and light trucks was 21.6 miles per gallon (FHWA, 2012). The average vehicle miles traveled in 2010 was 11,489 miles per year. In 2010, the ratio of carbon dioxide emissions to total greenhouse gas emissions (including carbon dioxide, methane, and nitrous oxide, all expressed as carbon dioxide equivalents) for passenger vehicles was 0.985 (USEPA, 2012b). The carbon dioxide emitted per gallon of motor gasoline burned is 8.92×10^{-3} metric tons, as calculated in the "Gallons of gasoline consumed" section.

The following methodology was used to determine annual greenhouse gas emissions per passenger vehicle: vehicle miles traveled (VMT) was driven by average gas mileage to determine gallons of gasoline consumed per vehicle per year. Gallons of gasoline consumed were multiplied by the carbon dioxide per gallon of gas to determine the carbon dioxide emitted per vehicle per year. Carbon dioxide emissions were then divided by the ratio of carbon dioxide emissions to total vehicle greenhouse gas emissions to account for vehicle methane and nitrous oxide emissions.

Calculation

Note: Due to rounding, performing the calculations given in the equations below may not return to the exact results shown.

8.92×10^{-3} metric tons CO_2/gal gasoline × 11,489 VMT car/truck average × 1/21.6 miles per gal car/truck average × 1 CO_2, CH_4, and N_2O/0.985 CO_2 = **4.8 metric tons CO_2e/vehicle/year**

Gallons of Gasoline Consumed

To obtain the number of grams of CO_2 emitted per gallon of gasoline combusted, the fuel's heat content per gallon is multiplied by the kg CO_2 per heat content of the fuel. The average heat content per gallon of gasoline is 0.125 MMBtu/gallon, and the average emission per gas heat content is 71.35 kg CO_2/MMBtu. The fraction oxidized to CO_2 is 100% (IPCC 2006).

Calculation

Note: Due to rounding, performing the calculations given in the equations below may not return to the exact results shown.

0.125 MMBtu/gal × 71.35 kg CO_2/MMBtu × 1 metric ton/1,000 kg = **8.92×10^{-3} metric tons CO_2/ gallon of gasoline**

Therms of Natural Gas

Carbon dioxide emissions per therm (the therm is a non-SI unit of heat energy equal to 100,000 BTU) are determined by multiplying heat content times the carbon coefficient times the fraction oxidized times the ratio of the molecular weight of carbon dioxide to carbon (44/12). The average heat content of natural gas is 0.1 MMBtu per therm (USEPA, 2012b). The average carbon coefficient of natural gas is 14.47 kg carbon per MMBtu (USEPA, 2012b). The fraction oxidized to CO_2 is 100% (IPCC, 2006).

Note: When using this equivalency, please remember that it represents the CO_2 equivalency for natural gas burned as a fuel, not natural gas released into the atmosphere. Direct methane emissions are 21 times more potent than CO_2 in terms of atmospheric warming.

Calculation

Note: Due to rounding, performing the calculations given in the equations below may not return to the exact results shown.

0.1 MMBtu/1 therm × 14.47 kg C/MMBtu × 44 g CO_2/12 g C × 1 metric ton/1,000 kg = **0.005 metric tons CO_2/therm**

Barrels of Oil Consumed

Carbon dioxide emissions per barrel of crude oil are determined by multiplying heat content times the carbon coefficient times the fraction oxidized times the ratio of the molecular weight of carbon dioxide to that of carbon (44/12). The average heat content of crude oil is 5.80 MMBtu per barrel (USEPA, 2012b). The average carbon coefficient of crude oil is 20.31 kg carbon per MMBtu (USEPA, 2012b). The fraction oxidized is 100% (IPCC, 2006).

Calculation

Note: Due to rounding, performing the calculations given in the equations below may not return to the exact results shown.

5.80 MMBtu/barrel × 20.31 kg C/MMBtu × 44 g CO_2/12 g C × 1 metric ton/1,000 kg = **0.43 metric tons CO_2/barrel**

Tanker Trucks Filled with Gasoline

Carbon dioxide emissions per barrel of gasoline are determined by multiplying the heat content times the carbon dioxide coefficient times the fraction oxidized times the molecular weight ratio of carbon dioxide to carbon (44/12). A barrel equals 42 gallons. A typical gasoline tanker truck contains 8,500 gallons. The average heat content of conventional motor gasoline is 0.125 MMBtu/gallon (USEPA, 2012b). The average carbon coefficient of motor gasoline is 71.35 kg CO_2 per 100% (IPCC, 2006).

Calculation

Note: Due to rounding, the calculations in the equations below may return different results than the exact results.

0.125 MMBtu/gallon × 71.35 kg CO_2/MMBtu × 1 metric ton/1,000 kg = 8.92×10^{-3} metric tons CO_2/gallon

8.92×10^{-3} metric tons CO_2/gallon × 8,500 gallons/tanker truck = **75.82 metric tons CO_2/tanker truck**

Number of Tree Seedlings Grown for Ten Years

Forest application of biosolids (an energy source as fertilizer for crops, plants, and trees, capturing and storing large amounts of carbon dioxide, and reducing the impact of global climate change) is extensively practiced in the Pacific Northwest. Thus, we need to be concerned about the carbon footprint involved. A medium-growth coniferous tree, planted in an urban setting and allowed to grow for ten years, sequesters 23.2 lbs. of carbon. This estimate is based on the following assumptions:

- Medium-growth coniferous trees are raised in a nursery for one year until they are 1 inch in diameter at 4.5 feet above the ground (the size of a tree purchased in a 15-gallon container).
- The nursery-grown trees are then planted in a suburban or urban setting; they are not densely planted.
- The calculation considers "survival factors" developed by the USEPA (1998). For example, after five years (one year in the nursery and four in the urban setting), the probability of survival is 68%; after ten years, the probability declines to 59%. Each year, the sequestration rate (in lbs. per tree) is multiplied by the survival factor to yield a probability-weighted sequestration rate. These values are summed over the ten years, beginning from the time of the plant, to derive the estimate of 23.2 lbs. of carbon per tree.

Please note the following caveats to these assumptions:

- While most trees take one year in a nursery to reach the seedling stage, trees grow under different conditions and trees of certain species may take longer: up to six years.
- Average survival rates in urban areas are based on broad assumptions and vary significantly depending on site conditions.
- Carbon sequestration depends on growth rate, which varies by location and other conditions.
- This method estimates only direct carbon sequestration and does not include the energy savings from buildings shaded by urban tree cover.

To convert to units of metric tons CO_2 per tree, multiply by the molecular weight of carbon dioxide divided by that of carbon (44/12) and the ratio of metric tons per pound (1/2,204.6).

23.2 lbs. C/tree × (44 units CO_2 ÷ 12 units C) × 1 metric ton ÷ 2,204.6 lbs. = **0.039 metric ton CO_2 per urban tree planted**

DID YOU KNOW?

Forest land in the United States includes land that is at least 10% stocked with trees of any size, or, in the case of stands dominated by certain Western woodland species for which stocking parameters are unavailable, at least 5% crown cover by trees of any size. Timberland is defined as unreserved productive forest land producing or capable of producing crops of industrial wood. Productivity is at a minimum rate of 20 cubic feet of industrial wood per acre per year. The remaining portion of forest land is classified as "reserved forest land," which is forest withdrawn from timber use by statute or regulation, or "other forest land," which includes forests on which timber is growing at a rate of less than 20 cubic feet per acre per year (Smith et al., 2010).

ACRES OF U.S. FORESTS STORING CARBON FOR ONE YEAR

Calculation

Note: Due to rounding, the calculations in the equations below may return different results than the exact results.

Acres of U.S. Forests Storing Carbon for One Year

Growing forests accumulate and store carbon. Trees remove CO_2 from the atmosphere through photosynthesis and store it as cellulose, lignin, and other compounds. The accumulation rate equals growth minus removals (i.e., harvest for producing paper and wood) minus decomposition. In most U.S. forests, growth exceeds removals and decomposition, so the amount of carbon stored nationally is increasing overall.

Calculation for U.S. Forests

The Inventory of U.S. Greenhouse Gas Emissions and Sinks: 1990-2010 (USEPA, 2012c) provides data on the net change in forest carbon stocks and forest area. The calculation does not include net changes in carbon attributed to harvested wood products.

Annual Net Change in Carbon Stocks per Area in Year n = (Carbon $\text{Stocks}_{(t+1)}$ − Carbon Stocks) ÷ Area of land remaining in the same land-use category

Step 1: Determine the carbon stock change between years by subtracting carbon stocks in year t from carbon stocks in the year (t+1). (This includes carbon stocks in the above-ground biomass, below-ground biomass, dead wood, litter, and soil organic carbon pools.)

Step 2: Determine the annual net change in carbon stocks (i.e., sequestration) per area by dividing the carbon stock change in U.S. forests from Step 1 by the total area of U.S. forests remaining in forests in year n+1 (i.e., the area of land that did not change land-use categories between the periods).

Applying these calculations to data developed by the USDA Forest Service for the *Inventory of U.S. Greenhouse Gas Emissions and Sinks: 1990-2010* yields a result of 150 metric tons of carbon per hectare (or 61 metric tons of carbon per acre; 1 hectare = 10,000 sq m or 100 m by 100 m; 2.47 acres) for the carbon stock density of U.S. forests in 2010, with an annual net change in carbon stock per area in 2010 of 0.82 metric tons of carbon sequestered per hectare per year (or 0.33 metric tons of carbon sequestered per acre per year). These values include carbon in the five forest pools, above-ground biomass, below-ground biomass, deadwood, litter, and soil organic carbon, and are based on state-level Forest Inventory and Analysis (FIA) data. Forest carbon stocks and carbon stock change are based on the stock difference methodology and algorithms described by Smith et al. (2010).

Conversion Factors for Carbon Sequestered Annually by 1 Acre of Average U.S. Forest

Note: Due to rounding, performing the calculations given in the equations below may not return to the exact results shown. In the following equation, negative values indicate carbon sequestration.

−0.33 metric ton C/acre/year (44 units CO_2 ÷ 12 units C) = **−1.22 metric tons CO_2 sequestered annually by one acre of average U.S. forest**

Note that this is an estimate for "average" U.S. forests in 2010, i.e., for U.S. forests in 2010. Geographical variations significantly influence the national estimates, so the values calculated here may not accurately reflect individual state regions. To estimate carbon sequestered for additional acres in one year, multiply the number of acres by 1.22 metric tons (mt) CO_2 acre/year. From 2000 to 2010 the average annual sequestration per area was 0.73 metric tons C hectare/year (or 0.30 metric tons C acre/year) in the United States, with a minimum value of 0.36 metric tons C hectare/year (or 0.15 metric tons C acre/year) in 2000, and a maximum value of 0.83 metric tons C hectare/year (or 0.34 metric tons C acre/year) in 2006.

DID YOU KNOW?

A comprehensive accounting of total greenhouse gas emissions includes more than just carbon dioxide emissions; it also provides methane, nitrous oxide, hydrocarbons, perfluorocarbons, sulfur hexafluoride, and nitrogen trifluoride. Manifestations from the COVID-19 pandemic included more than universal sickness and increased levels of annual mortality but also, due to its part, CO_2 emissions from fossil fuel combustion increased by 1% relative to the previous year. Interestingly, the continued rebound from the COVID-19 pandemic resulted in greenhouse gas emissions in 2022 at 17% below 2005 levels—this is the result of accounting for sequestration from the land sector (USEPA, 2024).

Acres of U.S. Forest Preserved from Conversion to Cropland

Based on data developed by the USDA Forest Service for the Inventory of U.S. Greenhouse Gas Emissions and Sinks: 1990–2010, the carbon stock density of U.S. forests in 2010 was 150 metric tons of carbon per hectare (or 61 metric tons of carbon per acre) (USEPA, 2012b). This estimate is composed of the five carbon pools: aboveground biomass (52 metric tons C/hectare), belowground biomass (10 metric tons C/hectare), dead wood (9 metric tons C/hectare), litter (17 metric tons/C hectare), and soil organic carbons (62 metric tons C/hectare).

The *Inventory of U.S. Greenhouse Gas Emissions and Sinks: 1990-2010* estimates soil carbon stock changes using U.S.-specific equations and data from the USDA Natural Resource Inventory and the Century Biogeochemical Model (USEPA, 2012b). When calculating carbon stock changes in biomass due to conversion from forest land to cropland, the IPCC guidelines indicate that the average carbon stock change is equal to the carbon stock change due to removal of biomass from the outgoing land use (i.e., forest land) plus the carbon stocks from one year of growth in the incoming land use (i.e., cropland), or the carbon in biomass immediately after the conversion minus the carbon in biomass before the conversion plus the carbon stocks from one year of growth in the incoming land use (i.e., cropland) (IPCC, 2006). The carbon stock in annual cropland biomass after one year is five metric tons C per hectare, and the carbon content of dry aboveground biomass is 45% (IPCC, 2006). Therefore, the carbon stock in cropland after one year of growth is estimated to be 2.25 metric tons C per hectare (or 0.91 metric tons C per acre).

The average reference soil carbon stock (for high-activity clay, low-activity clay, and sandy soils for all climate regions in the United States) is 40.83 metric tons C/hectare (USEPA, 2012b). Carbon stock change in soils is time-dependent, with a default period for transition between equilibrium soil organic carbon values of 20 years for mineral soils in cropland systems (IPCC, 2006). Consequently, it is assumed that the equilibrium mineral soil organic carbon change will be annualized over 20 years to represent the annual flux. The IPCC (2006) guidelines indicate that there is insufficient data to provide a default approach or parameters to estimate carbon stocks in perennial cropland (IPCC, 2006).

Calculations for Converting U.S. Forests to U.S. Cropland

Annual Change in Biomass Carbon Stocks on Land Converted to Other Land-Use Category

$$\mathbf{\Delta CB} = \Delta C_G + C_{conversion} - \Delta C_L$$

where

ΔCB = annual change in carbon stocks in biomass due to growth on land converted to another land-use category (i.e., 2.25 metric tons C/hectare).

ΔC_G = annual increase in carbon stocks in biomass due to growth on land converted to another land-use category (i.e., 2.25 metric tons C/hectare).

$C_{Conversion}$ = an initial change in carbon stocks in biomass on land converted to another land-use category. The sum of the carbon stocks aboveground, belowground, deadwood, and litter biomass (−88.47 metric tons C/hectare). Immediately after conversion from forest land to cropland, biomass is assumed to be zero, as the land is cleared of all vegetation before planting crops).

ΔC_L = annual decrease in biomass stocks due to losses from harvesting, fuel wood gathering, and disturbances on land converted to another land-use category (assumed to be zero).

Therefore, $\Delta CB = \Delta C_G + C_{Conversion} - \Delta C_L = -86.22$ metric tons C/hectare/year of biomass carbon stocks are lost when forest land is converted to cropland.

$$\Delta Change_{Mineral} = (SOC_O - SOC_{(O-T)}) \div D$$

where:

$\Delta C_{Mineral}$ = annual change in carbon stocks in mineral soils.

SOC_O = soil organic carbon stock in the last year of the inventory period (i.e., 40.83 mt/hectare).

$SOC_{(O\text{-}T)}$ = solid organic carbon stock at the beginning of the inventory period (i.e., 62 mt C/hectare).

D = time dependence of stock change factors, which is the default period for T transition between equilibrium SOC values (i.e., 20 years for cropland systems).

Therefore, $\Delta C_{Mineral}$ $(SPC_O - SOC_{(O\text{-}T)}) \div D = (40.83 - 62) \div 20 = -1.06$ metric tons C/hectare/year of soil organic C lost.

Consequently, the change in carbon density from converting forest land to cropland would be −86.22 metric tons of C/hectare/year of biomass plus −1.06 metric tons C/hectare/year of soil organic C, equaling a total loss of 87.28 metric tons c/hectare/year (or −35.32 metric tons C/acre/year). To convert to carbon dioxide, multiply the molecular weight of carbon dioxide to that of carbon (44/12) to yield a value of −320.01 metric tons CO_2 per hectare/year (or −129.51 metric tons CO_2 acre/year).

Conversion Factor for Carbon Sequestered Annually by 1 Acre of Forest Preserved from Conversion to Cropland

Note: Due to rounding, performing the calculations given in the equations below may not return to the exact results shown. Negative values indicate CO_2 that is NOT emitted.

−35.32 metric tons C/acre/year (44 units CO_2 ÷ 12 units C) = **−129.51 metric tons CO_2/acre/year**

To estimate CO_2 not emitted when an acre of forest is preserved from conversion to cropland, multiply the number of acres not converted by −129.51 mt CO_2/area/year. Note that this calculation method assumes that all forest biomass is oxidized during clearing (i.e., burned biomass remains as charcoal or ash). Also note that this estimate only includes mineral soil carbon stocks, as most forests in the contiguous United States are growing on mineral soils. In the case of mineral soil forests, soil carbon stocks could be replenished or even increased, depending on the starting stocks, how the agricultural lands are managed, and the time frame over which the lands are managed.

Propane Cylinders Used for Home Barbecues

Propane is 81.7% carbon. The fraction oxidized is 100% (IPCC, 2006; USEPA, 2012c). Carbon dioxide emissions per pound of propane were determined by multiplying the weight of propane in a cylinder by the carbon content percentage times the fraction oxidized times the ratio of the molecular weight of carbon dioxide to that of carbon (44/12). Propane cylinders vary in size; for this equivalency calculation, a typical cylinder for home use was assumed to contain 18 pounds of propane.

Calculation

Note: Due to rounding, the calculations in the equations below do not return the exact results.

18 pounds propane/1 cylinder × 0.817-pound C/pound propane × 0.4536 kilograms/pound × 44 kg CO_2/12 kg C × 1 metric ton/1,000 kg = **0.024 metric tons CO_2/cylinder**

Railcars of Coal Consumed

The average heat content of coal in 2009 was 27.56 MMBtu per metric ton. The average carbon coefficient of coal in 2009 was 25.34 kilograms per MMBtu (USEPA, 2011). The fraction oxidized is 100% (IPCC, 2006). Carbon dioxide emissions per ton of coal were determined by multiplying heat content times the carbon coefficient times the fraction oxidized times the ratio of the molecular weight of carbon dioxide to that of carbon (44/12). The amount of coal in an average railcar was assumed to be 100.19 short tons, or 90.89 metric tons (Hancock and Sreekanth, 2001).

Calculation

Note: Due to rounding, performing the calculations given in the equations below may not return the exact results shown.

27.56 MMBtu/metric ton coal × 25.34 kg C/MMBtu × 44g CO_2/12g C × 90.89 metric tons coal/railcar × 1 metric ton/1,000 kg = **232.74 metric tons CO_2/railcar**

Tons of Waste Recycled Instead of Landfilled

To develop the conversion factor for recycling rather than landfilling waste, emission factors from EPA's WAste Reduction Model (WARM) were used. These emission factors were developed following a life cycle assessment methodology using estimation techniques developed for national inventories of greenhouse gas emissions. According to WARM, the net emission reduction from recycling mixed recyclables (e.g., paper, metals, plastics), compared with a baseline in which the materials are landfilled, is 0.73 metric tons of carbon equivalent per short ton. This factor was then converted to metric tons of carbon dioxide equivalent by multiplying by 44/12, the molecular weight ratio of carbon dioxide to carbon.

Calculation

Note: Due to rounding, the calculation given in the equation below may not return the exact results shown.

0.73 metric tons of carbon equivalent/ton × 44 g CO_2/12 g C = **2.67 metric tons CO_2 equivalent/ton of waste recycled instead of landfilled**

Coal-fired Power Plant Emissions for One Year

In 2009, 457 power plants used coal to generate at least 95% of their electricity (USEPA, 2012a). These plants emitted 1,614,625,638.1 metric tons of CO_2 in 2009. Carbon dioxide emissions per power plant were calculated by dividing the total emissions from power plants whose primary fuel source was coal by the number of power plants.

Calculation

Note: Due to rounding, performing the calculations given in the equations below may not return the exact results shown.

1,614,625,638.1 metric tons of CO_2 × 1/457 power plants = **3,533,098 metric tons CO_2/power plant**

ENERGY STORAGE BY OTHER TERMS

When talking about energy storage, it seems a straightforward lingo, easily understood with little explanation required, right?

Wrong. Well, sort of, anyway. So, when we discuss energy storage, are we discussing and thinking about seasonal (stored in sunny summer), or are we referring to day storage (stored during daytime and reserved for nighttime), or momentary reactive energy, when, lasting for only seconds?

The bottom line: To understand the science of energy storage, one of the most essential tools in the energy storage professional's toolbox is to know the language (the jargon or lingo), terms, and definitions. Thus, the following jargon related to energy storage cost and performance metrics is presented. Note that the terms and definitions provided below are essential tools. In some cases, they are deliberately repetitive for retention and for explaining the broad scope of their meaning, but later, their meaning may differ slightly depending on their application.

- *Black Start Capable*: It is sometimes called bootstrapping in a self-starting process. It refers to the ability of technology to restore electric power from a complete blackout without relying on an external power source.
- *Calendar Life*[1] *(Years)*: It is the number of years a battery can be stored unused before it becomes useless. Stated differently, it is the number of years an energy storage system can be stored or minimally used while maintaining a high percentage of its initial capacity.
- *Capacity Charge*: The capacity charge, sometimes called the demand charge, is the portion of the electric service charge based on the amount of the customer's peak load (kW) within the specified billing period.
- *C Rate (C)*: It is a battery's charge/discharge rate normalized to its maximum capacity (further defined in the text and glossary).
- *Cycle Life*[2] *(Cycles)*: It is the number of complete charge/discharge cycles an energy storage system can complete while maintaining a high percentage of the initial capacity.
- *Cycles per Day (Number)*: It is the number of times the energy storage system charges and then discharges to a specific depth of discharge (usually 80%) within 24 hours.
- *Cycles per Year (Number)*: It is the number of times the energy storage system changes and then discharges to a certain depth of discharge (usually 80%) over a year.
- *Day-Ahead Markets*: These are forward markets for electricity to be supplied the following day. This market closes with acceptance by the independent system operator, power exchange, or scheduling coordinator of the final day-ahead schedule.
- *Degradation Factor (%)*: It refers to the amount of rated capacity of energy capacity lost over time, the components of the system experience wear and tear and/or chemical changes.
- *Depth of Discharge (%)*: It represents the ratio of discharged energy (kWh) to usable energy capacity (kWh).
- *Discharge Voltage Variability*: It is the variation in voltage magnitude caused by spikes, fluctuations, etc.
- *Duration (Seconds, Minutes, Hours, Days, etc.)*: It refers to the time when a storage system can discharge at its rated power capacity before depleting its energy capacity. For example, a storage system with 1 MV of rated power capacity and 5 MWh of energy capacity will have a storage duration of 5 hours.

- *Electric Utilities*: It encompasses all enterprises in the production and/or distribution of electricity for use by the public, such as investor-owned electric utility companies and government-owned electric utilities (municipal systems, federal agencies, state projects, and public power districts).
- *Energy Capacity (kWh)*: It is the maximum amount of stored energy the system can hold.
- *Energy Density (kWh/L)*: It is the amount of energy an energy storage system can store per unit volume occupied by the system.
- *Energy-to-Power Ratio (kWh/kW)*: It is the relationship between energy capacity and rated power capacity in a given application.
- *Footprint (Square Feet)*: It is the physical area a storage system requires.
- *Limited Oxygen Index (%)*: It is the minimum combination of oxygen expressed as a percent, in the air that will keep an electrolyte burning. The higher the limited Oxygen Index, the lower the risk of fire.
- *Load Management*: It is the application of measures to influence customers' use of electricity to modify the demand and load factor.
- *Maximum Operating Temperature (°C)*: It is the maximum temperature at which a storage system can effectively operate.
- *Operational Life (Years)*: It is the number of years an energy storage system can operate while maintaining its regular cycle rate for its given Use Case.
- *Percent Environmentally Sensitive Material*[3] *(%)*: It is the amount of environmentally sensitive or rare earth material (such as lithium) required to manufacture a given storage system.
- *Power Density (kW/L)*: It is the maximum available power per unit volume.
- *Ramp Rate (%/Second)*: It is the rate of change of power delivered to or absorbed by an energy storage over time, expressed in megawatts per second or as a percentage change in rated power over time (percent per second). It is the rate at which the power system load or generator output varies, or the limits to such rates due to mechanical or reliability issues.
- *Rate Base*: It is the portion of total assets (principally investments in plant and equipment) for regulated utilities, as defined by a regulatory body, upon which a utility can earn a return.
- *Rated Power Capacity (kW)*: It is the total possible instantaneous discharge capability of the storage system, or the maximum discharge rate the storage system can achieve starting from a fully charged state.
- *Reactive Power*: It is the portion of electricity that establishes and sustains the electric and magnetic fields of alternating current equipment.
- *Recyclability (%)*: It is the weight (%) of materials in a storage system that may be recycled for post-end-of-life use.
- *Response Time Constrained by Power Conversion System (Seconds)*: The time in seconds it takes an energy storage system to reach 100% of rated power during charge/discharge, constrained by the technical limits of its power conversion system.
- *Return on Debt*: A component of the carrying charge, return on debt is the revenue required to pay for using borrowed money. It is usually stated as a percentage and applies to unearned capital in a particular year. Numerically, it equals the cost of debt money times the debt ratio.
- *Return on Equity*: A component of the carrying charge, return on equity is the revenue required to pay for using equity money. It is commonly stated as a percentage and is applied to unrecovered capital in a particular year. Numerically, it equals the cost of equity money times the equity ratio.
- *Round-Trip Efficiency (%)*: The ratio of energy output (kWh) to energy input (kWh) of the storage system during one cycle. For battery technologies, these refer to DC/DC efficiencies, while mechanical-based systems are expressed in AC/AC terms.
- *Self-Extinguishing Time(s)*: It is a safety metric specific to storage systems with electrolytes that measures the time for an on-fire electrolyte to put itself out. The lower the Self-Extinguishing Time, the lower the fire risk.

- *Security Analysis*: The evaluation of a project, using several different assumptions, on the values of one or more uncertain variables.
- *Service Territory*: The area where a utility currently provides service to retail customers and specific areas adjacent to the utility's electric distribution lines or natural gas pipelines in cities and counties where the utility holds a franchise.
- *State of Charge (%)*: It represents the storage system's level of charge and ranges from completely discharged (0%) to fully charged (100%).
- *Theoretical Response Time (Seconds)*: In seconds, it takes an energy storage system to reach 100% of rated power during charge/discharge, or from an initial measurement taken when the system is at rest.
- *Time-of-Use Pricing*: It is a rate design imposing higher charges to customers during periods of the day when higher demand is experienced.
- *Trade-Offs*: They provide information on comparative values and possible substitutions among different outcomes of different natures.
- *Weight (kg)*: It describes how heavy the storage system is.

NOTES

1 Mongrid, K., Viswanathan, V., Balducci, P., Alam, J., Fotedar, V., Koritarov, V., and Hadjerioua, B. (2019). *Energy Storage Technology Cost Characterization Report* (Technical Report). PHNL-28866. Pacific Northwest National Laboratory.

2 Deliberate repetition is used for this definition in the sections applicable.

3 From DOE (2024). *OE Energy Storage Strategic plan*. Washington, DC: United States Department of Energy.

REFERENCES

FHWA (2012). *Highway Statistics 2010. Office of Highway Policy Information*. Table VM-1. Washington, DC: Federal Highway Administration.

Hancock, K., and Sreekanth, A. (2001). *Conversion of Weight of Freight to Number of Railcars. Transportation Research Board*. Paper 01-2056, 2001. Washington, DC: US Department of Transportation.

IPCC (2006). *2006 IPCC Guidelines for National Greenhouse Gas Inventories*. Geneva: Intergovernmental Panel on Climate Change.

Smith, J., Heath, L., and Nichols, M. (2010). *U.S. Forest Carbon Calculation Tool User's Guide: Forestland Carbon Stocks and Net Annual Stock Change*. General Technical Report NRS-13, Revised. Washington, DC: U.S. Department of Agriculture Forest Service, Northern Research Station.

USEPA (1998). *Method for Calculating Carbon Sequestration by Trees in Urban and Suburban Settings, Voluntary Reporting of Greenhouse Gases*. Washington, DC: U.S. Department of Energy, Energy Information Administration.

USEPA (2011). *Inventory of U.S. Greenhouse Gas Emissions and Sinks: Fast Facts 1990-2009. Conversion Factors to Energy Units (Heat Equivalents), Heat Contents and Carbon Coefficients of Various Fuel Types*. USEPA $430-F-11-007: Washington, DC: U.S. Environmental Protection Agency.

USEPA (2012a). *eGrid2012 Version 1.0, US. Annual Non-Baseload CO_2 Output Emission Rate, Year 2009, Year Data*. Washington, DC: U.S. Environmental Protection Agency.

USEPA (2012b). *Inventory of U.S. Greenhouse Gas Emissions and Sinks: 1990-2010*. Chapter 3 (Energy), Tables 3-12, 3-13, and 3-14. The US.EPA #430-R-12-001 (PDF) (481 pp., 16.6 MB). Washington, DC: U.S. Environmental Protection Agency.

USEPA (2012c). *Inventory of U.S. Greenhouse Gas Emissions and Sinks: 1990-2010, Annex 2 (Methodology for Estimating CO_2 Emissions from Fossil Fuel Combustion), P.A.-85*. The US.EPA #430-R-12-001 (PDF) (389 pp., 10.6 MB). Washington, DC: U.S. Environmental Protection Agency.

USEPA (2024). About the Inventory of Greenhouse Gas Emissions and Sinks. Accessed 10/09/24 @ https://epa.gov/ghgemissions/inventory-us-greenhouse-gas-emissions-and-sinks-1990-2200.

Prologue

During a Recent Rabbit and Grasshopper Conversation

At a recent meeting in the pristine wilderness in Western Washington State at a place where humans call it the Olympic National Park, Mr. Rabbitt and his friend Mr. Grasshopper were hidden together in the Hoh Rain Forest beneath a fall of moss standing on a bed of soft ferns next to a popular hiking trail just across from a shared camping spot for humans. They were quiet as eight hikers stood on the trail at first, observing the moss and Oregon oxalis painting the understory green amid Sitka spruce. The hikers were soaked despite their rain gear—staying dry is impossible in a place with over 200 inches of rain annually. Then, the hikers turned off to their right and entered the campsite. While removing their soaked clothing during a brief pause in the rain, Mr. Grasshopper and his friend Mr. Rabbit remained hidden. They were about to beat feet, literally and otherwise, toward the deep, dark surroundings when they paused as one of the female hikers stated, "Wow, we are one with Nature and her natural washing machine, so to speak."

This statement was like listening to something foreign to Mr. Rabbit and Mr. Grasshopper.

Mr. Grasshopper asked Mr. Rabbit, "What are those humans talking about?"

Mr. Rabbit, still listening to the human talk, thumped a leg against the wet ferns, shrugged his shoulders, and did not answer.

Another female individual stated, "Although I am drenched, it is of no concern. We are embracing nature and liberating ourselves from our everyday lives in the real world. We are free from the adverse impacts of various forms of energy polluting our usual environment."

For a moment, humans and their hidden observers could only hear the pit-pat, pat, pit-pat of raindrops striking against the lance-shaped blades of the copse of blue Piper's bellflowers. Mr. Grasshopper and Mr. Rabbit took shelter beneath the blades high on the stems, as their umbrellas from the wetness. They both looked at each other, shrugged their shoulders, and turned their attention back to the camping area where the humans were pitching their tents. They heard another older lady state, "I am so happy to be away from it all … this place is paradise, a wet one, for sure, but away from all the bull we put up out there in la-la land."

Mr. Rabbit and Mr. Grasshopper looked at each other and wondered what the humans were discussing.

One of the young male campers said, "Well, we certainly don't miss electricity out here, and I am not worried about paying a horrendous energy bill out here … but I am sure one of them terrible bills will be waiting for me … waiting for us all when we return to what Melody just called la-la land … which I agree with that."

Again, Mr. Grasshopper and Mr. Rabbit were all ears, so to speak.

A wailing sound rang through the forest. Mr. Rabbit and Mr. Grasshopper paid little attention to the bugling of the Roosevelt elk. In this place, this is a familiar sound. And for these two, it is a sound that is so reassuring to them … in their case, there is no need for CBD chill gummies.

"Yup, I like no need for electricity, and it's costs to worry about here," says another male camper. He says this before trying to power up his iPhone, but the phone battery is not cooperating, and also, in the dense forest, getting a signal from a distant tower is not possible.

"This is a real bummer, my friends. I like to stay informed about what is happening, but with a dead phone, this is impossible."

All the campers agreed, and one stated, "Well, hell's bells, and I ain't shipping out … or trekking out because we have no electricity. Let us get inside the tents, dry off, eat, and drink good wine."

Meanwhile, Mr. Rabbit and Mr. Grasshopper moved closer to the plant stems to get more cover from the flower blades from the now-pouring rain.

"Electricity … you know I often hear our camping visitors talk about electricity, especially when they try to use their flat-plate things or their phones," says Mr. Rabbit while peering through the rain at the six colorful wet tents in the camping area.

"Well, Mr. Rabbit, I have heard, over the season, you understand, the campers talking about electricity and that they do not miss it, but also, after a few days out here in Elk country, that they wished they had some electricity," Mr. Grasshopper said. "Them human types sure are a strange bunch."

Mr. Rabbit says, "You know you are correct … one time when I was eavesdropping on a couple who were sharing a tent mention electricity also … and one of the campers went so far as to state what she was missing being out here in our neighborhood … that my long-legged friend was a whole laundry list of missing items."

During this conversation, Mr. Rabbit told his attentive friend that everything in modern human life depends on electricity.

Mr. Rabbit paused momentarily and listened to the rain… so soothing, and licking the wet leaves was refreshing. Then, being one of those rabbits in the know, so to speak, he explained to his friend what humans need for electricity. He stated that lights, refrigerators and freezers, cell phones, air conditioning, heat, the Internet and computers, television, iPods, music, stores, ovens, microwaves, kitchen appliances, and clothes washer-dryers all need electricity.

It is essential to point out that Mr. Rabbit had no idea about the things he mentioned, but he had heard about them by overhearing the many visitors during the summer. He also listened to a few of them say they wished they could store more energy and take it camping with them.

Grasshopper told his friend Rabbit, "To fix human energy storage systems, population growth, unemployment, and many of their economic problems, what humankind needs right now, my friend, is innovation, innovation, innovation…."

After deliberate and well-practiced thumping of his foot, Rabbit replied to his friend, Grasshopper: "No, my long-legged friend … to fix all of humankind's energy storage and economic problems, what they need first is discovery, discovery, discovery and then invention, invention, invention … followed up by innovation, innovation, innovation … hmmmm, human leadership, brain power, common sense, (hmmm, common sense is not that common with them folks) anyway accountability would also help."

Grasshopper: hmmmmmmmmmm, buzzzzzzzz, etc., "Well, sometimes they ain't too smart … them humans…. All they need to do is ask us. We know how to economize … how to store the energy we need and to be efficient and produce usable energy and use it—well, at least for our use—and how to do the rest."

Rabbit replied: "Right on, Grasshopper!"

—A friend of all grasshoppers and rabbits, 2025

Part I

The Basics

1 The Grand Challenge and the 10-Hour Rule

Artificial intelligence (AI)-based energy storage advances benefits such as intelligent energy use, eco-friendliness, and suitability for use in many areas. Okay, that sounds good, but why is a book about energy storage needed? Is it not true that most people understand or know about energy storage via oil in the Strategic Petroleum Reserve and storage tanks, natural gas in underground storage reservoirs and pipelines, thermal energy in ice, and thermal mass/adobe? Others who might be less informed may ask, "What is energy storage? Why is it important? Why should we care?" Energy storage mediates between variable sources and variable loads. Note that without energy storage, energy generation must equal energy consumption. Energy storage works by moving energy through time. Energy generated at one time can be used at another time through storage. So, along with the other types of energy storage, electricity storage is another essential form of energy storage.[1]

1.1 WHEN THE LIGHTS GO OUT

When the lights go out, this does not mean that you, I, or anyone else necessarily lose electricity. No, it could mean or refer to that moment in ancient times when the fire in the cave went out. When navigating the sewers or tunnels of a large city, your flashlight may fail. Or, in later centuries, when the light goes out, it could mean that our candle has been extinguished. Lights out today could refer to lights out manufacturing when the methodology of fully automating the production of goods at factories and other industrial factories produces "whatever" with no human present on-site—AI-assisted operations. Today, in what we call the modern era, when the lights go out, it means that the electrical power system has shut down, failed, turned off, or inclement weather has shut it down.

If we were to reside in a cave and the interior lighting goes out, we would light our **backup** fire, though it may not be as straightforward as it seems. Or when your flashlight goes dead while navigating the sewer or tunnel, turn on your **backup** flashlight or light your torch if **backup** is available (Note: lighting a match in a sewer filled with hydrogen sulfide and methane gas is not recommended). If it is a matter of your candle blowing out or burning out, you light another candle (if you have a **backup** candle). Regarding lights out manufacturing, there is a need for a **backup energy source** system; if the power used to operate the autonomous factory goes out, you will need a **backup** power supply system.

A backup energy supply = source of available energy

We need an uninterruptible backup energy source similar to a UPS (uninterruptible power supply) commonly used with computer systems.

Do we have such **backup** energy systems?

Truthfully, we do have or are developing energy backup systems. Perfecting these developing reliable energy systems is the grand challenge of the here, now, and future. We need backup energy sources.

The preceding term, "backup," is highlighted for good reason. This is a book about energy storage, and when we need to store energy, it is because we need a backup, a constant supply—that simple—we need to continue the flow of energy, whether it is steam, water, gasoline, diesel, natural gas, nuclear, gravity, or transformation energy such as electricity. Moreover, a backup energy supply

DOI: 10.1201/9781003637851-2

is essential when using wind and solar power, which are variable—we can't tell the sun when to shine, nor can we order the wind to blow when we want or need it. Long-duration energy storage (LDES), which can provide electricity for 10 hours or more, is essential in this context. This has also been labeled by USDOE (2020) as the ***Energy Storage Grand Challenge***, which is a comprehensive program to accelerate the development, commercialization, and utilization of next-generation energy storage technologies. It is about finding, developing, and delivering reliable, readily available backup energy storage. When we energize a system, we expect energy to flow—it can only flow if it is stored and ready for use.

1.2 FORMS OF ENERGY

Currently, we know about several forms of energy.

- Potential energy
  - Chemical
  - Gravitational
  - Electrical
  - Temperature differential
  - Latent heat
  - Kinetic energy
  - Flywheels

The book discusses these forms of energy storage but concentrates on innovative energy storage technologies, including

- lithium batteries
- lead-acid batteries
- flow batteries
- zinc batteries
- sodium batteries
- pumped storage hydropower
- compressed air energy storage
- thermal energy storage
- supercapacitors
- flywheel stored energy
- hydrogen storage (fuel cells)
- gravity trains

Some of these storage technologies provide only short-term energy storage. In contrast, others can be long-term, such as converting power to gas using hydrogen and storing heat or cold between opposing seasons in deep aquifers or bedrock. These are known as thermal energy storage (TES) technologies, which heat or cool a storage medium and, when needed, deliver the stored thermal energy to meet heating or cooling needs. Whenever you wind your clock or raise the clock weights, this is the storage of potential energy. Compressed air storage stores potential energy from moving molecules; this technology is designed to decrease oil consumption in generating electric power. Then there is the battery storage that most people are familiar with, including vehicle starting batteries, iPhones, clocks, UPS devices, chainsaws, yard tools, E-bikes, hand tools, E-vehicles, laptops and tabletop computers, and other applications. Battery storage holds chemical energy rich in electrons, which can quickly convert into electricity—electrons flowing in a conductor. Hydroelectric dams also store energy in a reservoir as gravitational potential energy. This applies to pumped storage and the Advanced Rail Energy Storage (ARES) train system. When we think about renewable energy, we usually consider solar and wind power. Still, the ARES train system also produces green or renewable energy because of gravity. Ice or chilled water storage tanks store ice or chilled water (thermal energy) as latent heat at night to meet peak demand for cooling. Last, fossil fuels such as coal and gasoline store ancient energy derived from sunlight by organisms that later died and became buried, and, over time, were then converted into these fuels.

DID YOU KNOW?

A key point to remember in energy storage is that it converts kinetic or electrical energy to potential energy for its later use, such as when stored or backup energy is needed.

1.3 KEY ENERGY STORAGE TERMS

To understand the contents of this book, it is essential to recall Voltaire's words: "If you wish to converse with me, please define your terms." The key terms used in this book are listed and defined in the following; those terms not defined below are defined when presented:

- *Adiabatic:* It is a reversible thermodynamic process that occurs without gain or loss of heat and a change in entropy.
- *Alternating current (AC):* It is the electricity flow with changing polarity/voltage (alternates) between positive and negative.
- *Amp hours or ampere-hours:* It is a unit of measurement: the amount of current that flows over a given amount of time. For example, a battery rating of 5 Ah means it can enable 5 amperes of current flow for 1 hour.
- *Arbitrage:* It is the simultaneous purchase and sale of an asset to profit from a price difference. It is a trade that profits by exploiting price differences of identical or similar financial instruments in different markets or various forms. Arbitrage exists due to market inefficiencies: it provides a mechanism to ensure prices do not deviate significantly from fair value over extended periods.
- *Backup power:* It is the power source for ride-through or short-term power outages.
- *Base load:* It is the minimum amount of electric power delivered or required over a given period at a constant rate.
- *Base load generation:* It is the electricity generation designed/intended to operate constantly always.
- *Battery (accumulator):* In battery, two or more electric cells are connected electrically. In common usage, "battery" applies to a single cell, such as a household battery.
- *Black start service:* It is the ability of a generating unit to start without an outside electrical supply. Black start service is necessary to help ensure the reliable restoration of the grid following a blackout.
- *C rate:* It is the charge rate that, under ideal conditions, equals the energy storage capacity divided by 1 hour. 1C is the charge rate necessary to charge a battery in 1 hour.
- *Capacity:* It is the ability to produce/deploy electricity under specified conditions.
- *Capacity factor:* It is a value indicating the average percentage of full capacity used over a given period. For example, a generating facility that operates at an average of 70% of its expected full capacity over a measured period has a capacity factor of 0.7.
- *Capacity firming:* It provides energy to fill in when variable generation output is below the generator's rate power output. It is done to provide constant power output.
- *Charge:* It is the injecting energy to be stored into the electricity storage system.
- *Charge rate:* It is the rate at which electricity storage can be charged.

- *Constant-current charge:* It is a charging process during which electric current in a battery is constant. Charging is stopped when the battery is fully charged, and constant-voltage charging is used.
- *Constant-voltage charge:* For batteries, it is a charging process during which the voltage applied to the battery's terminals is constant.
- *Cutoff voltage:* It is the voltage at the end of applicable discharge.
- *Cycle:* It is the sequence of storage charging and discharging. It is also known as the charge-discharge cycle.
- *Cycle life:* It is the number of charge-discharge cycles after which electricity storage becomes inoperable or unusable for a given application.
- *Damping:* It is any effect that tends to reduce the amplitude (magnitude) of oscillations in a system.
- *Demand:* First, it is the rate at which electric energy is delivered to or by a system or part of a system, expressed as kilowatts or megawatts at a given instant or averaged over any designated interval. Second, it is the rate at which the customer is using energy.
- *Demand side:* It is of or related to the rate-payer or end-user electric demand, often said to be "on the customer side of the meter."
- *Depth of discharge:* It is the portion of energy discharged from a storage system relative to the amount of extractable stored energy.
- *Electrolyte:* In electrochemical batteries, the chemical medium provides the ion transport mechanism (conduct the electric current) between the positive and negative electrodes.
- *End user:* This is the person or entity that uses energy, as distinct from, for example, entities that engage in wholesale energy transactions or purchases made by a landlord to the distributor.
- *Energy:* It is the potential of a physical system to perform work.
- *Energy density:* It is the amount of energy a storage system can store per unit volume occupied by the system.
- *Float charging:* This is a method of maintaining a battery in a charged state by continuous, long-term constant-voltage charging at a level sufficient to balance its self-discharge rate.
- *Gigawatt:* It is a unit of power equal to 1 billion watts.
- *Gigawatt-hour:* This is a measure involving 1 billion watts being generated, transmitted, distributed, or used continuously for 1 hour.
- *Hydroelectric:* It is related to the production of electricity by waterpower.
- *Kilowatt-hour:* It is a unit of work or energy equal to that expended by 1 kilowatt in 1 hour or to 3.6 million joules (pronounced jewel).
- *Kinetic energy:* This is a form of energy an object has because of its motion. The kind of motion may be translation (motion along a path from one place to another), rotation about an axis, vibration, or any combination of motions. The total kinetic energy of a body or system is equal to the sum of the kinetic energies resulting from each type of motion. The kinetic energy of an object depends on its mass and velocity.
- *Megawatt:* It is a unit of power equal to 1 million watts.
- *Megawatt-hour:* It is defined as one thousand kilowatt-hours or 1 million-watt hours.
- *Off-peak:* These hours or other periods are defined by business practices, contracts, agreements, or guides as periods of lower electrical demand.
- *Power:* It is the rate at which energy is transferred, used, or transformed.
- *Power factor:* It is the ratio of real power to apparent power.
- *Primary cell (battery):* It is a cell or battery not designed to be recharged and discarded after delivering its charge.
- *Reactive power:* This is the portion of electricity that establishes and sustains alternating current equipment's electric and magnetic fields.
- *Real power:* This is the portion of electricity that supplies energy to the load.

- *Secondary battery:* It is a galvanic battery which, after discharge, may be restored to its charged state by passing an electric current through it in a direction opposite that of discharge.
- *Self-discharge:* It is the loss of helpful capacity of a storage system due to internal losses such as internal chemical action in a battery, frictional loss in a flywheel, or air loss from the storage reservoir.
- *Transmission:* An interconnected group of lines and associated equipment for the increasing movement or transfer of electric energy between points of supply and points at which it is transformed for delivery to customers or other electric systems is established.
- *Trickle charging:* It is an electrochemical cell or battery charging involving a continuous or intermittent constant-current supply that keeps the cell/battery fully charged. In contrast, the cell/battery is not being used (discharged).
- *Uninterruptible power supply (UPS):* a source of backup power—sometimes, including batteries and increasingly flywheel energy storage—that is used to ensure continuous availability of power in the event of an interruption of service.

DID YOU KNOW?

Renewable energy is energy generated partially or entirely from non-depleting energy sources for direct end use or electricity generation. Renewable energy definitions vary by state but usually include wind, solar, and geothermal energy. Some regions also consider low-impact or small hydro, biomass, biogas, and water-to-energy to be renewable energy sources.

1.4 THE DIGITAL TRANSFORMATION

Figure 1.1 shows humanity's steps in transforming energy from the time of cave-living to the time of so-called "modern" living.

Figure 1.1 also shows that transformative change in energy is ongoing. It means doing things differently. Actual transformative change is the change that becomes sweeping. Transformative change is like a railroad locomotive transiting an endless track. As the locomotive wheels along the endless track, it picks up passengers and boxcars with new technological developments. Again, all these developments are coupled to the locomotive as it continues along its endless track. One of the boxcars recently added is the digital metamorphosis—digital progress that continues to ride the endless rails.

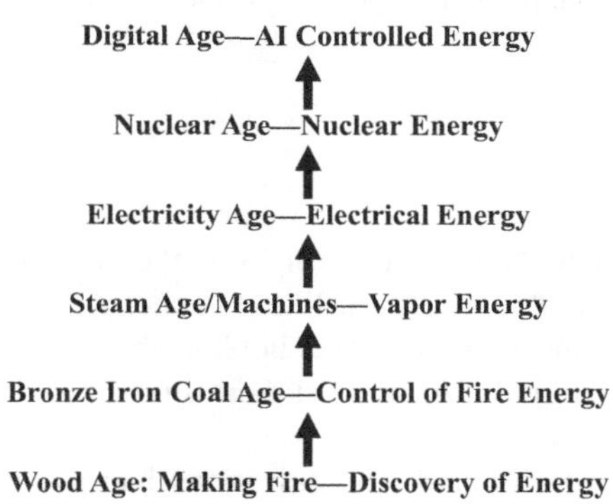

FIGURE 1.1 The main steps in the energy transformation process.

We must adapt as we ride the digital wave (aka the digital train). In this adaptation process, we need to learn the nuts and bolts and the applications of AI and its subsets, including machine learning (ML), deep learning (DL), natural language processing (NLP), expert systems, computer vision, and the Internet of Things (IoT). While traditional web technologies (i.e., the programming language, database, server administration, servers, markup language, style rules, and cloud-side codes) remain at the top of the ladder, in the majority of usage at present, the engine's cargo in digital service is a new wave of cloud-based services, coupled to the locomotive is the mobile Internet, Big Data, design thinking (DT), AI, robotics, advanced neural ML, augmented-reality technology (AR), and additive manufacturing, in that order, are moving (the movers) up that other train, that train of usage, rails of popularity and innovation.

DID YOU KNOW?

Natural language processing (NLP) focuses on understanding and generating human language. Expert systems use rules and knowledge to make decisions in specific domains. Computer vision involves analyzing visual information to understand and interpret images and videos.

1.5 THE BIG QUESTION

In my college-level classes in environmental engineering and associated disciplines, I always spend considerable time at the start of my varied lecturing sessions on AI. I do this in classes related not only to the science of energy, the science of energy storage, energy applications, water, air, soil, pollution, waste, climate change, assorted environmental health classes, with a few digressions thrown in here and there (commonly called "Sea Stories"). Again, I always begin my lectures on topics, subjects, disciplines, or whatever you want to call them, with a short discussion on AI. For example, at the beginning of each semester, at the start of the first class, with a new group of students, undergrads, and grad students in combined classes (ranging from class size of 45–150 students), I ask each student to take a scrap of paper with a pencil to write down not more than three words that in their opinion best describe AI—what it is; what it does; and so forth. I always receive a whole bunch of "and so forth" answers.

For example, consider the few responses listed below:

"It's Magic" (had a number of these responses)
"It's Today's World"
"We've been digitized"
"My job replacement"
"A communist plot" (more than one respondent provided this answer)
"The great repeater"
"AI break up"
"More data, please"
"Ability to learn"

The last response shown here, "ability to learn," was the only answer I sought. The other hundreds of responses are not printable in a work like this one. Anyway, because the nexus between the science of energy storage and AI is what this book is about, we need to look at the current AI components that make up the digital train and list their function:

- *Cloud-based services:* Traditionally, organizations had to purchase and maintain their physical servers. Now, cloud-based servers run in the cloud on infrastructure owned by the cloud service provider; they remove the need to buy, run, and manage physical servers.

- *Mobile Internet:* It is also known as *mobile web* for handheld devices, such as smartphones or feature phones (i.e., a pre-wireless smartphone) through a wireless mobile network.
- *Big Data Architecture:* This is a data management system's systemic design and structure.
- *The Internet of Things (IoT):* This is a network of physical objects embedded with sensors, software, and other technologies that connect and exchange data with other devices and systems via the Internet. IoT sensors are necessary for many applications. A few of these essential sensors and their purpose or function include the following:
  - *Pressure Sensors:* These sensors detect changes in a gas or liquid.
  - *Chemical and Gas Sensors:* Gas and chemical sensors are key devices for detecting and measuring unwanted ambient gases and chemicals, and generating warnings.
  - *Infrared Sensors:* These devices measure heat emissions from an object in applications such as remote controls and detecting environmental changes (temperature, motion, and proximity).
  - *Water Quality Sensors:* These devices collect real-time data on water quality. They measure pH, temperature, turbidity, and contaminant levels.
  - *Proximity Sensors:* These devices detect the presence or absence of objects near the sensor operates without physical contact.
  - *Temperature Sensors:* These devices measure the amount of heat generated from an area or an object.
  - *Smoke Sensors:* Most of us are familiar with smoke detectors, as they have protected our homes and businesses for a long time. However, these IoT sensors are much more sensitive than the standard smoke detector.
  - *Motion Sensors:* These devices are essential for ensuring security. The applications include automated sinks and toilet flushers, automatic door controls, energy management systems, and several others.
  - *Level Sensors:* These sensors detect the level of various substances, including energy operations. These sensors measure light and convert it into electrical signals. They are used in the auto, energy, healthcare and aerospace industries.
- *Digital Thinking (DT):* It is a nonlinear, iterative aspect or process that challenges assumptions, redefines complications, and creates groundbreaking models to be tested.
- *AI technology:* It enables computers and machines to simulate human learning, understanding, problem-solving, decision-making, originality, and independence.
- *Robotics:* It follows the digital train. Within environmental engineering and science, robotics focuses on robotic automation algorithms used in environmental functions such as performing vital tasks related to air pollution monitoring, water, soil, wildlife monitoring, and waste cleanup operations.
- *Advanced Neural ML:* The "name neural" is based on a model of how the human brain functions and comprises a collection of processing units called *nodes*. Advanced Neural ML is learning without definite instructions and is primarily used in DL.
- *Augmented-Reality Technology (AR):* AR is following in the digital train. AR is popular because it presents an enhanced real-world version via computer-generated digital information, including visual, sound, and other sensory elements.
- *Additive Manufacturing:* The final section of the digital train, not the caboose presented in this discussion, is known as additive manufacturing (aka 3D printing). In operation, additive manufacturing constructs a three-dimensional object from a CAD or digital 3D model.

Note: To be successful in AI-based energy storage, you must know the terminology, live it, write it, and use it in speech. Therefore, the keywords and abbreviations related to AI used in this book are presented and defined in the following section.

1.5.1 Key AI Terms and Abbreviations

An algorithm: is a specification of how to solve a class of problems by performing calculations, data processing, and automated reasoning tasks.

Artificial Intelligence (AI): intelligence produced by machines. AI is applied when a machine mimics cognitive functions that humans associate with learning and problem-solving. EO 14110 and the United States Code define artificial intelligence as

> … a machine-based system that can, for a given set of human-defined objectives, make predictions, recommendations, or decisions influencing real or virtual environments. Artificial Intelligence systems use machine- and human-based inputs to perceive real and virtual environments, abstract such perceptions into models through analysis in an automated manner, and use model inference to formulate options for information or action.
>
> *USDOE, 2008*

The reality is that there are various categories of AI; these categories overlap with each other to a point and include

Artificial General Intelligence (AGI): Artificial Intelligence, or human-level intelligence, aims to replicate human cognitive abilities across various tasks. It involves learning material or information and applying it to other domains.

- *Limited-memory Artificial Intelligence (LMAI):* It refers to a subset of AI narrow intelligence in which algorithms and networks can manage data exceeding the machine's memory capacity.
- *Artificial Narrow Intelligence (ANI):* This is AI, also known as **weak AI**, that performs one and only one task. ANI encompasses image recognition, autonomous vehicles, chatbots, spam filters, car engine diagnostics, handwriting recognition, and other applications.
- *Theory of Mind Artificial Intelligence:* This type of AI interacts socially and can discern and respond appropriately to people's beliefs, thoughts, emotions, expectations, and even facial expressions.
- *Reactive Artificial Intelligence (RAI):* Reactive AI is best known for implementing if-then rules to work through a set of inputs to determine what action to take or recommend. It can only react to current situations without memory of past experiences.
- *Self-aware Artificial Intelligence (SAAI):* This purely theoretical system can be aware of its existence and understand its thoughts.
- *Artificial Superintelligence (ASI):* This is a purely theoretical system in which the scope of intelligence surpasses that of humans.

Chatbot: An artificial intelligence that conducts conversation via auditory or word-based methods.

Generative AI: refers to models or algorithms that generate new output, such as text, photos, videos, data code, and 3D representations, from vast amounts of data they were trained on. Understanding the differences between AI and Generative AI is crucial in today's technology-driven world, as these two concepts, while married at the hip, so to speak, offer unique capabilities and applications in several different fields. Traditional AI systems are designed to understand, interpret, and make decisions based on the data they have been supplied. They can answer questions, recommend purchasing products, and drive motor vehicles autonomously. This is accomplished through machine learning algorithms that allow these systems to gain knowledge from experience and improve over time. On the other hand, Generative AI goes beyond understanding and interpreting data. It generates entirely new content from the very beginning. It can create human-like text through advanced deep learning, compose music, and create images indistinguishable from those made by humans.

Generative Pre-trained Transformers (GPT): denotes a family of AI models. These models, primarily larger language models such as those developed by OpenAI (founded in 2015 in San Francisco by Elon Musk and others), are based on a transformer architecture capable of understanding and generating human-like language.

Machine Learning (ML): The scientific study of statistical models and algorithms computer systems use to perform specific tasks.

AE	auto recorder
AI	artificial intelligence
ANN	artificial neural network
CFD	computational fluid dynamics
CUSUM	cumulative sum
DBN	deep belief network
DL	deep learning
DRL	deep reinforcement learning
GPU	graphical processing unit
GRU	gated recurrent unit
LSTM	long short-term memory
MLP	multi-layer perception
ML	machine learning
NN	neural network
TL	transfer learning

1.6 ENERGY USE

Have you ever spent time pondering energy storage systems? No, or yes? No comes into play whenever we use some type of energy system for our personal or workplace use and simply throw the switch, push a button, or turn a key. There is no issue—no problem—energy is readily available and used. However, if we throw a switch, push a button, or turn a key and nothing happens, we may wonder where the problem is—where is the energy?

The truth is that there are so many uses of energy that we do not often think about them until they are not operable or available. So, what are some of the uses of energy? There are numerous potential applications for energy, and the following provides a partial list of several current uses.

- *Agriculture:* Energy is used in farming operations for irrigation, machinery, and crop processing
- *Environmental conservation:* Energy supports efforts in renewable energy generation
- *Industrial manufacturing:* Energy is used for operating machinery, powering assembly lines
- *Mining and extraction:* Energy is used in mining operations for drilling and excavation
- *Lighting:* Energy is essential for lighting applications
- *Transportation infrastructure:* Energy is required to construct and maintain transportation infrastructure, including bridges, tunnels, and highways.

- *Water supply and treatment:* Energy powers pumps, filtration systems, and treatment processes in providing clean water for residential and industrial use
- *Automotive:* Energy powers the production of vehicles
- *Chemical industry:* Energy is required for chemical manufacturing processes
- *Financial services:* Energy powers banks, financial institutions, and data centers
- *Firefighting:* Energy is used in firefighting equipment
- *Forest industry:* Energy is used in sawmills, paper mills, and wood process
- *Marine transportation:* Energy powers ships, boats, and marine vessels for cargo hauling and fishing
- *Oil and gas industry:* Energy is required for the oil and gas sector's drilling, extraction, and refining processes
- *Military operations:* Energy is used in military activities
- *Space exploration:* Energy is used in spacecraft propulsion systems and life support
- *Surveillance systems:* Energy powers surveillance cameras and other security systems
- *Weather monitoring:* Energy is used in weather monitoring systems
- *Water transportation:* Energy powers ships, boats, and vessels in commercial shipping
- *Biofuels production:* Energy is used in the production of biofuels
- *Carbon capture and storage:* It involves using energy to capture, compress, and store carbon dioxide emissions to reduce pollution and greenhouse gas levels
- *HVAC systems:* Energy is used in heating, ventilation, and air-conditioning
- *Public transportation:* Energy powers buses, trains, and other public systems
- *Renewable energy production:* Energy is harnessed from renewable sources for electricity generation
- *Artificial Intelligence:* Energy is used in AI systems.

1.7 THE 10-HOUR RULE

The 10-hour rule is all about LDES. Energy storage duration is usually expressed in the number of hours a storage device can provide continuous output at its rated capacity. In literature, LDES is defined as little as 2 hours to as much as multiple days or even months.

There is a need for a standard LDES that can be accepted and used by most, if not all. Duration of at least 10 hours is increasingly recognized as a standard definition in academic and general education.

It is difficult—if not impossible—to reconcile all of the different definitions into one standard definition and arrive at a single numerical value for duration that helps to make the LDES value for duration a universal standard. The duration required for storage to offer high-capacity credit can vary widely.

The term "long-duration energy storage" is a standard vernacular term, but without a clear definition; thus, we need a consensus definition of the long-duration value to establish its ability to provide resource adequacy.

1.8 THE FLY IN THE OINTMENT

We are only starting to employ AI in energy storage and other applications. What will its full impact, influence, and effect be?

I'm not sure if Mr. Rabbit or Mr. Grasshopper can help us with that, but it all boils down to one word: Awareness, Awareness, Awareness.

For example, we need to be aware that AI is not the end that will end all... meaning it is not the last digital word or pixel. AI is evolving, and where it stands in its overall growth is not known at present because progress is ongoing. What AI is today is not what it will be in the future.

We know of a few things that need to be focused on, including AI's challenges and the lessons learned from our experience using AI for whatever purpose it is or was intended.

So, let's examine what we know now from our limited use of AI to date (remember, AI is a work in progress). I am unsure if anyone knows where we are headed with this so-called endpoint technology. Does AI have an endpoint? The jury is still out on that one.

- We must be aware of bogus, erroneous data that distorts AI applications. The point?
  Data models are only as good as the data they use. We must question and understand the data. Otherwise, the output may be flawed. It is all about predisposition and preconception (aka bias). This is the old "Garbage in, Garbage out" scenario.
- *Bias:* Whatever is made by humans contains biases. In AI, these biases become common in machine learning and algorithms.

XXXXXXXXXXXXXXXXXXXXXXXXXXXXXXXXXXXXXXXXXXXX

Sidenote: Having served for 26 years in the US Navy, I have picked up or fashioned a few dozen "sea stories" that I often relate to anyone who will listen. I know of some instances related to the biases in this AI bias issue. One of my favorite *non-sea stories* I often relate to my students or anyone else is about a statement by one of my former upper-level graduate college students in one of my environmental health and engineering classes. When I lectured on AI biases that affect the outcome of specific results, I pointed out that we must be careful because AI is "taught" by humans, and humans are the source of many biases. Well, as is often the case, my classes always contained some extroverted students who like to make their opinions known to all (I encourage this behavior because it is a chance for all, including their instructor, to learn—is this not why we teach and attend school at any level?). This student (there were others) stated that AI and its subsets might have learned bias from other digital machines and not directly from humans.

Hmmmmmmmmmm Buzz (as Mr. Rabbit and Mr. Grasshopper might respond), I looked at the student and then performed one of my classic eyeballs sweeps of the entire classroom: I asked, "Who programmed the machines that train other digital machines?"

At first, the silence was deafening, and then I received bombastic comments from all directions by many different students, including the resounding statement (in one form or several others): "Humans program or teach the digital machines or programs."

I agree with that. It is encouraging to find out that most of my students have IQs in the upper ranges.

I will explain to all those in attendance that any of the corpora is applicable. The corpora, the plural form of corpus, is an extensive collection of structured texts (including books, articles, and websites) for machine learning training taught by humans. Therefore, it is or can be biased.

Okay, okay, now what?

The question that always arises is about the biases themselves. Specifically, what biases infect and distort AI learning?

Over time, I learned how to answer this question; I expected it and hoped it would be asked.

It is always asked.

Anyway, when the question comes up, I take some time and show a prepared slide from the overhead projector on bias and the different types of cognitive bias. The projected slides display the following[2]:

- *Algorithm bias:* It occurs when misinformation is part and parcel of the algorithm.
- *Ambiguity bias:* It occurs when we prefer known probabilities over unknown ones.
- *Anchoring bias:* It arises when excessive reliance is placed on initial data encountered.
- *Availability bias:* It occurs when we overestimate the importance of information readily available.
- *Bandwagon bias:* It occurs when we do something because others are doing it.
- *Blind spot bias:* This bias happens when people identify biases in others but fail to recognize them in themselves.

- *Cognitive bias:* It occurs because AI requires input from a human thought process.
- *Confirmation bias:* It occurs when AI relies too much on pre-existing trends in data. Stated differently occurs when favoring information that confirms existing beliefs.
- *Dunning-Kruger Effect:* This is often described as human hubris, illusory superiority, narcissism, the Pygmalion effect, a superiority complex, or self-serving bias. Dunning (2011) describes it as overestimating competence in a specific area. One of my former students labeled it dumb as a brick, but more intelligent than the mortar. Needless to say, this statement puzzled all who heard it, including me.
- *Highlight bias:* Believing an event was predictable after it has occurred.
- *Measurement bias:* It results when incomplete data is used.
- *Self-serving bias:* It occurs when positive events are attributed to oneself and adverse events to external factors.
- *Sunk cost fallacy:* It is the tendency for investors to continue investing in a project based on the cumulative prior investment (time, $, resources) rather than on the current and future benefits.

XXXXXXXXXXXXXXXXXXXXXXXXXXXXXXXXXXXXXXXXXXXXXXXXXXXXXXX

Getting back to the challenges facing AI implementation and use:

- One of the significant challenges to ensuring AI input is not biased is using various tools and technologies to mitigate bias in AI. One of the essential tools is to pick the proper AI installation and operation team, meaning diverse personnel should make up the team to ensure that a diverse range of people compose the team. Groupthink should not be the order of business. Another option is to use fairness toolkits. Finally, ensure robust ethics and strong governance are used.
- Stay away from predictions. Remember that AI learning models can predict conclusions based on input data, but this does not indicate that the data used causes the conclusion.

  Consider this analogy: Imagine you are camping out in a tent during the night, a few miles from Mt. St. Helens in Washington State. This particular area is commonly known as the stomping grounds of the 9-foot-tall and 600-lb creature (of sorts), Big Foot (aka Sasquatch, Yeti, Skookum, Fouke Monster, and just Jimmy in the Curmudgeon books). Sometime after midnight and before sunrise, you first sense and then hear the presence of something or someone outside your tent. However, detecting sounds or movements outside your tent does not necessarily indicate an imminent visit from someone. Predicting an outcome does not inherently provide information about the factors that cause it.
- Discrimination is a dirty four-letter word. Note that AI algorithms have no concept of dirty four-letter words unless you program them. Note also that an AI algorithm can be unfair in its predictions.
- AI subsets can recognize interference rather than subject matter.
- When we query using AI, we may automatically assume certainty. This is an allusion. Remember that death and taxes are certain only.

1.9 THE BOTTOM LINE

Shelter shields us from weather and predators, crucial for cave dwellers and spelunkers. And no matter where we are, in a cave, tent, or home, we also need a supply of breathable air, water, and food, and for extended stays, we may also need reliable energy for cooking, heating, and light. A backup energy supply, such as a torch or flashlight in the cave, is essential; moreover, a backup light source might be necessary for transiting a long tunnel or a sewer system. Also, if we depend on candlelight as our primary light source, it is good practice to have some backup energy source for lighting and other uses. The energy available from throwing a knife switch or a toggle switch, setting a thermostat, pushing a button, or lighting a fire must be available—some type or form of energy is necessary. Having some form of energy available requires us to have a source of energy. AI use in storing energy is valuable, but it is also important to note that AI has limitations. This book is all about the storage of various energy sources, that is, energy storage is the capture of energy produced at one time for use later, and this definition is the real **bottom line**.

NOTES

1 From F.R. Spellman (2025). *The Science of Energy.* Boca Raton, FL: CRC Press.

2 Adapted from *A List of Common Cognitive Biases—Clearer Thinking.* Accessed 5/19/25 @ https:www.clearerthinking.org.

REFERENCES

Dunning, D. (2011). *Chapter Five: The Dunning-Kreger Effect: On Being Ignorant of One's Ignorance.* Advances in Experimental Psychology. Vol. 44, Academic Press, pp. 247–296.

USDOE (2008). *The Smart Grid Introduction.* Washington, DC: U.S. Department of Energy.

USDOE (2020). *Energy Storage Grand Challenge.* Washington, DC: U.S. Department of Energy.

2 AI and the Connected Mindset

AI-based energy storage balances load during peak and off-peak times based on weather, time, season, and/or the day.

2.1 ENERGY STORAGE IS A MAJOR ELEMENT

For several years, the technology used to produce, generate, and develop effective systems to distribute energy has been the primary goal of energy technology. Historically, the prioritization of energy production and distribution has often overshadowed the importance of energy storage for immediate use. Still, the need for battery power, stored energy, and the development and use of renewable energy sources, such as wind and solar technology, which are variable in their ability to supply energy, have ratcheted up the need for backup energy storage technologies beyond batteries.

Energy storage has become just as important an element in energy supply as energy production and distribution. This is the case because energy storage is the key to unlocking the future of the power sector, which can be designed to be more flexible, dependable, and is predictable in terms of power availability when needed—advances in renewable energy technologies and their variabilities have made energy storage for at least a 10-hour backup supply important. Also, with the increased development and use of short-duration, rapid-response energy storage, and longer-duration applications, energy can be economically shifted to periods of high seasonal demand—think hot, miserable, scorching summer heat waves and the need for energy for cooling. Indications that new storage technologies will continue to emerge are becoming increasingly evident and vital.

2.2 ALONG CAME AI

Advancement of artificial intelligence (AI) with human actions or activities comes down to an issue of attitude, outlook, approach, belief, and conviction that humans and AI can meld into a workable enterprise or mission that can accomplish great things for our benefit that we are still discovering, imagining, and hoping for. This certainly applies to energy storage for instantaneous use—this is the old flip the switch, and the lights come on scenario. The advancement of energy storage technology with AI requires an across-the-board strategy combining varied and complex data sources, including scientific research, numerical models, new data, and environmental considerations. It all concerns the need for AI technologies to implement domain-aware algorithms. These domain-aware algorithms leverage context-specific procedures and limits to attain advanced accuracy and efficiency. This provides critical insight that guides the machine learning (ML) process and ensures that the models developed are applicable and operative in their intended application.

Presented in simple, understandable language, this section discusses the human/AI mindset, the connection, and concerns about applying AI to accelerate the development and deployment of storage technologies in our current and evolving energy infrastructure. The capabilities of multimodal AI foundation models should not be ignored. The possibility of enabling the discovery of new patterns and the prediction of overall impacts, physically and environmentally, should not be understated.

It is interesting to note the parallel between the 1970s environmental movement, the incorporation of personal computers, and the programs that run them, which have blossomed to the present extent. The movement toward decarbonizing our energy systems and the growing demand for renewable energy have provided the impetus for developing energy storage systems that are environmentally friendly and able to incorporate AI to accelerate the development of new storage technologies.

DOI: 10.1201/9781003637851-3

Because these new energy technologies are variable in providing output, AI-supported energy storage technologies reliably control and efficiently optimize multi-scale storage networks. Energy storage for renewable energy sources is a gap-filling measure to provide backup energy when the sun goes down and the wind is not blowing.

In Chapter 1, the USDOE characterizes the development of energy storage systems as the "Grand Challenge." It should be acknowledged that this Grand Challenge is multifaceted. One significant facet of the challenge is developing reliable, efficient, and universally applicable energy storage systems. This is no minor issue because backup gap-filling energy storage technology is filled with gaps. For example, given the growing demand for renewable energy, there is an immediate need to rapidly accelerate the development of affordable, efficient, and secure energy storage solutions for the future. But again, there are gaps in the advancement of energy storage systems, mainly because of the lack of coordination among those entities involved in developing and producing energy storage systems. This is where the digital and AI world can be and is becoming increasingly the White Knights of energy storage.

Let us move on.

Are these gaps in coordination between entities involved in developing and implementing advanced energy storage systems purposely made? There is such a pressing need to establish energy storage systems for renewable energy technologies that many developers have not shared technical information because of the speed of development and the all-consuming nature of the work. Moreover, many developers may have proprietary requirements, such as the fact that many technical developments are considered privately owned and not to be shared. However, many developers would readily share information because such advancements help all concerned. The goal is to realize the vision of ubiquitous and reliable energy storage. AI is the tool needed to incorporate data quickly and to distribute the data to a wide range of recipients.

Another facet of the Grand Challenge is the multi-scale and multi-physics problems in designing energy storage systems. Solving these problems is dependent on understanding the activation barrier (i.e., a sort of energetic hurdle that a reaction must overcome), molecular-scale transport (i.e., the movement of molecules from one place to another) thermodynamics, and kinetics of response, as it does on full-scale design and all scales in between. Designing, developing, and deploying a new storage solution involves many interdependent yet often isolated technical advances. Note that these steps do not occur overnight; they take time, and because of this, one or more may fail, and this will or might impact all by reversing any progress made. Consider, for example, that one of the entities develops a crucial new innovation that works like a dream on the micro-scale. Still, with time, it might display behavior to the point of being toxic and possibly degrade beyond practical use; that is, it proves very hard to manufacture at scale. The biggest problem with some of these specific challenges is the overarching technical problem of slow transfer of knowledge and requirements between different aspects of the problem. Again, proper implementation of AI and/or its subsections can address and mitigate these issues.

Accelerating science related to various compounds, such as lithium and other solid-state electrolytes, is critical for ensuring the development of energy storage. Energy storage technology needs ongoing research and new materials that offer higher energy density, extended lifespans, natural abundance, and environmental friendliness. Many innovators have realized that incorporating digital twins (DTs) technology is essential, whereby a digital replica of a physical object, system, or process is contextualized in a digital version of its environment. The challenge is devising computationally accurate and cheap DTs to simulate and analyze many new materials before synthesis. Experience has demonstrated that DTs can significantly accelerate the discovery and testing of materials with higher energy density, longer lifespans, and environmental sustainability. The challenge is developing DTs that can emulate production processes. Producing virtual replicas of manufacturing systems enables automation and optimization and can offer promising methods to implement and evaluate rigorous testing protocols. These DTs are beneficial in that they can be adapted for new and related systems.

2.3 EFFICIENCY

Along with the challenge and need for rapid development of energy storage technology, efficiency in energy storage deployment, operations, and control is essential and is one of the challenges we face. Efficiency is necessary to decarbonize the electric grid and transition to net zero successfully. We must address the national energy storage system design, optimization, deployment, operations, and control to accomplish this. The range of applications includes enhancing grid stability and resilience and enabling the decarbonization of buildings, transportation, and industrial systems.

Energy storage systems must be designed and located with efficiency and reliability in mind. Storage is also needed to support mobile transportation and stationary systems for the electric grid, industries, and buildings.

2.4 COMMON SENSE ENERGY STORAGE

Common sense dictates that energy storage solutions should be deployed in an equitable and accessible manner to ensure that advancements in energy storage benefit diverse markets and communities. Ethical and accessible deployment will also help ensure that the benefits of renewable energy are distributed. A broad workforce for energy storage should be considered.

2.5 THE BOTTOM LINE

AI tools and approaches can optimize manufacturing processes for energy storage devices. The adoption of AI in automation technologies can improve production efficiency. AI has the potential to optimize every step of the various manufacturing processes. AI can also help develop modular and flexible manufacturing systems responsive to evolving technologies and market demands.

3 Fundamentals of Electricity

Using AI, a professionally designed energy storage system aids in balancing load during peak and off-peak times based on the day, time, season, or weather. Why does this energy book place so much emphasis on electricity? An accurate and applicable discussion about energy storage cannot be made if electrical energy production is not part and parcel of the discussion. This is the case because electrical energy is harnessed from both conventional sources, such as hydrocarbon, and renewable sources, such as solar, wind, hydro, and geothermal power, or electricity generation—it is all about electron flow: electricity.

3.1 BEYOND THE ALGORITHM

All around us, technology is evolving at warp speed—yesterday's science fiction has become today's daily routine, and today's impossibilities will be tomorrow's conveniences. Tomorrow is today, and our major convenience is the digital and AI worlds. A key point to remember is that without electricity, there are no digital or AI worlds.

Possessing some rudimentary or fundamental knowledge of electricity is vital to understanding energy storage systems, even at the margins. This is especially important for renewable energy sources, such as waterpower (aka, the water battery), solar, and wind, which transform energy into electricity. Focusing on electricity fundamentals and production is important because we can produce electricity if the hydrocarbon well runs dry. So, when hydrocarbon energy is replaced with renewable energy, the primary transformation of the energy source to electricity is and will be even more common than it is today.

This book is about the science of energy storage, and the AI connection cannot overlook the primary energy or power source used in most locations throughout the globe: electricity. Electricity is produced in Nature via lightning. At present, we have not determined how to harness lightning-produced electricity or how to store it. Keep in mind that storing energy is no easy task. Then, there is the naturally produced static electricity. Static electricity is produced when an imbalance between positive and negative charges within a material occurs when electrons (the negatively charged particles in an atom) move from one material to another. Electricity is the flow of electrons in a conductor, usually copper or aluminum, but in electronics, conductors are often silver or gold. Again, it is not easy to tap static electricity for practical power use, and although it can be stored in certain materials, it isn't easy to use. For practical electrical energy use, it is essential to point out that humans have developed electromagnetic generation, solar generation, and wind generation, which points out a necessary characteristic of the electricity we use. Electricity is a transformative source of energy and power, meaning it takes other forms of energy, such as mechanical generators, wind power, or solar energy, to produce electricity.

DID YOU KNOW?

Direct use of energy includes:

- electricity
- gasoline
- natural gas
- agriculture—for irrigation, machinery, and processing agricultural products
- heating, ventilation, and air conditioning (HVAC)
- battery charging

DOI: 10.1201/9781003637851-4

3.2 PRINCIPLES OF BASIC ELECTRICITY[1]

People living and working in modern societies have little difficulty in recognizing that electrical equipment is everywhere and (if one pays attention to his or her surroundings) is easy to spot. Despite its great importance in our daily lives, few of us stop to think about what life would be like without electricity—no computers, iPhones, dishwashers, clothes dryers, lights, and so on, with too many apps to list. The point is, we take electricity for granted, like air and water; we tend to take electricity for granted. But again, we use electricity for many daily jobs—from lighting, heating, and cooling our homes to powering our televisions and computers. Then there is the workplace. Can we do work without electricity? For example, the typical industrial workplace is outfitted with electrical equipment that

- generates electricity (a generator or emergency generator)
- stores electricity (batteries)
- change electricity from one level (voltage or current) to another (transformers)
- transport or transmit and distribute electricity throughout the plant site (wiring distribution systems)
- measures electricity (meters)
- converts electricity into other forms of energy (rotating shafts–mechanical energy, heat energy, light energy, chemical energy, or radio energy)
- protects other electrical equipment (fuses, circuit breakers, or relays)
- operates and controls other electrical equipment (motor controllers)
- converts some condition or occurrence into an electric signal (sensors)
- converts some measured variable to a representative electrical signal (transducers or transmitters)

3.2.1 NATURE OF ELECTRICITY

The word *electricity* is derived from the Greek word "electron" (meaning AMBER). Amber is a translucent (semitransparent) yellowish fossilized mineral resin. The ancient Greeks used "electric force" to refer to the mysterious forces of attraction and repulsion exhibited by amber when rubbed with cloth. They did not understand the nature of this force. They could not answer the question, "What is electricity?" This question remains unanswered. Today, we often attempt to answer this question by describing the effect, not the force. The standard answer given in physics is that electricity is "the force that moves electrons," which is about the same as defining a sail as "that force which moves a sailboat."

At present, little more is known than the ancient Greeks knew about the fundamental nature of electricity, but we have made tremendous strides in harnessing and using it. As with many other unknown (or unexplainable) phenomena, elaborate theories concerning the nature and behavior of electricity have been advanced and have gained wide acceptance because of their apparent truth—and because they work.

Scientists have determined that electricity behaves constantly and predictably in given situations or when subjected to given conditions. Faraday, Ohm, Lenz, and Kirchhoff have described the predictable characteristics of electricity and electric current in specific rules, often called laws. Thus, though electricity has never been clearly defined, its predictable nature and easily used energy form have made it one of the most widely used power sources in modern times.

The bottom line: You can "learn" about electricity by learning the rules or laws that govern its behavior and by understanding the methods of producing, controlling, and using it. Thus, this learning about electricity can be accomplished without ever having determined its fundamental identity.

You are scratching your head, puzzled.

I understand the main question running through your brain now: "This is a section in the text about the physics of electricity, and the author can't even explain what electricity is?"

That is correct; we cannot. The point is that no one can definitively define electricity. Electricity is one of those subject areas where the old saying "we don't know what we don't know about it" fits perfectly.

Again, a few theories about electricity have so far stood the test of extensive analysis and much time (relatively speaking, of course). One of the oldest and most generally accepted theories concerning electric current flow (or electricity) is the *electron theory.*

The electron theory states that electricity or current flow results from the flow of free electrons in a conductor. Thus, electricity is the flow of free electrons, or simply electron flow. This is how we define electricity in this text: Electricity is the flow of free electrons.

Electrons are extremely tiny particles of matter. To understand electrons and precisely what is meant by "electron flow," we must examine a discussion about the structure of matter.

3.2.2 The Structure of Matter

Matter is anything that has mass and occupies space. To study any matter's fundamental structure or composition, it must be reduced to its essential components. All matter is made of *molecules,* or combinations of *atoms* (Greek: not able to be divided), bound together to produce a given substance, such as salt, glass, or water. For example, if you keep dividing water into smaller and smaller drops, you will eventually arrive at the smallest particle that is still water. That particle is the molecule, which is defined as the *slightest bit of a substance that retains the characteristics of that substance.*

A molecule of water (H_2O) comprises one atom of oxygen and two atoms of hydrogen. If the water molecules were further subdivided, only unrelated oxygen and hydrogen atoms would remain, and the water would no longer exist. Thus, the molecule is the smallest particle to which a substance can be reduced and still be called by the same name. This applies to all substances—solids, liquids, and gases.

Important point: Molecules are made up of atoms, which are bound together to produce a given substance.

In various combinations, atoms comprise subatomic particles of *electrons, protons,* and *neutrons.* These particles differ in weight (a proton is much heavier than an electron) and charge. We are not concerned with the weights of particles in this text, but the *charge* is significant in electricity. The electron is the fundamental negative charge (–) of electricity. Electrons revolve about the nucleus or center of the atom in paths of concentric *orbits* or shells. The proton is the fundamental positive (+) charge of electricity. Protons are found in the nucleus. The number of protons within an atom's nucleus specifies that atom's atomic number. For example, the helium atom has two protons in its nucleus, so the atomic number is 2. The neutron, the fundamental neutral charge of electricity, is also found in the nucleus.

Most of the atom's weight is in the nucleus's protons and neutrons. Whirling around the nucleus are one or more negatively charged electrons. Usually, there is one proton for each electron in the entire atom, so the net positive charge of the nucleus is balanced by the net negative charge of the electrons rotating around the nucleus (see Figure 3.1).

Important point: Most batteries are marked with + and – or even with the abbreviations POS (positive) and NEG (negative). The concept of a positive or negative polarity and its importance in electricity will become clear later. However, for the moment, you need to remember that an electron has a negative charge and that a proton has a positive charge.

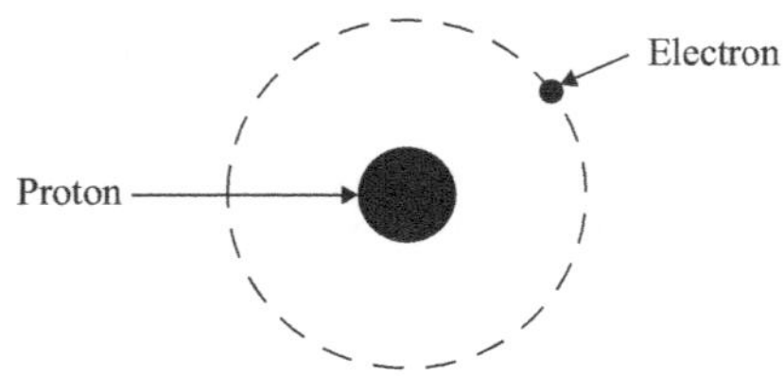

FIGURE 3.1 One proton and one electron = electrically neutral.

Again, the number of protons is usually the same as the number of electrons in an atom. This is an important point because this relationship determines the kind of element (the atom is the smallest particle that makes up an element; an element retains its characteristics when subdivided into atoms) in question. Figure 3.2 shows a simplified drawing of several atoms of different materials based on the conception of electrons orbiting about the nucleus. For example, hydrogen has a nucleus consisting of one proton, around which rotates one electron. The helium atom has a nucleus containing two protons and two neutrons, with two electrons encircling the nucleus. These elements are electrically neutral (or balanced) because each has an equal number of electrons and protons. Since each electron's negative (–) charge is equal in magnitude to each proton's positive (+) charge, the two opposite charges cancel.

A balanced (neutral or stable) atom has a certain amount of energy equal to the sum of the energy of its electrons. Electrons, in turn, have different energies called *energy levels*. The energy level of an electron is proportional to its distance from the nucleus. Therefore, the energy levels of electrons in shells are farther from the nucleus than those in shells nearer the nucleus.

When an electric force is applied to a conducting medium, such as a copper wire, electrons in the outer orbits of the copper atoms are forced out of orbit (i.e., liberating or freeing electrons) and impelled along the wire. This electrical force, which forces electrons out of orbit, can be produced in several ways, such as by moving a conductor through a magnetic field, by friction, when a glass rod is rubbed with cloth (silk), or by chemical action, as in a battery.

When the electrons are forced from their orbits, they are called *free electrons*. Some of the electrons of certain metallic atoms are so loosely bound to the nucleus that they are relatively free to

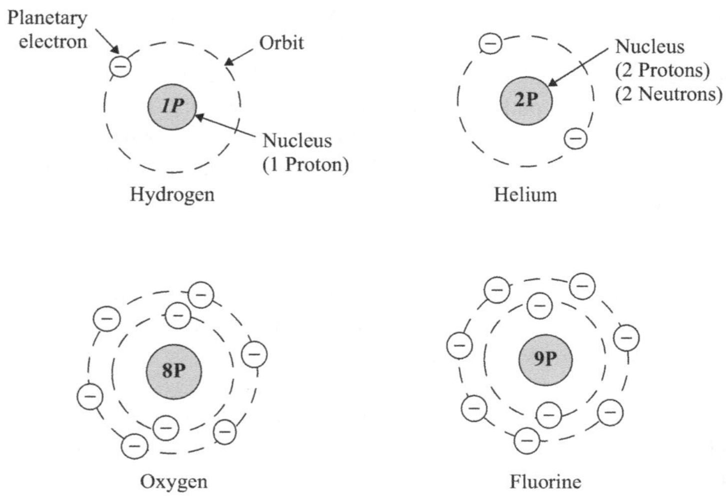

FIGURE 3.2 Atomic structure of a few elements.

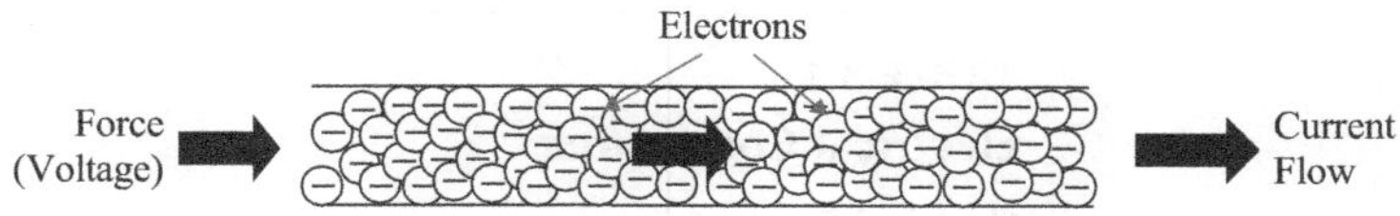

FIGURE 3.3 Electron flow in a copper wire.

move from atom to atom. These free electrons constitute the flow of an electric current in electrical conductors.

Important point: When an electric force is applied to a copper wire, free electrons are displaced from the copper atoms and move along the wire, producing an electric current, as shown in Figure 3.3.

If the internal energy of an atom is raised above its normal state, the atom is said to be *excited*. Excitation may be produced by causing the atoms to collide with particles propelled by an electric force, as shown in Figure 3.3. In effect, energy is transferred from the electric source to the atom. The excess energy an atom absorbs may become sufficient to cause loosely bound outer electrons (as shown in Figure 3.3) to leave the atom against the force that acts to hold them within.

Important point: An atom that has lost or gained one or more electrons is said to be *ionized*. If the atom loses electrons, it becomes positively charged and is referred to as a *positive ion*. Conversely, if the atom gains electrons, it becomes negatively charged and is referred to as a *negative ion*.

3.2.3 Conductors, Semiconductors, and Insulators

Electric current moves quickly through some materials but with incredible difficulty through others. Substances that permit the free movement of many electrons are called conductors. Copper is the most widely used electrical conductor because of its high conductivity (how good a conductor the material is) and cost-effectiveness.

Electrical energy is transferred through a copper or other metal conductor by moving free electrons that migrate from atom to atom inside the conductor (see Figure 3.3). Each electron moves a truly short distance to the neighboring atom, which replaces one or more electrons by forcing them out of their orbits. The replaced electrons repeat the process in other nearby atoms until the movement is transmitted throughout the entire length of the conductor. A good conductor is said to have a low opposition, or *resistance*, to the electron (current) flow.

Note: If many electrons flow through a material with only a small force (voltage) applied, we call that material a **conductor**. A convenient way to understand the purpose and function of an electrical conductor is to think of or visualize a pipe or conveyor of water or some other liquid or gaseous substance. As a liquid or gas is conveyed or conducted through a pipe, an electron or electricity is transferred through a conductor. Regarding the force applied, it is a pump that pushes the water, and a voltage that pushes the electrons. It is important to remember that in an electrical circuit, voltage does not move; instead, it is the force that makes electrons move.

Table 3.1 lists many of the metals commonly used as electric conductors. The best conductors appear at the top of the list, with the poorer ones shown last.

Note: The movement of each electron (e.g., in copper wire) takes a minimal amount of time and is almost instantaneous. This is a crucial point to remember when events in an electrical circuit seem to co-occur.

TABLE 3.1
Electrical Conductors

Electrical Conductors
Silver
Copper
Gold
Aluminum
Zinc
Brass
Iron
Tin
Mercury

While it is true that electron motion is known to exist in all matter, some substances, such as rubber, glass, and dry wood, have very few free electrons. In these materials, large amounts of energy must be expended to break the electrons loose from the influence of the nucleus. Substances containing very few free electrons are called *insulators*. Insulators are essential in electrical work because they prevent the current from being diverted from the wires, which helps prevent electrocutions and fires.

Note: If the voltage is large enough, even the best insulators—rubber, plastic, and wood—will break down and allow their electrons to flow.

Table 3.2 lists some materials we often use as insulators in electrical circuits. The list is in a top-to-bottom sequence to show their decreasing ability to withstand high voltages without conducting.

A material that is neither a good conductor nor a good insulator is called a *semiconductor*. Silicon and germanium are substances that fall into this category. Because of their peculiar crystalline structure, these materials may, under certain conditions, function as conductors and, under other conditions, as insulators. As the temperature rises, however, a limited number of electrons become available for conduction.

TABLE 3.2
Materials That Withstand Voltage without Conducting

Common Insulators
Rubber
Mica
Wax or Paraffin
Porcelain
Bakelite
Plastics
Glass
Fiberglass
Dry wood
Air

3.2.4 Static Electricity

Electricity at rest is often referred to as *static electricity*. More specifically, when two bodies of matter have unequal charges and are near one another, an electric force is exerted between them because of their unequal charges. However, since they are not in contact, their charges cannot equalize. The existence of such an electric force, where current cannot flow, is *static electricity*.

However, static electricity at rest will flow if given the opportunity—this is the case because static electricity is an imbalance of negative and positive charges. An example of this phenomenon is often experienced when one walks across a dry carpet and then touches a doorknob—a slight shock is usually felt, and a spark at the fingertips is likely noticed. Another familiar example is "static cling." For example, whenever we rub an air-filled balloon against the hair on our heads and then place the balloon against a wall, the balloon sticks to the wall, defying gravity because of static cling. In the workplace, static electricity is prevented from being built up by adequately bonding or grounding equipment to the ground or earth.

3.2.5 Charged Bodies

The fundamental law of charged bodies states: "Like charges repel each other, and unlike charges attract each other." Positive and negative charges, opposite or unlike, tend to move toward each other, attracting each other. In contrast, like bodies tend to repel each other. Electrons repel each other because of their negative charges (like charges repel), and protons repel each other because of their positive charges. Figure 3.4 demonstrates the law of charged bodies.

It is essential to point out another significant part of the fundamental law of charged bodies—the *force of attraction or repulsion between two magnetic poles decreases rapidly as the poles are separated.* More specifically, the force of attraction or repulsion varies directly as the product of the separate pole strengths and inversely as the square of the distance separating the magnetic poles, provided the poles are small enough to be considered points.

Let's look at an example:

> If you increase the distance between the magnet's two north poles from 2 feet to 4 feet, the force of repulsion between them decreases to one-fourth of its original value. If either pole strength is doubled, the distance remaining the same, the force between the poles will be doubled.

3.2.6 Coulomb's Law

Coulomb's Law states that the amount of attracting or repelling force that acts between two electrically charged bodies in free space depends on two things:

a. Their charges
b. The distance between them.

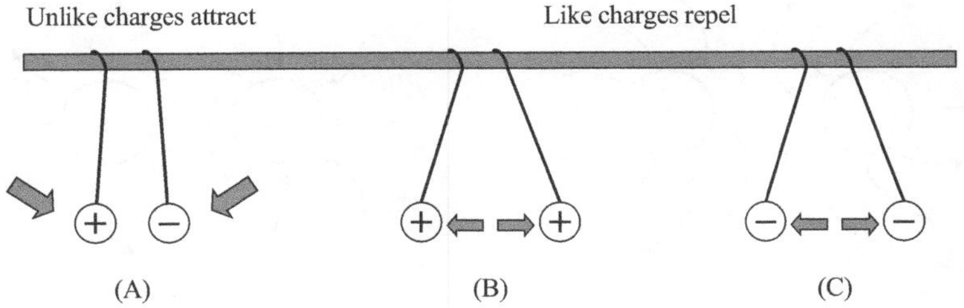

FIGURE 3.4 Reaction between two charged bodies. The opposite charge in (A) attracts. The like charges in (B) and (C) repel each other.

Specifically, Coulomb's Law states: "Charged bodies attract or repel each other with a force that is directly proportional to the product of their charges and is inversely proportional to the square of the distance between them."

Note: The number of electrons determines the magnitude of a body's electric charge compared to the number of protons within the body. The symbol for the magnitude of electric charge is Q, expressed in units of *coulombs (C)*. A charge of one positive coulomb means a body contains a charge of 6.25×10^{18}. A charge of one negative coulomb, $-Q$, means a body includes a charge of 6.25×10^{18} more electrons than protons.

3.2.7 Electrostatic Fields

The fundamental characteristic of an electric charge is its ability to exert force. The space between and around charged bodies in which their influence is felt is called an *electric field of force*. The electric field is terminated on material objects and extends between positive and negative charges. This region of force can consist of air, glass, paper, or a vacuum. This region of force is referred to as an *electrostatic field*.

When two objects of opposite polarity are brought near each other, the electrostatic field is concentrated in the area between them. The field is represented by lines referred to as *electrostatic lines of force*. These imaginary lines are used merely to describe the direction and strength of the field. To avoid confusion, the positive lines of force are always shown as leaving charge; for a negative charge, they are shown as entering. Figure 3.5 illustrates using lines to represent the field of charged bodies.

Note: A charged object will retain its charge temporarily if there is no immediate transfer of electrons to or from it. In this condition, the charge is said to be at rest. Remember, electricity at rest is called static electricity.

3.2.8 Magnetism

Most electrical equipment depends directly or indirectly upon magnetism. Magnetism is a phenomenon associated with magnetic fields; that is, it can attract substances such as iron, steel, nickel, or cobalt (metals known as magnetic materials). Correspondingly, a substance is said to be a magnet if it has the property of magnetism. For example, a piece of iron can be magnetized and is thus a magnet.

When magnetized, the piece of iron (note: we will assume a piece of flat bar 6 inches long × 1-inch-wide × 0.5 inches thick; a bar magnet—see Figure 3.6) will have two points opposite each other, which most readily attract other iron pieces. The points of maximum attraction (one on each

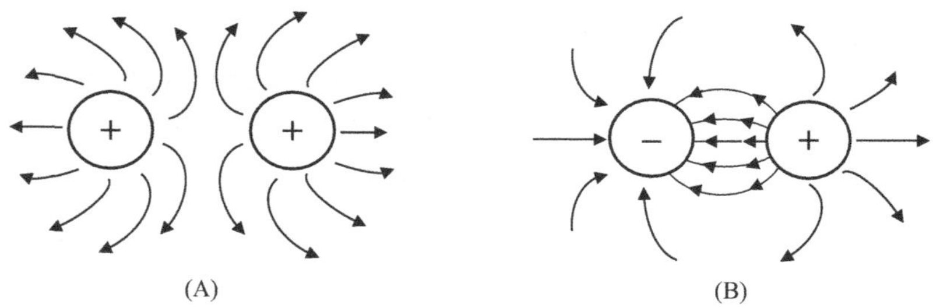

FIGURE 3.5 Electrostatic lines of force. Figure (A) represents the repulsion of like-charged bodies and their associated fields. Figure (B) represents the attraction between unlike-charged bodies and their related fields.

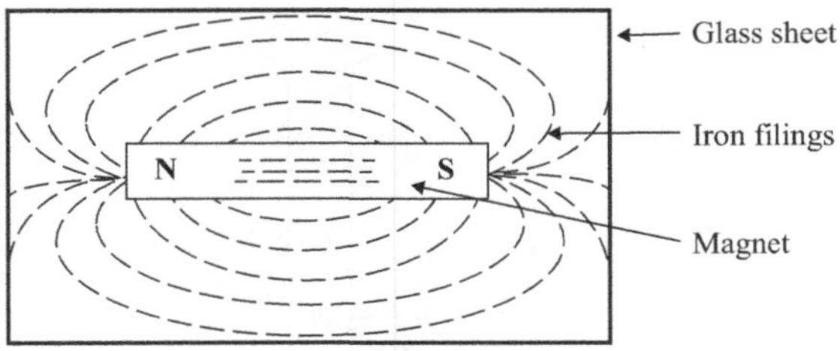

FIGURE 3.6 The magnetic field around a bar magnet. If the glass sheet is tapped gently, the filings will move into a definite pattern that describes the field of force around the magnet.

end) are called the *magnetic poles* of the magnet: the north (N) pole and the south (S) pole. Just as electric charges repel each other and opposite charges attract each other, magnetic poles repel each other, and unlike poles attract each other. Although invisible to the naked eye, its force can be shown to exist by sprinkling small iron filings on a glass covering a bar magnet, as shown in Figure 3.6.

Figure 3.7 shows how the field looks without iron filings. It is shown as lines of force [known as magnetic flux or flux lines; the symbol for magnetic flux is the Greek lowercase letter (phi)] in the field, repelled away from the magnet's north pole and attracted to its south pole.

Note: A *magnetic circuit* is a complete path through which magnetic lines of force may be established under the influence of a magnetizing force. Most magnetic circuits contain magnetic materials to contain the magnetic flux. These circuits are like the *electric circuit* (important point), which is a complete path through which current is caused to flow under the influence of an electromotive force.

There are three types or groups of magnets:

a. *Natural magnets:* They are found in the natural state as a mineral (an iron compound) called magnetite.
b. *Permanent magnets:* They are artificial magnets hardened steel or some alloy such as Alinco bars that have been permanently magnetized. Most people are familiar with the permanent horseshoe magnet; the red U-shaped magnet is the universal symbol of magnets, recognized worldwide (see Figure 3.8).
c. *Electromagnets:* They are artificial magnets composed of soft-iron cores around wound coils of insulated wire. When an electric current flows through the coil, the core becomes magnetized. When the current ceases to flow, the core loses its magnetism.

3.2.8.1 Magnetic Materials

Natural magnets are no longer used (they have no practical value) in electrical circuitry because more powerful and conveniently shaped permanent magnets can be produced artificially. Commercial magnets are made from special steels and alloys, which are magnetic materials.

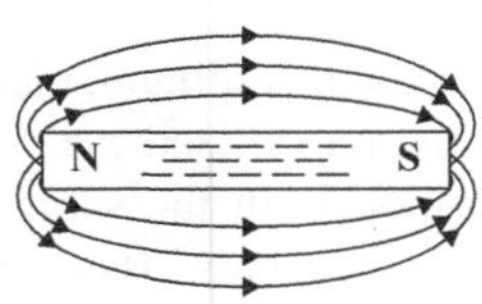

FIGURE 3.7 Magnetic field of force around a bar magnet, indicated by lines of force.

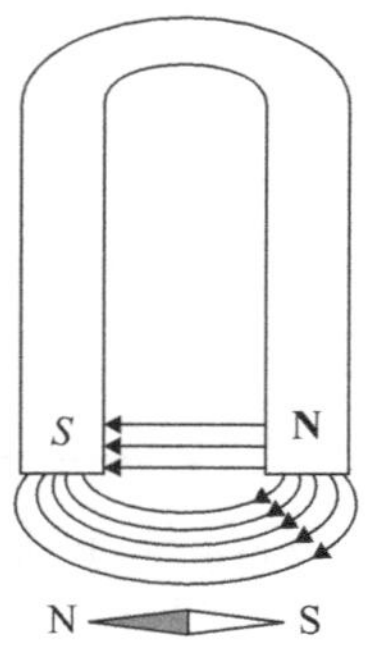

FIGURE 3.8 Horseshoe magnet.

Magnetic materials are attracted or repelled by a magnet and can be magnetized. Iron, steel, and alloy bars are the most common magnetic materials. These materials can be magnetized by inserting the material (in bar form) into a coil of insulated wire and passing a heavy direct current through the coil. The same material may also be magnetized if stroked with a bar magnet. It will then have the same magnetic property as the magnet used to induce the magnetism—namely, there will be two poles of attraction, one at either end. This process produces a permanent magnet by induction—that is, the magnetism is induced in the bar by the influence of the stroking magnet.

Note: Permanent magnets are hard magnetic materials (hard steel or alloys) that retain their magnetism when the magnetizing field is removed. A temporary magnet cannot retain a magnetized state when the magnetizing field is removed.

Even though they are classified as permanent magnets, it is essential to point out that hardened steel and certain alloys are relatively tricky to magnetize and are said to have low permeability because the magnetic lines of force do not easily permeate or distribute themselves readily through the steel.

Note: Permeability refers to the ability of a magnetic material to concentrate magnetic flux. Any material that is easily magnetized has high permeability. A measure of permeability for different materials compared to air or vacuum is called *relative* permeability, symbolized by μ or (mu).

Once hard steel and other alloys are magnetized, they retain a large part of their magnetic strength and are called *permanent magnets*. Conversely, easy-to-magnetize materials—such as soft iron and annealed silicon steel—are said to have a high permeability. Such materials retain only a tiny part of their magnetism after removing the magnetizing force and are called *temporary magnets*.

The magnetism that remains in a temporary magnet after the magnetizing force is removed is called *residual magnetism*.

Early magnetic studies classified magnetic materials merely as magnetic and nonmagnetic based on the strong magnetic properties of iron. However, since weak magnetic materials can be necessary in some applications, present studies classify materials into one of three groups: paramagnetic, diamagnetic, and ferromagnetic.

a. *Paramagnetic materials:* These include aluminum, platinum, manganese, and chromium—materials that become only slightly magnetized even under the influence of a strong magnetic field. This slight magnetization is in the same direction as the magnetizing field. Relative permeability is somewhat more than 1 (i.e., considered nonmagnetic materials).
b. *Diamagnetic materials:* These include bismuth, antimony, copper, zinc, mercury, gold, and silver—materials that can also be slightly magnetized when under the influence

of a powerful field. Relative permeability is less than 1 (i.e., considered nonmagnetic materials).

c. *Ferromagnetic materials:* These include iron, steel, nickel, cobalt, and commercial alloys, the most important group for applications in electricity and electronics. Ferromagnetic materials are easy to magnetize and have high permeability, ranging from 50 to 3,000.

3.2.8.2 Magnetic Earth

The Earth is a giant magnet, and its magnetism produces the magnetic field surrounding it. Most people would have no problem understanding or at least accepting this statement. However, if told that the Earth's north magnetic pole is its south magnetic pole and the south magnetic pole is its north magnetic pole, they might not accept or understand this statement. But, in terms of magnets, it is true.

Figure 3.9 indicates the Earth's magnetic polarities. The geographic poles are also shown at each end of the Earth's rotation axis. The magnetic axis does not coincide with the geographic axis; therefore, the magnetic and geographic poles are not at the same place on the Earth's surface.

Recalling that magnetic lines of force are assumed to emanate from the magnet's north pole and enter the south pole as closed loops. Because the Earth is a magnet, lines of force emanate from its north magnetic pole and enter the south magnetic pole as closed loops. A compass needle aligns itself so that the Earth's lines of force enter at its south pole and leave at its north pole. Because the needle's north pole is defined as the end that points in a northerly direction, it follows that the magnetic pole near the north geographic pole is a south magnetic pole, and vice versa.

3.2.9 Difference in Potential

Because of the force of its electrostatic field, an electric charge can move another charge by attraction or repulsion. The force that causes free electrons to move in a conductor as an electric current may be referred to as follows:

- Electromotive force (emf)
- Voltage
- Difference in potential

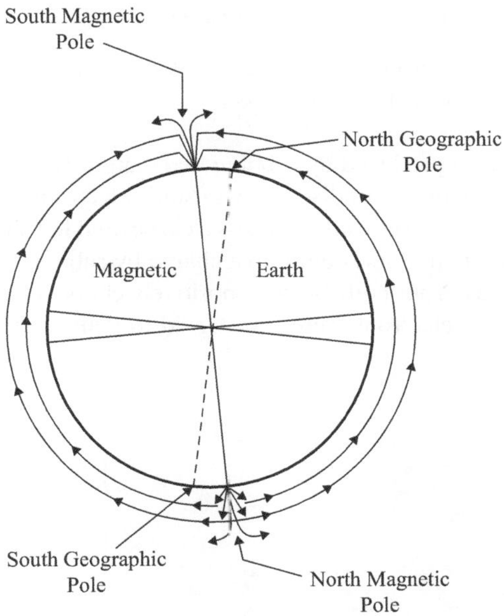

FIGURE 3.9 Earth's magnetic poles.

When a difference in potential exists between two charged bodies connected by a wire (conductor), electrons (current) flow along the conductor. This flows from the negatively charged body to the positively charged body until the two charges are equalized, and the potential difference no longer exists.

Note: The basic unit of potential difference is the *volt* (V). The symbol for the potential difference is *V*, indicating the ability to force electrons (current flow) to move. Because the volt unit is used, a potential difference is called *voltage.*

3.2.10 Water Analogy

In attempting to train individuals in the concepts of basic electricity, especially regarding the difference of potential (voltage), current, and resistance relationships in a simple electrical circuit, it has been common practice to use what is referred to as the water analogy. We use the water analogy later to explain (in a simple, straightforward fashion) voltage, current, and resistance and their relationships in more detail. Still, for now, we use analogy to explain the basic concept of electricity: the difference between potential and voltage. Because a difference in potential causes current flow (against resistance), it is essential to understand this concept before the concept of current flow and resistance is explained.

Consider the water tanks in Figure 3.10—two connected by a pipe and valve. First, the valve is closed, and all the water is in Tank A. Thus, the water pressure across the valve is at maximum. When the valve is opened, the water flows through the pipe from A to B until the water level becomes the same in both tanks. The water then stops flowing in the pipe because there is no longer a difference in water pressure (difference in potential) between the two tanks.

Just as the flow of water through the pipe in Figure 3.10 is directly proportional to the difference in water level in the two tanks, current flow through an electric circuit is directly proportional to the difference in potential across the circuit.

Important point: A fundamental law of current electricity states that the current is directly proportional to the applied voltage; if the voltage increases, the current increases. If the voltage decreases, the current decreases.

3.2.11 Principal Methods of Producing a Voltage

There are many ways to produce electromotive force or voltage, some more widely used than others. The following list the seven most common methods.

a. *Friction:* It is the voltage produced by rubbing two materials together (static electricity or electrostatic force). Remember our discussion of static electricity? Let's refresh our memories. For example, have you ever walked across a carpet and felt shocked when you touched a metal doorknob? Your shoe soles built up a charge by rubbing on the carpet, which was transferred to your body. Your body became positively charged, and when you touched the zero-charged doorknob, electrons were transferred to your body until you and the doorknob had equal charges.

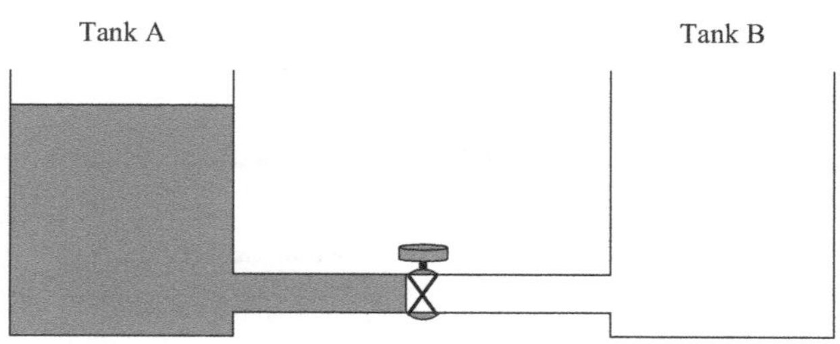

FIGURE 3.10 Water analogy of the electric difference of potential.

b. *Pressure (piezoelectricity):* It is the voltage produced by squeezing or applying pressure to crystals of certain substances (e.g., crystals like quartz or Rochelle salts or certain ceramics like barium titanate). When pressure is applied to such substances, electrons can be driven out of orbit in the direction of the force. Electrons leave one side of the material and accumulate on the other, building up positive and negative charges on opposite sides. When the pressure is released, the electrons return to their orbits. Some materials react to bending pressure, while others respond to twisting pressure. This voltage generation is known as the *piezoelectric effect (using* Rochelle salts to generate electricity is called pyroelectricity). If external wires are connected while pressure and voltage are present, electrons will flow, and current will be produced. If the pressure is constant, the current will flow until the potential difference is equalized. When the force is removed, the material is decompressed and immediately causes an electric force in the opposite direction. The power capacity of these materials is minimal. However, these materials are handy because of their extreme sensitivity to changes in mechanical force. One example is the crystal phonograph cartridge containing a Rochelle salt crystal (potassium sodium tartrate tetrahydrate). A phonograph needle is attached to the crystal. As the needle moves in the grooves of a record, it swings from side to side, applying compression and decompression to the crystal. This mechanical motion applied to the crystal generates a voltage signal to reproduce sound.

c. *Heat (thermoelectricity):* It is the voltage produced by heating the joint (junction) where two unlike metals are joined. Some materials readily give up their electrons, and others readily accept electrons. For example, electrons can transfer when two dissimilar metals, like copper and zinc, join. Electrons will leave the copper atoms and enter the zinc atoms. The zinc gets a surplus of electrons and becomes negatively charged. The copper loses electrons and takes on a positive charge. This creates a voltage potential across the junction of the two metals. The heat energy of normal room temperature is enough to make them release and gain electrons, causing a measurable voltage potential. As more heat energy is applied to the junction, more electrons are released, and the voltage potential becomes greater. When heat is removed, the junction cools, the charges dissipate, and the voltage potential decreases. This process is called thermoelectricity. A device like this is generally referred to as a thermocouple.

   The thermocouple voltage in a thermocouple depends upon the heat energy applied to the junction of the two dissimilar metals. Thermocouples are widely used to measure temperature and as heat-sensing devices in automatic temperature-controlled equipment.

   Thermocouple power capacities are minimal compared to other sources but are more significant than those of crystals. A thermocouple can be subjected to higher temperatures than ordinary mercury or alcohol thermometers.

d. *Light (photoelectricity):* It is the voltage produced by light (photons) striking photosensitive (light-sensitive) substances. When the photons in a light beam hit the surface of a material, they release their energy and transfer it to the atomic electrons of the material. This energy transfer may dislodge electrons from their orbits around the surface of the substance. The photosensitive (light-sensitive) material becomes positively charged upon losing electrons, creating an electric force.

   This phenomenon is called the photoelectric effect and has wide applications in electronics, such as photoelectric cells, photovoltaic cells, optical couplers, and television camera tubes. Three uses of the photoelectric effect are described below.

   - *Photovoltaic:* Light energy in two plates joined together causes one plate to release electrons to the other, building up opposite charges, like a battery.
   - *Photoemission:* The photon energy from a light beam could cause a surface to release electrons in a vacuum tube. A plate would then collect the electrons.
   - *Photoconduction:* Light energy applied to some typically poor materials produces free electrons, making the materials better conductors.

e. *Chemical action:* It is the voltage produced by a chemical reaction in a battery cell. For example, the voltaic chemical cell wherein a chemical reaction produces and maintains opposite charges on two dissimilar metals that serve as positive and negative terminals. The metals are in contact with an electrolyte solution. Connections between more than one of these cells will produce a battery.
f. *Magnetism:* It is the voltage produced in a conductor when the conductor moves through a magnetic field, or a magnetic field moves through the conductor in such a manner as to cut the magnetic lines of force of the field. A generator is a machine that converts mechanical energy into electrical energy using the magnetic induction principle. This is one of the most useful and widely employed applications for producing vast electric power.
g. *Thermionic emission:* A thermionic energy converter consists of two electrodes placed near one another in a vacuum. One electrode is typically called the cathode or emitter, and the other is called the anode or plate. Ordinarily, electrons in the cathode are prevented from escaping from the surface by a potential energy barrier. When an electron starts to move from the surface, it induces a corresponding positive charge in the material, which tends to pull it back to the surface. To escape the electron, one must somehow acquire enough energy to overcome this energy barrier. Almost none of the electrons can gain enough energy to escape at ordinary temperatures. However, thermal motion dramatically increases the electron energy when the cathode is hot. At sufficiently high temperatures, a considerable number of electrons can escape. The liberation of electrons from a hot surface is called *thermionic emission.*

   The electrons that have escaped from the hot cathode form a cloud of negative charges called a space charge near it. If a battery maintains the plate positive for the cathode, the electrons in the cloud are attracted to it. If the potential difference between the electrodes is maintained, there will be a steady current flow from the cathode to the plate.

   The simplest example of a thermionic device is a vacuum tube diode, in which the only electrodes are the cathode and plate, or anode. The diode can convert alternating current (AC) flow to pulsating direct current (DC) flow.

In the study of the basic electricity related to renewable energy production, we are most concerned with magnetism (generators powered by hydropower, for example), light (photoelectricity produced by solar cells), and chemistry (chemical energy converted to electricity in batteries) to produce voltage. Friction has few practical applications, though we discussed it earlier in studying static electricity. Pressure and heat have valuable applications, but we do not need to consider them in this text. Magnetism used in generators, solar light-produced electricity, and the chemistry involved with storing electricity in batteries, on the other hand, are, as mentioned, the principal sources of voltage and are discussed at length in this text.

3.2.12 Electric Current

The movement or flow of electrons is called current. The electrons must be moved by a potential difference or pressure (voltage) to produce current.

Note: The terms current, current flow, electron flow, or electron current, etc., may be used to describe the same phenomenon.

For our purposes in this text, electron flow, or current, in an electric circuit moves from a region of less negative potential to an area of more positive potential—from negative to positive.

Note: Current is represented by the letter I. Its basic unit of measurement is the *ampere, or amp (A).* One ampere of current is defined as the movement of one coulomb past a conductor's point during 1 second.

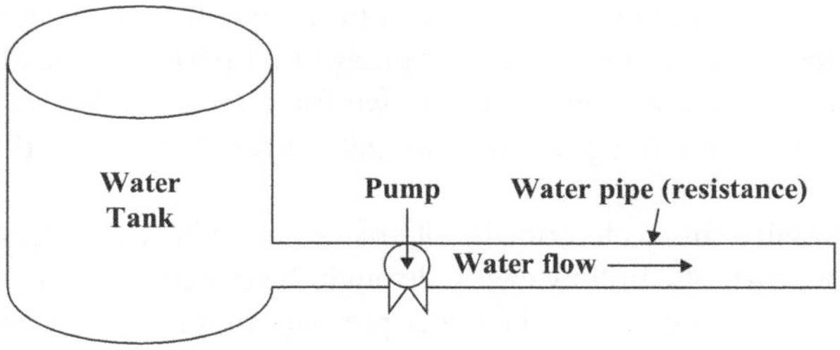

FIGURE 3.11 Water analogy: Current flow.

Recall that we used the water analogy to help us understand the potential differences. We can also use the water analogy to help us understand current flow through a simple electric circuit. Consider Figure 3.11, which shows a water tank connected via a pipe to a pump with a discharge pipe. If the water tank contains an amount of water above the level of the pipe opening to the pump, the water exerts pressure (a difference in potential) against the pump. When sufficient water is available for pumping, water flows through the pipe against the resistance of the pump and pipe. The analogy should be clear with an electric circuit. If a difference of potential exists, current will flow in the circuit.

Another simple way of looking at this analogy is to consider Figure 3.12, where the water tank has been replaced with a generator, the pipe with a conductor (wire), and water flow with the flow of electric current.

Again, Figures 3.11 and 3.12 illustrate that the electrons must be moved by a potential difference to produce current.

Electric current is classified into two general types:

- direct current (D-C)
- alternating current (A-C)

Direct current moves through a conductor or circuit in one direction only. *Alternating current* periodically reverses direction.

3.2.13 Resistance

Earlier, it was pointed out that free electrons, or electric current, could move quickly through a good conductor, such as copper, but an insulator, such as glass, was an obstacle to current flow. In the water analogy shown in Figure 3.11 and the simple electric circuit shown in Figure 3.12, resistance is indicated by either the pipe or the conductor.

Every material offers some resistance to the flow of electric current. Good conductors like copper, silver, and aluminum provide little resistance. Poor conductors, like glass, wood, and paper, resist current flow.

Note: The current flow in each circuit depends on voltage and resistance.

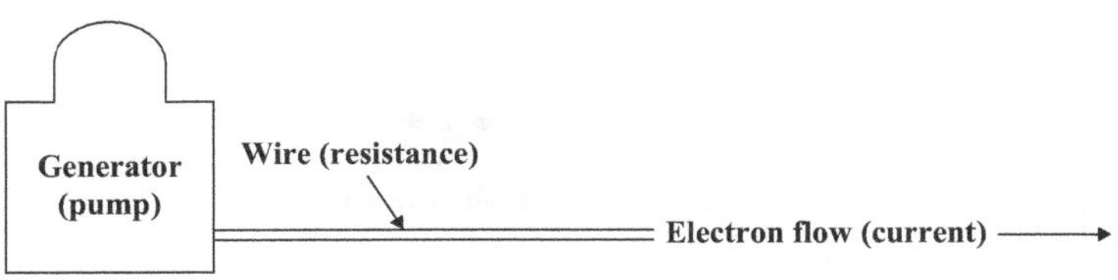

FIGURE 3.12 Simple electric circuit with current flow.

Note: The letter R represents resistance. The basic unit in which resistance is measured is the *ohm* (Ω). 1 Ω is the resistance of a circuit element or circuit that permits a steady current of 1 amp (1 C per second) to flow when a constant electromotive force (emf) of 1 V is applied to the circuit. Manufactured circuit parts containing definite amounts of resistance are called **resistors**.

The size and type of wires in an electric circuit are chosen to keep the electrical resistance as low as possible. In this way, current can flow easily through the conductors, as water flows through the pipe between the tanks in Figure 3.11. If the water pressure remains constant, the water flow in the pipe will depend on how far the valve is open. The smaller the opening, the greater the opposition (resistance) to the flow; the smaller the flow rate in gallons per second.

In the electric circuit shown in Figure 3.12, the larger the diameter of the wire, the lower its electrical resistance (opposition) to the flow of current through it. In the water analogy, pipe friction opposes the water flow between the tanks. This friction is like electrical resistance. The resistance of the pipe to the flow of water through it depends upon (1) the length of the pipe, (2) the diameter of the pipe, and (3) the nature of the inside walls (rough or smooth). Similarly, the electrical resistance of the conductors depends upon (1) the length of the wires, (2) the diameter of the wires, and (3) the material of the wires (copper, silver, etc.).

It is important to note that temperature also affects the resistance of electrical conductors to some extent. The resistance increases with temperature in most conductors (copper, aluminum, etc.). Carbon is an exception. In carbon, the resistance decreases as the temperature increases.

Important note: Electricity is a study frequently explained in terms of opposites. The term that is precisely the opposite of resistance is *conductance*. Conductance (*G*) is the ability of a material to pass electrons. The SI-derived unit of conductance is the Siemens. The commonly used unit of conductance is the *Mho,* which is ohm spelled backward. Resistance and conductance have a reciprocal relationship; one is the inverse of the other. If the resistance of a material is known, dividing its value by one will give its conductance. Similarly, if the conductance is known, dividing its value by one will provide its resistance.

3.3 D-C CIRCUITS

An electric circuit includes an **energy source** [source of electromotive force (emf) or voltage; that is, a battery or generator], a conductor (wire), a load, and a means of control (see Figure 3.13). The energy source could be a battery, as in Figure 3.13, or some other means of producing a voltage. The **load** that dissipates the energy could be a lamp, a resistor, or some other device (or devices) that does valuable work, such as an electric toaster, a power drill, a radio, a soldering iron, a laptop computer, or a battery pack for storage of energy as a backup to a renewable energy source such as solar or

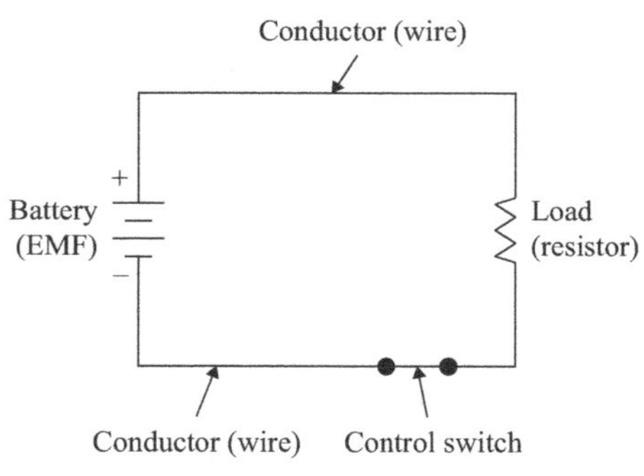

FIGURE 3.13 Simple closed circuit.

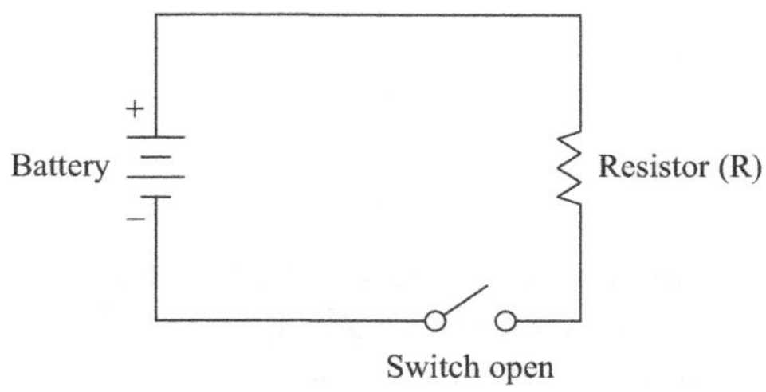

FIGURE 3.14 Open circuit.

wind. Again, **conductors** are wires that offer low resistance to current; they connect all the loads in the circuit to the voltage source. No electrical device dissipates energy unless current flows through it. Because conductors, or wires, are not perfect, because of their resistance, they heat up (dissipate energy) and are part of the load. For simplicity, we assume connecting wires have no resistance to avoid tedious calculations of their low resistance values. **Control devices** might be switches, variable resistors, circuit breakers, fuses, or relays.

A complete pathway for current flow, or **closed circuit** (Figure 3.13), is an unbroken path for current from the emf, through a load, and back to the source. A circuit is called **open** (see Figure 3.14) if a break in the circuit (e.g., an open switch) does not provide a complete path for current.

Important point: Current flows from the negative (−) terminal of the battery, shown in Figures 3.13 and 3.14, through the load to the positive (+) battery terminal and continues by going through the battery from the positive (+) terminal to the negative (−) terminal. If this pathway is unbroken, it is a closed circuit, and current will flow. However, if the path is broken at ANY point, it is an open circuit, and no current flows (see Figure 3.14).

A fuse is placed directly into the circuit to protect it (see Figure 3.15). A fuse will open the circuit whenever a dangerously large current starts to flow (i.e., a short circuit condition occurs caused by an accidental connection between two points in a circuit that offers little resistance). A fuse will permit currents smaller than the fuse value to flow, but will melt and break or open the circuit if a more significant current flows.

3.3.1 Schematic Representation

The simple circuits shown in Figures 3.13–3.15 are displayed schematically. A *schematic diagram* (usually shortened to "schematic") is a simplified drawing representing the **electrical**, not the physical, situation in a circuit. The symbols used in schematic diagrams are the electrician's "shorthand";

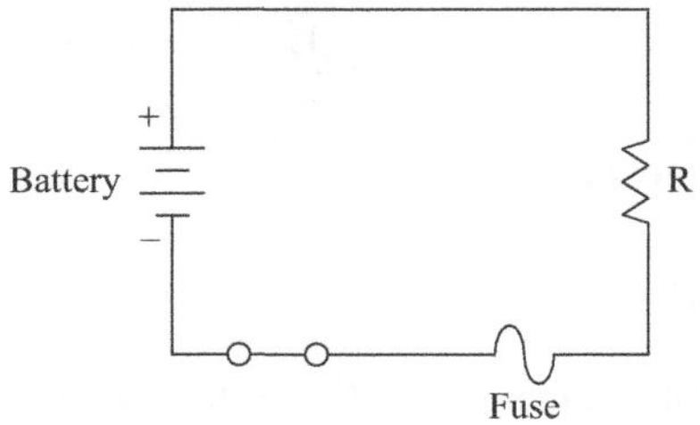

FIGURE 3.15 A simple fused circuit.

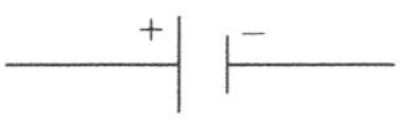

FIGURE 3.16 Schematic symbol for a battery.

they make the diagrams easier to draw and understand. Consider the symbol in Figure 3.16 used to represent a battery power supply. The symbol is relatively straightforward, but it is also essential. For example, by convention, the shorter line in the symbol for a battery represents the **negative** terminal. It is important to remember this because it is sometimes necessary to note the direction of the current flow, which is from negative to positive, when you examine the schematic. The battery symbol shown in Figure 3.16 has a single cell, so only one short and one long line are used. The number of lines used to represent a battery varies (and they are not necessarily equivalent to the number of cells), but they are always in pairs, with long and short lines alternating. In the circuit shown in Figure 3.15, the current would flow counterclockwise, in the opposite direction that a clock's hands move. If the long and short lines of the battery symbol (symbol shown in Figure 3.16) were reversed, the current in the circuit shown in Figure 3.15 would flow clockwise, that is, in the direction of a clock's hands.

Important note: In studies of electricity and electronics, many circuits, mainly composed of specially designed resistive components, are analyzed. As previously stated, these components are called **resistors**. The resistive component will be a physical resistor throughout the remaining fundamental circuit analysis. However, the resistive component could be any one of several electrical devices.

Remember that in the simple circuits shown in the figures, we have only illustrated and discussed a few symbols used in schematics to represent circuit components. (Other symbols will be introduced as we need them.)

Remembering that a closed loop of wire (conductor) is not necessarily a circuit is also essential. A source of voltage must be included to make it an electric circuit. In any electric circuit where electrons move around a closed loop, current, voltage, and resistance are present. The physical pathway for current flow is the circuit. By knowing any two of the three quantities, such as voltage and current, the third (resistance) may be determined. This is done mathematically using Ohm's Law, which is the foundation of electrical theory.

3.4 OHM'S LAW

Ohm's Law defines the relationship between an electric circuit's current, voltage, and resistance. It can be expressed mathematically in three ways.

a. The *current* (I) in a circuit is equal to the voltage applied to the circuit divided by the resistance of the circuit. Stated another way, the current in a circuit is DIRECTLY proportional to the applied voltage and INVERSELY proportional to the circuit resistance. Ohm's Law may be expressed as an equation:

$$I = \frac{E}{R} \tag{3.1}$$

where I = current in amperes (amps)
E = Voltage in volts
R = resistance in ohm

b. The *resistance* (R) of a circuit is equal to the voltage applied to the circuit divided by the current in the circuit:

$$R = \frac{E}{I} \tag{3.2}$$

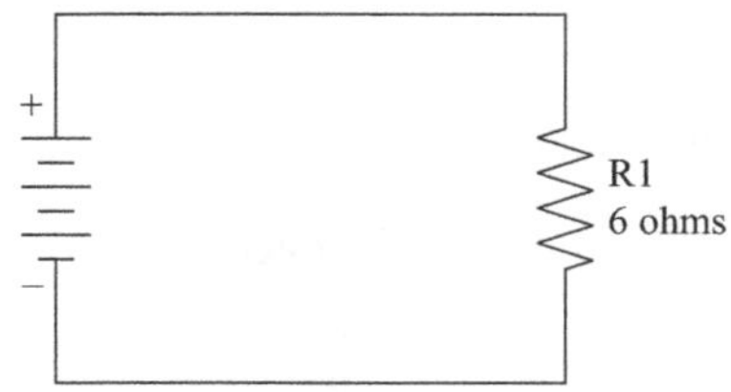

FIGURE 3.17 Determining current in a simple circuit.

c. The applied *voltage* (E) to a circuit is equal to the product of the current and the resistance of the circuit:

$$E = I \times R = IR \tag{3.3}$$

The third may be easily found if any two of the quantities in Equations (3.1–3.3) are known. Let's look at an example.

EXAMPLE 3.1

Problem:

Figure 3.17 shows a circuit with a resistance of 6 Ω and a voltage source of 3 V. How much current flows in the circuit?

Given: E = 3 V
R = 6 Ω
I = ?

SOLUTION:

$$I = \frac{E}{R}$$

$$I = \frac{3}{6}$$

$$I = 0.5 \text{ amp}$$

The key Point: Circuit current is directly proportional to the applied voltage and will change by the same factor as the voltage.

An Ohm's Law circle can help students memorize and practice the equations effectively (see Figure 3.18).

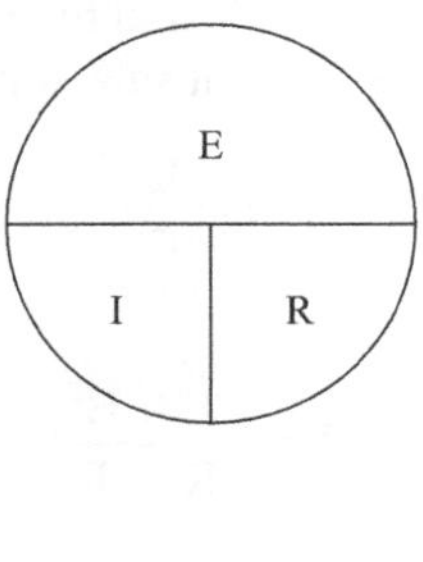

FIGURE 3.18 Ohm's Law circle.

EXAMPLE 3.2

Problem: Find I when E = 120 V and R = 40 Ω.

SOLUTION:

$$I = \frac{E}{R}$$

$$= \frac{120}{40} = 3\,A$$

Note: In the previous examples, we have demonstrated how the Ohm's Law circle can help us solve simple voltage, current, and amperage problems. However, the beginning student is cautioned not to rely wholly on using this circle when transposing simple formulas, but rather to supplement his/her knowledge of the algebraic method. Algebra is essential in solving electrical problems, and knowing how to use it should not be under-emphasized or bypassed.

3.5 ELECTRIC POWER

Power, whether electrical or mechanical, pertains to the rate at which work is being done, so the power consumption in your plant is related to current flow. A large electric motor or air dryer consumes more power (and draws more current) in each length of time than, for example, an indicating light on a motor controller. **Work** is done whenever a force causes motion. Work is done if a mechanical force is used to lift or move a weight. However, force exerted WITHOUT causing motion, such as the force of a compressed spring acting between two fixed objects, does not constitute work.

Key point: Power is the rate at which work is done.

The electric power P used in any part of a circuit equals the current I in that part multiplied by the voltage V across that part of the circuit. In equation form,

$$P = EI \tag{3.4}$$

where P = power, Watts (W)
E = Voltage, V
I = current, A

If we know the current I and the resistance R but not the voltage V, we can find the power P by using Ohm's Law for voltage, so that substituting

$$E = IR \tag{3.5}$$

into

$$P = IR \times I = I^2R \tag{3.6}$$

In the same manner, if we know the voltage V and the resistance R but not the current I, we can find the P by using Ohm's Law for current, so that substituting

$$I = \frac{E}{E}$$

into

$$P = E\frac{E}{R} - \frac{E^2}{R} \tag{3.7}$$

Key point: We can calculate the third if we know two quantities.

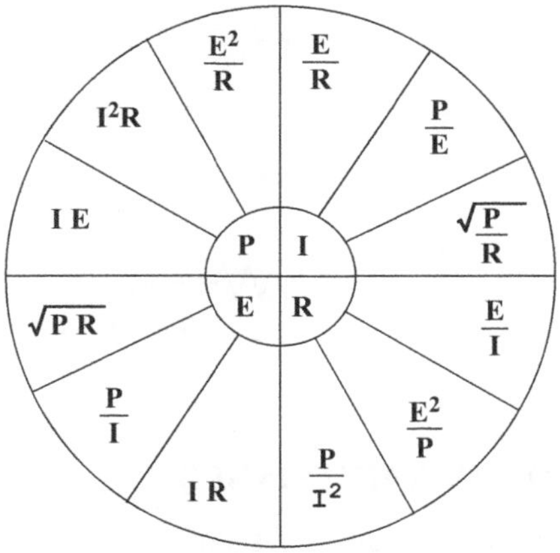

FIGURE 3.19 Ohm's law circle—Summary of basic formulas.

EXAMPLE 3.3

Problem:

The current through a 200-Ω resistor in a circuit is 0.25 A. Find the resistor's power rating.

SOLUTION:

Given that I and R are known, refer to Equation (13.6) to determine P.

$$P = I^2R = (0.25)^2(200) = 0.0625(200) = 12.5\,W$$

Important point: To prevent a resistor from burning out, the power rating of any resistor used in a circuit should be twice the wattage calculated by the power equation. Figure 3.19 shows a summary of basic Ohm's Law formulas.

Adjacent to each quantity are three segments. Note that the fundamental quantity is expressed in terms of two other basic quantities in each segment, and no two segments are alike.

3.6 ELECTRIC ENERGY

Energy (the mechanical definition) is the ability to do work (energy and time are essentially the same and are expressed in identical units). Energy is expended when work is done because it takes energy to maintain a force when that force acts through a distance. The total energy expended to do a certain amount of work equals the working force multiplied by the distance through which the force moved to do the job.

In electricity, the total energy expenditure equals the rate at which work is done multiplied by the time the rate is measured. Energy is equal to power P multiplied by time t. Kilowatt-hours (kWh) are units commonly used for large amounts of electric energy or work. Kilowatt-hours are calculated by multiplying power in kilowatts (kW) by time in hours (h).

$$kWh = kW \times h \tag{3.8}$$

EXAMPLE 3.4

Problem:

How much energy is delivered in 4 hours by a generator supplying 12 kW?

SOLUTION:

$$\begin{aligned} kWh &= kW \times h \\ &= 12(4) = 48 \end{aligned} \tag{3.9}$$

$$\text{Energy delivered} = 48\,kWh$$

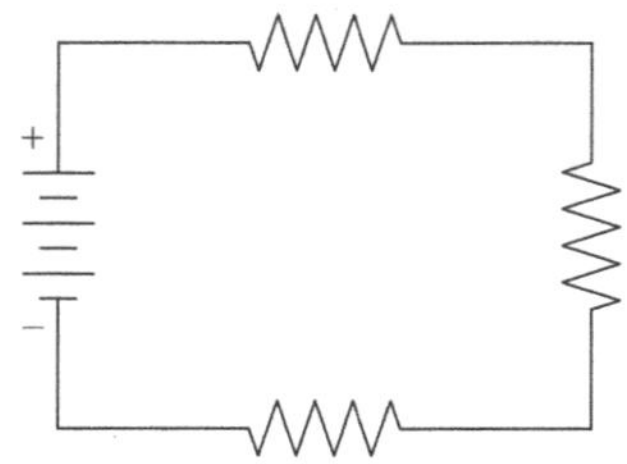

FIGURE 3.20 Series circuit.

3.7 SERIES D-C CIRCUIT CHARACTERISTICS

As previously mentioned, an electric circuit consists of a voltage source, the necessary connecting conductors, and the effective load.

If the circuit is arranged so that the electrons have only ONE possible path, the circuit is called a *Series Circuit.* Therefore, a series circuit is defined as a circuit that contains only one path for current flow. Figure 3.20 shows a series circuit having several loads (resistors).

Key point: A **series circuit** has only one path for the current to flow.

3.7.1 D-C Resistance

Referring to Figure 3.20, the current in a series circuit must flow through each resistor inserted into the circuit to complete its electrical path. Thus, each additional resistor offers added resistance. In a series circuit, the total circuit resistance (R_T) equals the sum of the individual resistances. As an equation:

$$R_T = R_1 + R_2 + R_3 \ldots R_n \tag{3.10}$$

where

R_T = total resistance, Ω

R_1, R_2, R_3 = resistance in series, Ω

R_n = any number of additional resistors in the equation

EXAMPLE 3.5

Problem:

Three resistors of 10 Ω, 12 Ω, and 25 Ω are connected in series across a battery whose emf is 110 V (Figure 3.21). What is the total resistance?

SOLUTION:

$$\text{Given}: R_1 = 10\,\Omega$$
$$R_2 = 12\,\Omega$$
$$R_3 = 25\,\Omega$$
$$R_T = ?$$

$$R_T = R_1 + R_2 + R_3$$
$$R_T = 10 + 12 + 25$$
$$R_T = 47\,\Omega$$

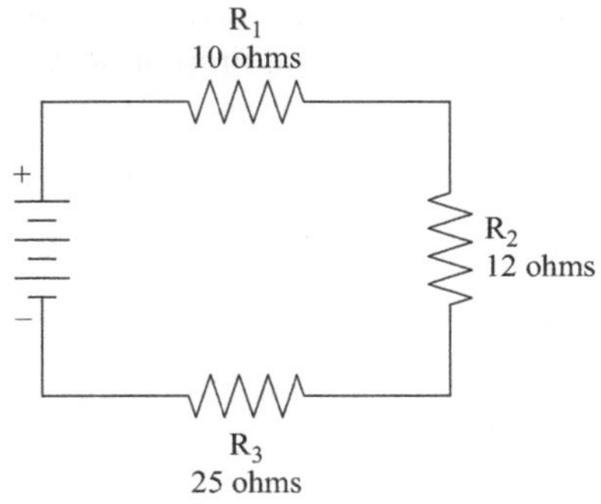

FIGURE 3.21 Solving total resistance in a series circuit.

Transposition can be used in some circuit applications where the total resistance is known, but a circuit resistor's value must be determined.

Key point: When resistances are connected in series, the total resistance in the circuit is equal to the sum of the resistances of all the parts of the circuit.

3.7.2 D-C Current

Because there is but one path for current in a series circuit, the same *current* (I) must flow through each part of the circuit. Thus, to determine the current throughout a series circuit, only the current through one of the parts needs to be known.

Key point: In a series circuit, the same current flows in every part of the circuit. Do not add the currents in each part of the circuit to obtain I.

3.7.3 D-C Voltage

The *voltage* drop across the resistor in the basic D-C circuit is the total voltage across the circuit and is equal to the applied voltage. The total voltage across a series circuit is equal to the applied voltage but consists of two or more individual voltage drops. In this circuit, a source potential (E_T) of 30 V is impressed across a series circuit consisting of two 6-Ω resistors. The circuit's total resistance equals the sum of the two individual resistances, or 12 Ω. Using Ohm's Law, the circuit current may be calculated as follows:

$$I = \frac{E_T}{R_T}$$

$$I = \frac{30}{12}$$

$$I = 2.5 \text{ amps}$$

Knowing the value of the resistors to be 6 Ω each and the D-C current through the resistors to be 2.5 amps, the voltage drops across the resistors can be calculated. The voltage (E_1) across R_1 is therefore:

$$E_1 = IR_1$$

$$E_1 = 2.5 \text{ amps} \times 6\ \Omega$$

$$E_1 = 15 \text{ V}$$

Since R_2 is the same as R_1 and carries the same current, the voltage drop across R_2 equals 15 V. Adding these two 15-V drops together gives a total drop of 30 V, equal to the applied voltage. For a series circuit, then:

$$E_T = E_1 + E_2 + E_3 \ldots E_n \tag{3.11}$$

where
E_T = total voltage, V
E_1 = voltage across resistance R_1, V
E_2 = voltage across resistance R_2, V
E_3 = voltage across resistance R_3, V

3.7.4 D-C Power

Each resistor in a D-C series circuit consumes *power*. This power is dissipated in the form of heat. Because this power must come from the source, the total power must equal the power consumed by circuit resistance. In a series circuit, the total power equals the **sum** of the powers the individual resistors dissipate. Total power (P_T) is thus equal to:

$$P_T = P_1 + P_2 + P_3 \ldots P_n \tag{3.12}$$

where
P_T = total power, W
P_1 = power used in the first part, W
P_2 = power used in the second part, W
P_3 = power used in the third part, W
P_n = power used in *n*th part, W

Key point: We found that Ohm's Law can be used to calculate total values in a D-C series circuit and individual parts. Similarly, the formula for power may be used to calculate total values.

$$P_T = IE_T \tag{3.13}$$

To this point, we have covered many critical factors governing the operation of basic D-C series circuits. We have laid a strong foundation to build on in preparation for the more advanced circuit theory that follows. A summary of the critical factors governing the operation of a D-C series circuit is listed as follows:

a. The same current flows through each part of a series circuit.
b. The total resistance of a series circuit is equal to the sum of the individual resistances.
c. The total voltage across a series circuit equals the sum of the voltage drops.
d. The voltage drop across a resistor in a series circuit is proportional to the size of the resistor.
e. The total power dissipated in a series circuit equals the sum of individual dissipation.

Important note: When applying Ohm's Law to a DC series circuit, consider whether the values used are component or total. When the information available enables the use of Ohm's Law to find total resistance, total voltage, and total current, total values must be inserted into the formula.

To find total resistance:

$$R_T = \frac{E_T}{I_T}$$

To find the total voltage:

$$E_T = I_T \times R_T$$

To find the total current:

$$I_T - \frac{E_T}{R_T}$$

3.8 KIRCHHOFF'S VOLTAGE LAW

Kirchhoff's voltage law states that the voltage applied to a closed circuit equals the sum of the voltage drops in that circuit. It should be obvious that this fact was used in the study of series circuits to this point. It was expressed as follows:

Voltage applied = sum of voltage drops

$$E_A = E_1 + E_2 + E_3$$

where E_A is the applied voltage and E_1, E_2, and E_3 are voltage drops.

Another way of stating Kirchhoff's Law is that the algebraic sum of the instantaneous EMFs and voltage drops around any closed circuit is zero.

Kirchhoff's Law can solve circuit problems, which would be difficult and often impossible with only knowledge of Ohm's Law. When Kirchhoff's Law is properly applied, an equation can be set up for a closed loop, and the unknown circuit values may be calculated.

Kirchhoff's Law can be used to solve multiple-source circuit problems. This method uses the same procedure for multiple-source circuits as for single-source circuits.

3.9 GROUND

The ground is used to denote a common electrical point of zero potential. The reference point of a circuit is always considered to be at zero potential. Because the Earth (ground) is said to have zero potential. In Figure 3.22, point A is the zero reference or ground and is symbolized as such. Point C is 60 V positive, and point B is 20 V positive with respect to ground.

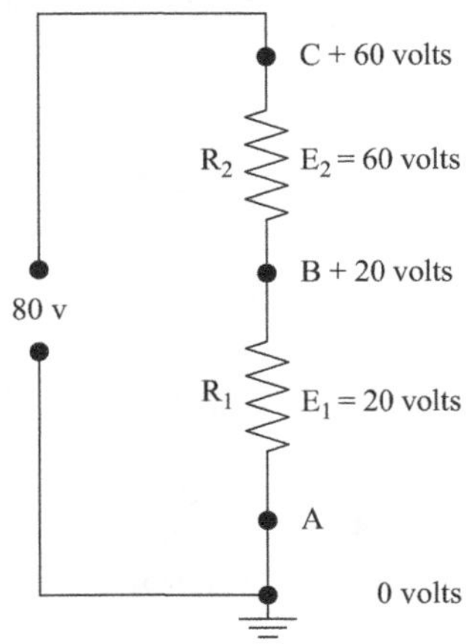

FIGURE 3.22 Use of ground symbol.

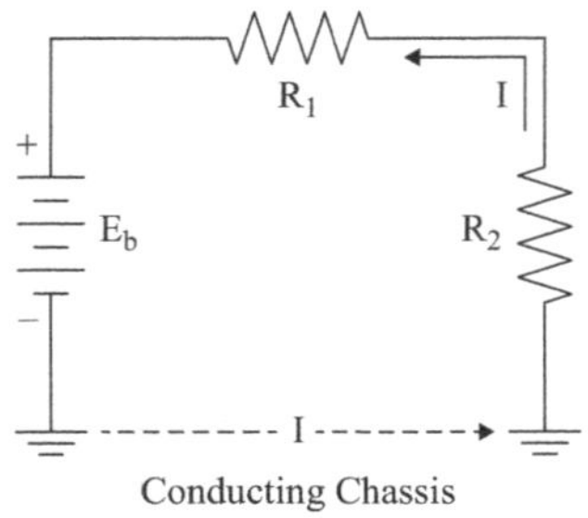

FIGURE 3.23 Ground used as a conductor.

The common ground for many electrical/electronic equipment is the metal chassis. The value of ground is noted when considering its contribution to the economy, simplification of schematics, and ease of measurement. When completing each electrical circuit, common points of a circuit at zero potential are connected directly to the metal chassis, thereby eliminating a large amount of connecting wire. An example of a grounded circuit is illustrated in Figure 3.23.

Note: Most voltage measurements used to check proper circuit operation in electronic equipment are taken with respect to the ground. One meter lead is attached to the ground, and the other is moved to various test points.

3.10 OPEN AND SHORT CIRCUITS

A circuit is *open* if a break does not provide a complete path for DC or AC. Figure 3.24 shows an open circuit because the fuse is blown.

A fuse is placed directly into the circuit to protect the circuit (see Figure 3.25). A fuse will open the circuit whenever a dangerously large current flows. A fuse will permit currents smaller than the fuse value to flow, but will melt and break or open the circuit if a more significant current flows. A dangerously large current will flow when a "short circuit" occurs. A short circuit is usually caused by an accidental connection between two points in a circuit that offers little resistance and passes an abnormal amount of current (see Figure 3.25). A short circuit often occurs because of improper wiring or broken insulation.

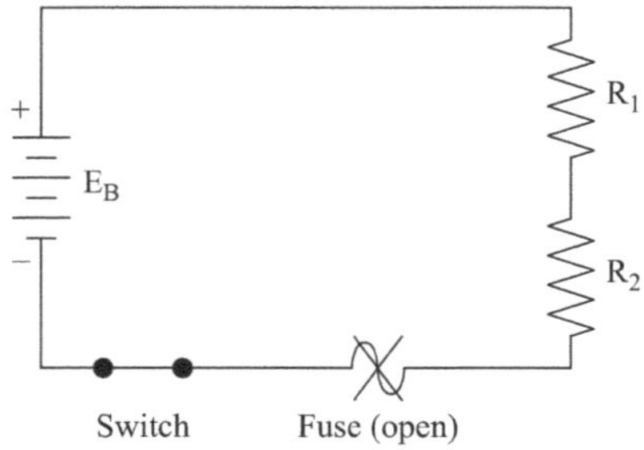

FIGURE 3.24 Open circuit—Fuse blown.

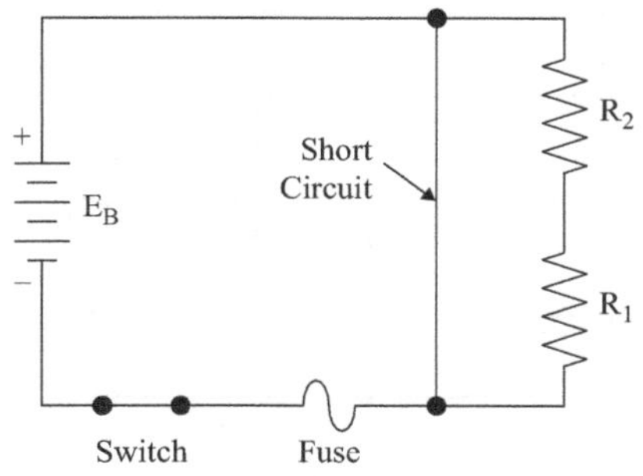

FIGURE 3.25 Short circuit.

3.11 PARALLEL D-C CIRCUITS

The principles we applied to solving simple D-C series circuit calculations to determine voltage, current, and resistance reactions can also be used in D-C parallel and series-parallel circuits.

A *parallel circuit* connects two or more components across the same voltage source (see Figure 3.26). Recall that there is only one path for current flow in a series circuit. As additional loads (resistors etc.) are added to the circuit, the total resistance increases, and the total current decreases. This does not apply in a **parallel** circuit. Each load (or branch) is connected directly across the voltage source in a parallel circuit. In Figure 3.26, commencing at the voltage source (E_b) and tracing counterclockwise around the circuit, two complete and separate paths can be identified where current can flow. One path goes from the source through resistance R_1 and returns to the source; the other path goes through resistance R_2 and returns to the source.

3.11.1 Voltage in Parallel Circuits

Recall that in a series circuit, the source voltage divides proportionately across each resistor in the circuit. In a parallel circuit (see Figure 3.27), the same voltage is present across all the resistors of a parallel group. This voltage is equal to the applied voltage (E_b) and can be expressed in equation form as:

$$E_b = E_{R1} = E_{R2} = E_{Rn} \tag{3.14}$$

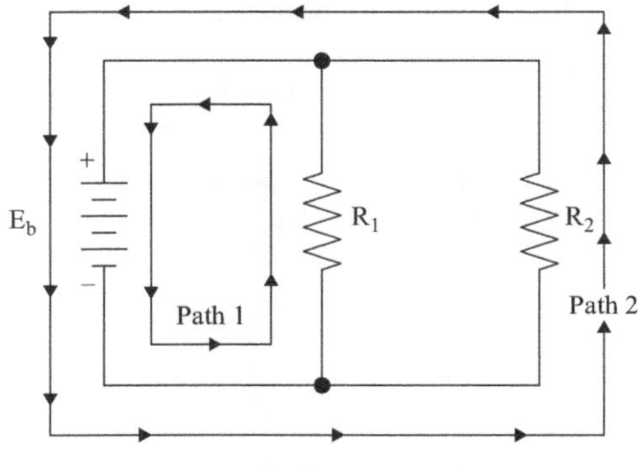

FIGURE 3.26 Basic parallel circuit.

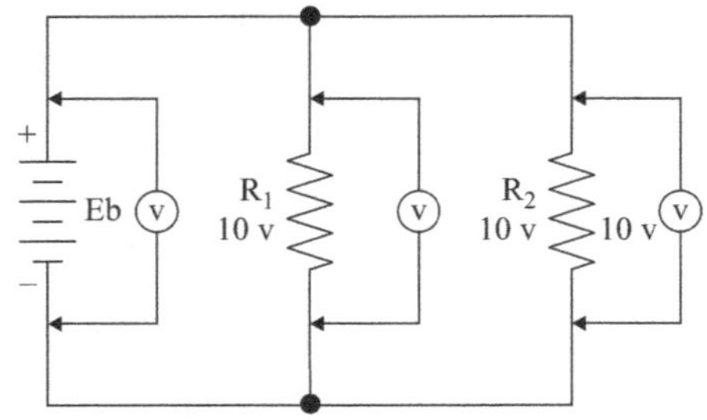

FIGURE 3.27 Voltage comparison in a parallel circuit.

We can verify Equation 3.17 by taking voltage measurements across the resistors of a parallel circuit, as illustrated in Figure 3.27. Notice that each voltmeter indicates the same amount of voltage; that is, the voltage across each resistor is the same as the applied voltage.

Key point: In a parallel circuit, the voltage remains the same.

3.11.2 Current in Parallel Circuits

Important point: Ohm's Law states that the current in a circuit is inversely proportional to the circuit resistance. This fact, essential as a fundamental building block of electrical theory, also helps explain current flow in parallel circuits.

In a series circuit, a single current flows. Its value is determined in part by the total resistance of the circuit. However, the source current in a parallel circuit is divided among the available paths according to the value of the resistors in the circuit. Ohm's Law remains unchanged. For a given voltage, current varies inversely with resistance.

The behavior of current in a parallel circuit is best illustrated by example. The example we use is Figure 3.28. The resistors R_1, R_2, and R_3 are parallel to each other and the battery. Each parallel path is then a branch with its current. When the total current I_T leaves the voltage source E, part I_1 of the current I_T will flow through R_1, part I_2 will flow through R_2, and the remainder I_3 will flow R_3. The branches I_1, I_2, and I_3 can be different. However, if a voltmeter (used for measuring the voltage of a circuit) is connected across R_1, R_2, and R_3, the respective voltages E_1, E_2, and E_3 will be equal. Therefore,

$$E = E_1 = E_2 = E_3$$

The total current I_T is equal to the sum of all branch currents.

$$I_T = I_1 = I_2 = I_3 \tag{3.15}$$

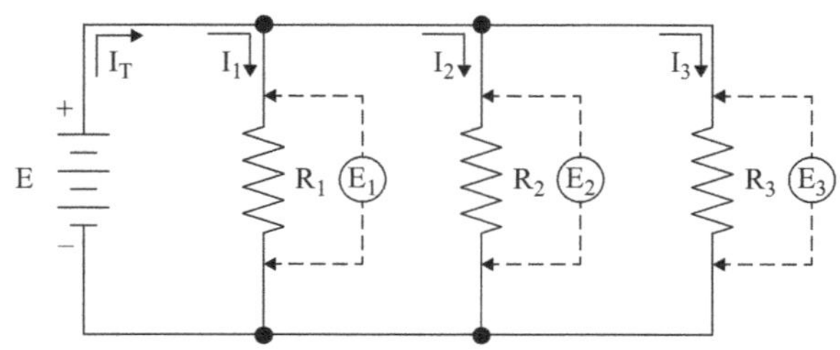

FIGURE 3.28 Parallel circuit.

This formula applies to any number of parallel branches, whether the resistances are equal or unequal.

By Ohm's Law, each branch current equals the applied voltage divided by the resistance between the two points where the voltage is applied. Hence, we have the following equations:

$$\text{Branch 1:} \quad I_1 = \frac{E_1}{R_1} = \frac{E}{R_1}$$

$$\text{Branch 2:} \quad I_2 = \frac{E_2}{R_2} = \frac{E}{R_2}$$

$$\text{Branch 3:} \quad I_3 = \frac{E_3}{R_3} = \frac{V}{R_3}$$

With the same applied voltage, any branch with less resistance allows more current through it than a branch with higher resistance.

EXAMPLE 3.6

Problem:

Two resistors, each drawing 2 amps, and a third resistor drawing 1 amp, are connected in parallel across a 100-V line (see Figure 3.29). What is the total current?

SOLUTION:

The formula for total current is

$$\begin{aligned} I_T &= I_1 + I_2 + I_3 \\ &= 2 + 2 + 1 \\ &= 5\,A \end{aligned}$$

The total current is 5 A.

3.11.3 Parallel Circuits and Kirchhoff's Current Law

The division of current in a parallel network follows a definite pattern. This pattern is described by *Kirchhoff's Current Law*, which is stated as follows:

The algebraic sum of the currents entering and leaving any junction of conductors equals zero. This can be stated mathematically as

$$I_a + I_b + \ldots\ldots + I_n = 0 \tag{3.16}$$

where I_a, I_b +, etc., are the currents entering and leaving the junction. Currents entering the junction are assumed to be positive, and currents leaving the junction are considered negative.

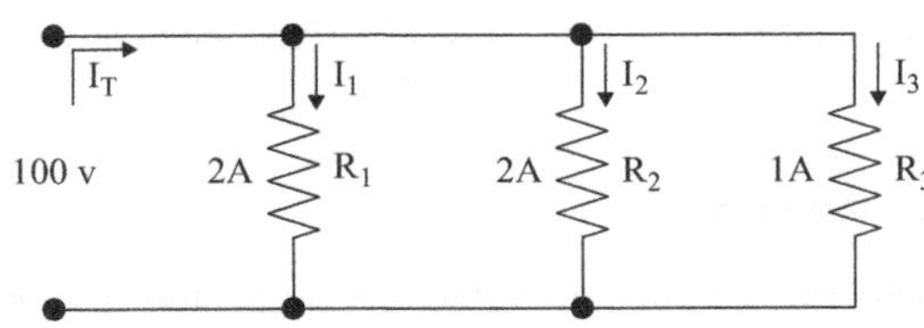

FIGURE 3.29 (for Example 3.6).

3.11.4 Parallel Resistance

Unlike series circuits, where total resistance (R_T) is the sum of the individual resistances, in a parallel circuit, the total resistance is not the sum of the individual resistances.

Ohm's Law can be applied to determine the total resistance in a parallel circuit. We use the equation.

$$R = \frac{R}{I}$$

or

$$R_T = \frac{E_S}{I_t}$$

R_T is the total resistance of all the parallel branches across the voltage source E_S, and I_T is the sum of all the branch currents.

Key point: When two equal-value resistors are connected in parallel, they present a total resistance equivalent to a single resistor of one-half the value of either of the original resistors.

What about parallel circuits containing resistance of unequal value? How is equivalent resistance determined? Example 3.7 demonstrates how this is accomplished.

EXAMPLE 3.7

Problem:

Given:

$$R_1 = 3\ \Omega$$

$$R_2 = 6\ \Omega$$

$$E_a = 30\ V$$

Known:

$$I_1 = 10 \text{ amps}$$

$$I_2 = 5 \text{ amps}$$

$$I_t = 15 \text{ amps}$$

Determine:

$$R_{eq} = ?$$

$$R_{eq} = \frac{E_a}{I_t}$$

$$R_{eq} = \frac{30}{15} = 2\ \Omega$$

Key point: In Example 3.7, the equivalent resistance of 2 Ω is less than the value of either branch resistor. Remember, in parallel circuits, the equivalent resistance will always be smaller than the resistance of any branch.

3.11.5 The Reciprocal Method

When circuits in which resistors of unequal value are connected in parallel are encountered, the equivalent resistance may be computed using the *reciprocal method*.

Note: A *reciprocal* is an inverted fraction; the reciprocal of the fraction 3/4, for example, is 4/3. We consider a whole number to be a fraction with one as the denominator, so the reciprocal of a whole number is that number divided by 1. For example, the reciprocal of R_t is $1/R_t$.

The formula gives equivalent resistance in parallel.

$$\frac{1}{R_T} = \frac{1}{R_1} + \frac{1}{R_2} + \frac{1}{R_3} + \ldots + \frac{1}{R_n} \quad (3.17)$$

where
R_T is the total resistance in parallel, and R_1, R_2, R_3, and R_n are the branch resistances.

EXAMPLE 3.8

Problem:
Find the total resistance of a 2-Ω, a 4-Ω, and an 8-Ω resistor in parallel

SOLUTION:

Write the formula for three resistors in parallel.

$$\frac{1}{R_T} = \frac{1}{R_1} + \frac{1}{R_2} + \frac{1}{R_3}$$

Substituting the resistance values.

$$\frac{1}{R_T} = \frac{1}{2} + \frac{1}{4} + \frac{1}{8}$$

Add fractions.

$$\frac{1}{R_T} = \frac{4}{8} + \frac{2}{8} + \frac{1}{8} = \frac{7}{8}$$

Invert both sides of the equation to solve for R_T.

$$R_T = \frac{8}{7} 1.14\ \Omega$$

Note: When resistances are connected in parallel, the total resistance is always **less** than the smallest resistance of any single branch.

3.11.6 Product over the Sum Method

To find the total resistance of two unequal parallel resistors, multiply their resistances and divide by their sum.

$$R_T = \frac{R_1 \times R_2}{R_1 + R_2} \quad (3.18)$$

where

R_T is the total resistance in parallel, and R_1 and R_2 are the two resistors in parallel.

EXAMPLE 3.9

Problem:

What is the equivalent resistance of a 20-Ω and a 30-Ω resistor connected in parallel?

SOLUTION:

$$\text{Given}: R_1 = 20$$
$$R_2 = 30$$
$$R_T = \frac{R_1 \times R_2}{R_1 + R_2}$$
$$R_T = \frac{20 \times 30}{20 + 30}$$
$$R_T = 12\,\Omega$$

It is often necessary to study basic electricity to resolve a complex circuit into a more straightforward form. Complex circuits with individual resistors can be reduced to a basic equivalent circuit containing the source and total resistance. This process is called reduction to an equivalent circuit.

3.11.7 Power in Parallel Circuits

As in the series circuit, the total *power* consumed in a parallel circuit is equal to the sum of the individual resistors.

Note: Because power dissipation in resistors consists of heat loss, power dissipation is additive regardless of how the resistors are connected to the circuit.

$$P_T = P_1 + P_2 + P_3 + \ldots + P_n \tag{3.19}$$

where

P_T is the total power, and P_1, P_2, P_3, and P_n are the branch powers.

The equation can also calculate the total power

$$P_T = EI_T \tag{3.20}$$

where

P_T is the total power, E is the voltage source across all parallel branches, and I_T is the total current.

The power dissipated in each branch is equal to EI and equal to V^2/R.

Note: In parallel and series arrangements, the sum of the individual values of power dissipated in the circuit equals the total power generated by the source. The circuit arrangements cannot change the fact that all power in the circuit comes from the source.

EXAMPLE 3.10

Total power consumed by the circuit can be calculated as follows:

$$P_{R1} = E_b \times I_{R1}$$
$$P_{R1} = 50 \times 5$$
$$P_{R1} = 250\ \text{W}$$
$$P_{R2} = E_b \times I_{R2}$$
$$P_{R2} = 50 \times 2$$
$$P_{R2} = 100\ \text{W}$$
$$P_{R3} = E_b \times I_{R3}$$
$$P_{R3} = 50 \times 1$$
$$P_{R3} = 50\ \text{W}$$
$$P_T = P_1 + P_2 + P_3$$
$$P_T = 250 + 100 + 50$$
$$P_T = 400\ \text{W}$$

Because, in the example shown, the total current is known, we could determine the total power by the following method:

$$P_T = E_b \times I_T$$
$$P_T = 50\ \text{V} \times 8\text{a}$$
$$P_T = 400\ \text{W}$$

Problems involving determining resistance, voltage, current, and power in a parallel circuit are solved simply as in a series circuit. The procedure is the same–(1) draw a circuit diagram, (2) state the values given and the values to be found, (3) state the applicable equations, and (4) substitute the given values and solve for the unknown.

Along with following the problem-solving procedure above, it is also important to remember and apply the rules for solving parallel D-C circuits. These rules are:

a. The same voltage exists across each parallel circuit branch and is equal to the source voltage.
b. The current through a branch of a parallel network is inversely proportional to the branch's resistance.
c. The total current of a parallel circuit is equal to the sum of the currents of the individual branches of the circuit.
d. The total resistance of a parallel circuit is equal to the reciprocal of the sum of the reciprocals of the individual resistances of the circuit.
e. The total power consumed in a parallel circuit is equal to the sum of the power consumption of the individual resistances.

3.12 SERIES-PARALLEL CIRCUITS

To this point, we have discussed series and parallel D-C circuits. However, the maintenance operator will seldom encounter a circuit that consists solely of either type of circuit. Most circuits consist of both series and parallel elements. A circuit of this type is referred to as a *series-parallel circuit* or as a combination circuit. The solution of a series-parallel (combination) circuit is simply a matter of applying the laws and rules discussed before this point.

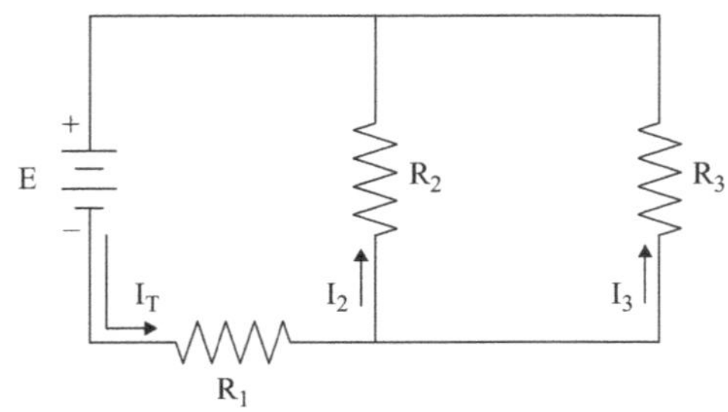

FIGURE 3.30 Series-parallel circuit.

At least three resistors are required to form a series-parallel circuit. An example of a series-parallel circuit is shown in Figure 3.30, where two parallel resistors, R_2 and R_3, are connected in series with the resistor R_1 and the voltage source E. In a circuit of this type, the current I_T divides after it flows through R_1; part flows through R_2, and part flows through R_3. Then, the current joins at the junction of the two resistors and flows back to the positive terminal of the voltage source (E), which goes through the voltage source to the positive terminal.

In solving values in a series-parallel circuit (current, voltage, and resistance), follow the rules that apply to a series circuit for the series part of the circuit and the rules that apply to a parallel circuit for the parallel part of the circuit. Solving series-parallel circuits is simplified if all parallel and series groups are first reduced to single equivalent resistances and the circuits redrawn in simplified form. Recall, the redrawn circuit is called an **equivalent circuit**.

Note: There are no general formulas for solving series-parallel circuits because there are so many different forms of these circuits.

Important note: The total current in the series-parallel circuit depends on the effective resistance of the parallel portion **and** the other resistances.

3.13 CONDUCTORS

Recall that we pointed out earlier that electric current moves quickly through some materials but with more incredible difficulty through others. Three good electrical conductors are copper, silver, and aluminum (generally, we can say that most metals are good conductors). At present, copper is the material of choice used in electrical conductors. Under special conditions, certain gases are also used as conductors (e.g., neon gas, mercury vapor, and sodium vapor are used in various lamps).

The function of the wire conductor is to connect a source of applied voltage to a load resistance with a minimum IR voltage drop in the conductor so that most of the applied voltage can produce current in the load resistance. Ideally, a conductor must have a very low resistance (e.g., a typical value for a conductor—copper—is less than 1 Ω per 10 ft).

Because all electrical circuits utilize conductors of one type or another, this section discusses the basic features and electrical characteristics of the most common types of conductors.

A conductor's standard (or unit size) has been established to compare the resistance and size of one conductor with another. The unit of linear measurement used is (regarding the diameter of a piece of wire) the **mil** (0.001 of an inch). A convenient unit of wire length used is the **foot**. Thus, in most cases, the standard unit of size is the **mil-foot** (i.e., a wire will have unit size if it has a diameter of 1 mil and a length of 1 ft). The resistance in ohms of a unit conductor or a given substance is called the **resistivity** (or specific resistance) of the substance.

For further convenience, **gauge** numbers are also used to compare the diameter of wires. The B and S (Browne and Sharpe) gauge was used in the past; now, the most commonly used gauge is the **American Wire Gauge** (AWG).

The *square mil* is a convenient unit of cross-sectional area for square or rectangular conductors.

EXAMPLE 3.11

Problem:

Find the cross-sectional area of a large rectangular conductor 5/8 inch thick and 5 inches wide.

SOLUTION:

The thickness may be expressed in mils as 0.625 × 1,000 = 625 mils and the width as 5 × 1,000 = 5,000 mils. The cross-sectional area is calculated as 625 multiplied by 5,000, resulting in 3,125,000 square mils.

The *circular mil* is the standard unit of wire cross-sectional area used in most wire tables. Expressing these diameters in mils is convenient to avoid decimals (because most wires used to conduct electricity may be only a tiny fraction of an inch). For example, the diameter of a wire is described as 25 mils instead of 0.025 inches. A circular mil is the area of a circle having a diameter of 1 mil. The area in circular mils of a round conductor is obtained by squaring the diameter measured in mils. Thus, wires having a diameter of 25 mils have an area of 25^2 or 625 circular mils. By way of comparison, the basic formula for the area of a circle is

$$A = \pi R^2 \tag{3.21}$$

and in this example, the area in square inches is

$$A = \pi R^2 = 3.14(0.0125)^2 = 0.00049 \text{sq.in.}$$

If D is the diameter of the wire in mils, the area of the wire in square mils can be determined.

$$A = \pi(D / 2)^2 \tag{3.22}$$

which translates to

$$= 3.14 / 4D^2$$
$$= 0.785D^2 \text{ sq.mils}$$

Thus, a wire one mil in diameter has an area of

$$A = 0.785 \times 1^2 = 0.785 \text{sq.mils},$$

which is equivalent to 1 circular mil. The cross-sectional area of a wire in circular mils is therefore determined as

$$A = \frac{0.785D_2}{0.785} = D^2 \text{ circular mil},$$

where D is the diameter in mils, the constant π/4 is eliminated from the calculation.

Note that in comparing square and round conductors, the circular mil is a smaller unit of area than the square mil, and therefore, there are more circular mils than square mils in any given area.

Important point: To determine the circular-mil area when the square-mil area is given, divide the area into square mils by 0.785. Conversely, to determine the square-mil area when the circular-mil area is provided, multiply the area in circular mils by 0.785.

EXAMPLE 3.12

Problem:

A No. 12 wire has a diameter of 80.81 mils. What is (1) its area in circular mils and (2) square mils?

SOLUTION:

1. $A = D^2 = 80.81^2 = 6{,}530$ circular mils
2. $A = 0.785 \times 6{,}530 = 5{,}126$ square mils

Note: Many electric cables are composed of stranded wires. The strands are usually single wires twisted together in sufficient numbers to make up the necessary cross-sectional area of the cable. The total area in circular mils is determined by multiplying the area of one strand in circular mils by the number of strands in the cable.

3.13.1 Circular-Mil-Foot

A *circular-mil foot* is a unit of volume. Specifically, it is a unit conductor 1 ft long with a cross-sectional area of 1 circular mil. The circular mil foot helps compare wires of different metals because it is considered a unit of conductor. Because it is regarded as a conductor unit, the circular mil foot helps compare wires of different metals. For example, a basis for comparing the resistivity of various substances may be made by determining the resistance of a circular mil-foot of each substance.

Note: It is sometimes more convenient to employ a different unit of volume when working with certain substances. Accordingly, unit Volume may also be taken as the centimeter cube. The inch cube may also be used. The unit of volume employed is given in tables of specific resistances.

3.13.2 Resistivity

All materials differ in their atomic structure and, therefore, in their ability to resist the flow of an electric current. The measure of the ability of a specific material to resist the flow of electricity is called its *Resistivity*, or specific resistance—the resistance in ohms offered by unit volume (the circular-mil-foot) of a substance to the flow of electric current. Resistivity is the reciprocal of conductivity (i.e., the ease with which current flows in a conductor). A substance with high resistivity will have low conductivity, and vice versa.

The resistance of a given length, for any conductor, depends upon the resistivity of the material, the size of the wire, and the cross-sectional area of the wire according to the equation.

$$R = \rho \frac{L}{A} \tag{3.23}$$

where R = resistance of the conductor, Ω

L = length of the wire, ft

A = cross-sectional area of the wire, CM

ρ = specific resistance or resistivity, CM × Ω/ft

The factor ρ (Greek letter rho, pronounced "roe") permits different materials to be compared for resistance according to their nature without regard to different lengths or areas. Higher values mean more resistance.

Key point: The resistivity of a substance is the resistance of a unit volume of that substance.

TABLE 3.3
Resistivity (Specific Resistance)

Substance	Specific Resistance @ 20° (CM ft Ω)
Silver	9.8
Copper (drawn)	10.37
Gold	14.7
Aluminum	17.02
Tungsten	33.2
Brass	42.1
Steel (soft)	95.8
Nichrome	660.0

Many tables of resistivity are based on the resistance in ohms of a volume of the substance 1 ft long and 1 circular mil in cross-sectional area. The temperature at which the resistance measurement is made is also specified. If the kind of metal from which the conductor is made is known, the resistivity of the metal may be obtained from a table. The resistivity, or specific resistance, of some common substances is given in Table 3.3.

Note: Since silver, copper, gold, and aluminum have the lowest resistivity values, they are the best conductors. Tungsten and iron have much higher resistivity.

EXAMPLE 3.13

Problem:

What is the resistance of 1,000 ft of copper wire with a cross-sectional area of 10,400 circular mils (No. 10 wire) and a wire temperature of 20°C?

SOLUTION:

The resistivity (specific resistance), from Table 3.4, is 10.37. Substituting the known values in the preceding Equation (3.23), the resistance, R, is determined as

$$R = \rho \frac{L}{A} = 10.37 \times \frac{1,000}{10,400} = 1\,\Omega\text{, approximately}$$

3.13.3 Wire Measurement

Wires are manufactured in sizes numbered according to a table known as the American Wire Gauge (AWG). Table 3.4 lists the standard wire sizes that correspond to the AWG. The gauge numbers specify the size of round wire in terms of its diameter and cross-sectional area. Note the following:

a. As the gauge numbers increase from 1 to 40, the diameter and circular area decrease. Higher gauge numbers mean smaller wire sizes. Thus, No. 12 is a smaller wire than No. 4.
b. The circular area doubles for every three-gauge size. For example, No. 12 wire has about twice the area of No. 15.
c. The higher the gauge number and the smaller the wire, the better the resistance of the wire for any given length. Therefore, for 1,000 ft of wire, No. 12 has a resistance of 1.62, while No. 4 has 0.253 Ω.

TABLE 3.4
Copper Wire Table

Gage #	Diameter	Circular mils	Ohms/1,000 ft @ 25°C
1	289.0	83,700.0	0.126
2	258.0	66,400.0	0.159
3	229.0	52,600.0	0.201
4	204.0	41,700.0	0.253
5	182.0	33,100.0	0.319
6	162.0	26,300.0	0.403
7	144.0	20,800.0	0.508
8	128.0	16,500.0	0.641
9	114.0	13,100.0	0.808
10	102.0	10,400.0	1.02
11	91.0	8,230.0	1.28
12	81.0	6,530.0	1.62
13	72.0	5,180.0	2.04
14	64.0	4,110.0	2.58
15	57.0	3,260.0	3.25
16	51.0	2,580.0	4.09
17	45.0	2,050.0	5.16
18	40.0	1,620.0	6.51
19	36.0	1,290.0	8.21
20	32.0	1,020.0	10.4
21	28.5	810.0	13.1
22	25.3	642.0	16.5
23	22.6	509.0	20.8
24	20.1	404.0	26.4
25	17.9	320.0	33.0
26	15.9	254.0	41.6
27	14.2	202.0	52.5
28	12.6	160.0	66.2
29	11.3	127.0	83.4
30	10.0	101.0	105.0
31	8.9	79.7	133.0
32	8.0	63.2	167.0
33	7.1	50.1	211.0
34	6.3	39.8	266.0
35	5.6	31.5	335.0
36	5.0	25.0	423.0
37	4.5	19.8	533.0
38	4.0	15.7	673.0
39	3.5	12.5	848.0
40	3.1	9.9	1,070.0

3.13.4 Factors Governing Selection of Wire Size

When selecting the wire size to transmit and distribute electric power, several factors must be considered. These factors include the allowable power loss in the line, the permissible voltage drop in the line, the current-carrying capacity of the line, and the ambient temperatures in which the wire is to be used.

a. *Allowable power loss (I^2R) in the line:* This loss represents electrical energy converted into heat. Using large conductors will reduce resistance and, therefore, the I^2R loss. However, large conductors are heavier and require more substantial support; thus, they are more expensive initially than small ones.

b. *Permissible voltage drop (IR drop) in the line:* If the source maintains a constant voltage at the input to the line, any variation in the load on the line will cause a variation in line current and a consequent variation in the IR drop in the line. A wide variation in the IR drop in the line causes poor voltage regulation at the load.
c. *The current-carrying capacity of the line:* When the current is drawn through the line, heat is generated. The temperature of the line will rise until the heat radiated, or otherwise dissipated, is equal to the heat generated by the passage of current through the line. If the conductor is insulated, the heat generated in the conductor is not so readily removed as it would be if the conductor were not insulated.
d. *Conductors installed where the ambient temperature is relatively high:* When installed in such surroundings, the heat generated by external sources constitutes an appreciable part of the total conductor heating. Due allowance must be made for the influence of external heating on the allowable conductor current, and each case has its specific limitations.

3.14 ELECTROMAGNETISM

Earlier, we discussed the fundamental theories concerning simple magnets and magnetism. Those discussions dealt with forms of magnetism that were not related to electricity—permanent magnets, for instance. Furthermore, only a brief mention was made of those forms of magnetism related to electricity, producing electricity with magnetism, for example.

Anatomy and physiology are so closely related that medical students cannot study one at a time without involving the other. A similar relationship holds for the electrical field; magnetism and basic electricity are so closely related that one cannot be studied at length without involving the other. This close fundamental relationship is continually borne out in the study of generators, transformers, battery packs, and motors. To be proficient in electricity, we must become familiar with the general relationships that exist between magnetism and electricity as follows:

a. Electric current flow will always produce some form of magnetism.
b. Magnetism is the most commonly used means for producing or using electricity.
c. Magnetic influences cause the peculiar behavior of electricity under certain conditions.

3.14.1 Magnetic Field around a Single Conductor

In 1819, Hans Christian Oersted, a Danish scientist, discovered that a field of magnetic force exists around a single wire conductor carrying an electric current. In Figure 3.31, a wire is

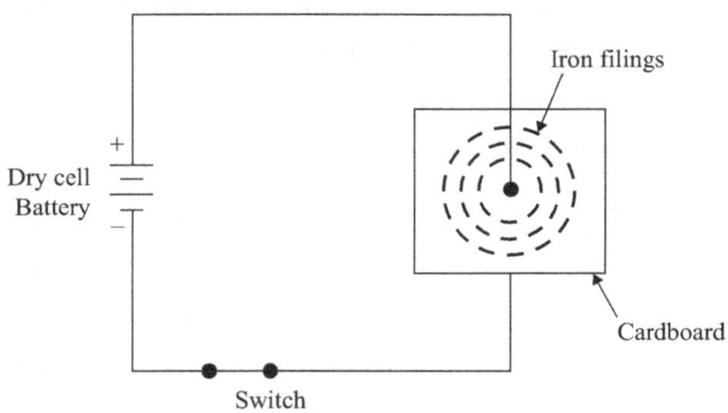

FIGURE 3.31 A circular pattern of magnetic force exists around a wire carrying an electric current.

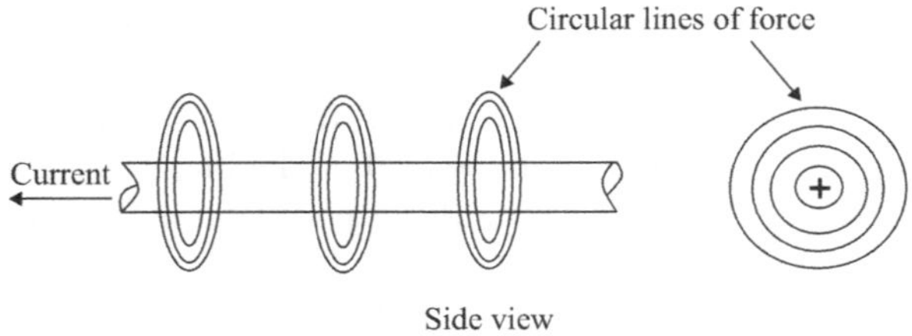

FIGURE 3.32 The circular force fields around a wire carrying a current are in planes perpendicular to the wire.

passed through a piece of cardboard and connected through a switch to a dry cell. With the switch open (no current flowing), if we sprinkle iron filings on the cardboard and tap it gently, the filings will fall back haphazardly. If we close the switch, the current will flow in the wire. If we tap the cardboard again, the magnetic effect of the current in the wire will cause the filings to fall back into a definite pattern of concentric circles with the wire as the center of the circles. Every section of the wire has this field of force around it in a plane perpendicular to the wire, as shown in Figure 3.32

The ability of the magnetic field to attract bits of iron (as demonstrated in Figure 3.33) depends on the number of lines of force present. The strength of the magnetic field around a wire carrying a current depends on the current, since it is the current that produces the field. The greater the current, the greater the strength of the field. A large current will create many lines of force extending far from the wire, while a small current will produce only a few lines close to the wire, as shown in Figure 3.33.

The relation between the direction of the magnetic lines of force around a conductor and the direction of current flow along the conductor may be determined using the **left-hand rule for a conductor**. If the conductor is grasped in the left hand with the thumb extended in the direction of electron flow (– to +), the fingers will point in the direction of the magnetic lines of force. This is the same direction that the north pole of a compass would point if the compass were placed in the magnetic field.

Important note: Arrows are used in electrical diagrams to denote the direction of current flow along the length of the wire. A unique view of the arrow is used, where cross sections of wire are shown. A cross-sectional view of a conductor that is carrying current toward the observer is illustrated in Figure 3.34(A). The direction of the current is indicated by a dot, which represents the head of the arrow. A conductor carrying current away from the observer is illustrated in Figure 3.34(B). The direction of the current is indicated by a cross, which represents the arrow's tail.

When two parallel conductors carry current in the same direction, the magnetic fields tend to encircle both conductors, drawing them together with a force of attraction, as shown

FIGURE 3.33 The strength of the magnetic field around a wire carrying a current depends on the amount of current.

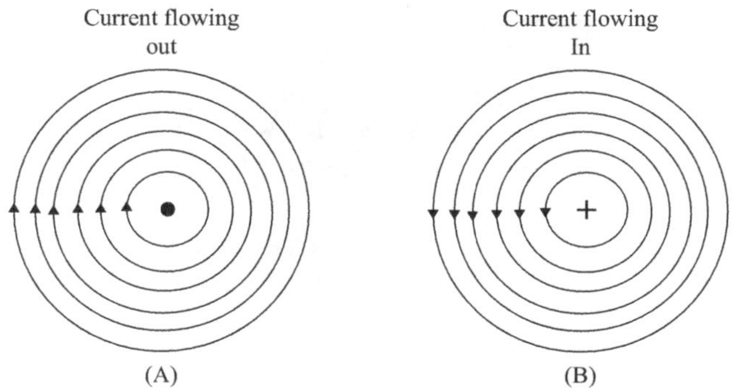

FIGURE 3.34 Magnetic field around a current-carrying conductor.

in Figure 3.35(A). Two parallel conductors carrying currents in opposite directions are shown in Figure 3.35(B). The field around one conductor is opposite to the field around the other. The resulting lines of force are crowded together in the space between the wires and tend to push the wires apart. Therefore, two parallel adjacent conductors carrying currents in the same direction attract each other, and two parallel conductors carrying currents in opposite directions repel each other.

3.14.2 Magnetic Field of a Coil

The magnetic field around a current-carrying wire exists at all points along its length. Bending the current-carrying wire into a single loop has two results. First, the magnetic field consists of more dense concentric circles in a plane perpendicular to the wire, the number of lines is the same as for

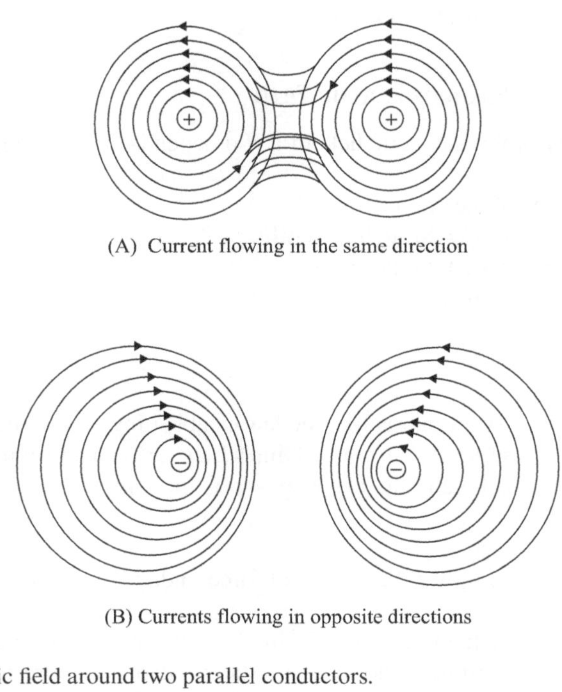

FIGURE 3.35 Magnetic field around two parallel conductors.

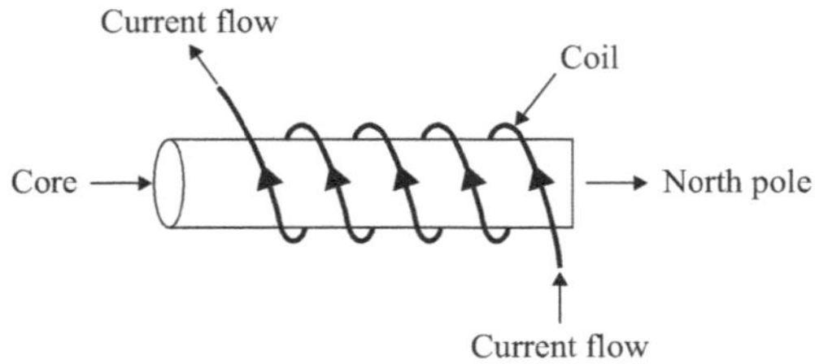

FIGURE 3.36 Current-carrying coil.

the straight conductor. Second, all the lines inside the loop are in the same direction. When this straight wire is wound around a core, as is shown in Figure 3.36, it becomes a coil, and the magnetic field assumes a different shape. When current is passed through the coiled conductor, the magnetic field of each turn of wire links with the fields of adjacent turns. The combined influence of all the turns produces a two-pole field like a simple bar magnet. One end of the coil will be a north pole, and the other will be a south pole.

In Figure 3.34, it was shown that the direction of the magnetic field around a straight conductor depends on the direction of current flow through that conductor. Thus, a reversal of current flow through a conductor causes a reversal in the direction of the produced magnetic field. It follows that a reversal of the current flow through a coil also causes a reversal of its two-pole field. This is true because that field is the product of the linkage between the individual turns of wire on the coil. Therefore, if the field of each turn is reversed, the total field (coils' field) is also reversed.

When the direction of electron flow through a coil is known, its polarity may be determined using the **left-hand rule for coils**. This rule is illustrated in Figure 3.36 and is stated as follows: Grasping the coil in the left hand, with the fingers "wrapped around" in the direction of electron flow, the thumb will point toward the north pole.

3.14.3 Strength of an Electromagnetic Field

The strength or intensity of the magnetic field of a coil depends on several factors.

- The <u>number of turns</u> of the conductor.
- The <u>amount of current flow</u> through the coil.
- The ratio of the coil's length to its width.
- The type of material in the core.

3.14.4 Magnetic Units

The current flow in the electric circuit is like the law establishing flux in the magnetic circuit.

The *magnetic flux* (phi) is like the current in Ohm's Law formula and comprises the total number of lines of force in the magnetic circuit. The Maxwell is the unit of flux—that is, 1 line of force is equal to 1 Maxwell.

Note: The Maxwell is often simply called a line of force, induction, or line.

The *strength* of a magnetic field in a coil of wire depends on how much current flows in the coil turns; the more current, the stronger the magnetic field. Also, the more turns, the more concentrated the lines of force are. The *force* that produces the flux in the magnetic circuit (comparable to electromotive force in Ohm's Law) is known as *magnetomotive force* or mmf. The practical unit of magnetomotive force is the **ampere-turn** (At). In equation form,

$$F = \text{ampere} - \text{turns} = NI \tag{3.24}$$

where
F = magnetomotive force, At
N = number of turns
I = current, A

EXAMPLE 3.14

Problem:
Calculate the ampere-turns for a coil with 2,000 turns and a 5-Ma current.

SOLUTION:

Use Equation (3.21) and substitute N = 2,000 and I = 5 × 10–³ A.

$$NI = 2000(5 \times 10^{-3}) = 10\,At$$

The *intensity* of magnetizing force per unit of length is designated as H and sometimes expressed as Gilberts per centimeter. Expressed as an equation,

$$H = \frac{NI}{L} \tag{3.25}$$

where
H = magnetic field intensity, ampere-turns per meter (At/m)
NI = ampere-turns, At
L = length between poles of the coil, m

Note: Equation (3.25) is for a solenoid. *H* is the intensity of an air core. With an iron core, *H* is the intensity through the entire core, and *L* is the length or distance between poles of the iron core.

3.14.5 Properties of Magnetic Materials

This section discusses two important properties of magnetic materials: permeability and hysteresis.

3.14.5.1 Permeability

When the core of an electromagnet is made of annealed sheet steel, it produces a stronger magnet than if a cast iron core is used. This is the case because the magnetizing force of the coil more readily acts upon annealed sheet steel than the hard cast iron. Simply put, soft sheet steel is said to have better *permeability* because of its greater ease with which magnetic lines are established.

Recalling that permeability is the relative ease with which a substance conducts magnetic lines of force. The permeability of air is arbitrarily set at 1. The permeability of other substances is the ratio of their ability to conduct magnetic lines to that of air. The permeability of nonmagnetic materials, such as aluminum, copper, wood, and brass, is unity, or the same as for air.

Important note: The permeability of magnetic materials varies with the degree of magnetization, being more negligible for high flux density values. *Reluctance*—analogous to resistance, which is the opposition to the production of flux in a material—is inversely proportional to permeability.

Iron has high permeability and, therefore, low reluctance. Air has low permeability and, hence, high reluctance.

3.14.6 Hysteresis

When the current in a coil of wire reverses thousands of times per second, a considerable energy loss can occur. This energy loss is caused by *hysteresis*, which means "lagging behind." That is, the magnetic flux in an iron core lags behind the increases or decreases of the magnetizing force.

3.14.7 Electromagnets

An *electromagnet* is composed of a coil of wire wound around a normally soft iron core because of its high permeability and low hysteresis. When direct current flows through the coil, the core becomes magnetized with the same polarity that the coil would have without the core. If the current is reversed, the coil's and core's polarity are reversed.

Electromagnets are essential in electricity because magnetism can be "turned on" or "turned off" at will. A good example is the starter solenoid (an electromagnet) in automobiles and power boats. In a car or boat, an electromagnet is part of a relay that connects the battery to the induction coil, generating the high voltage needed to start the engine. The starter solenoid isolates this high voltage from the ignition switch. When no current flows in the coil, it is an "air core," but a movable soft-iron core does two things when the coil is energized—first, the magnetic flux increases because the soft-iron core is more permeable than the air core. Second, the flux is more concentrated. All this concentration of magnetic lines of force in the soft-iron core results in a perfect magnet when current flows in the coil. But soft iron loses its magnetism quickly when the current is shut off. The effect of soft iron is the same whether it is movable, as in some solenoids, or permanently installed in the coil. An electromagnet consists of a coil and a core; it becomes a magnet when current flows through the coil.

The ability to control the action of magnetic force makes an electromagnet particularly useful in many circuit applications, many of which are discussed in this manual.

3.15 A-C THEORY

Because voltage is induced in a conductor when lines of force are cut, the induced electromotive force (emf) amount depends on the number of lines cut in a unit of time. To induce an emf of 1 V, a conductor must cut 100,000,000 lines of force per second. To obtain this significant number of "cuttings," the conductor is formed into a loop and rotated on an axis at great speed (see Figure 3.37). The two sides of the loop become individual conductors in series, each side cutting lines of force and inducing twice the voltage that a single conductor would induce. In commercial generators, the

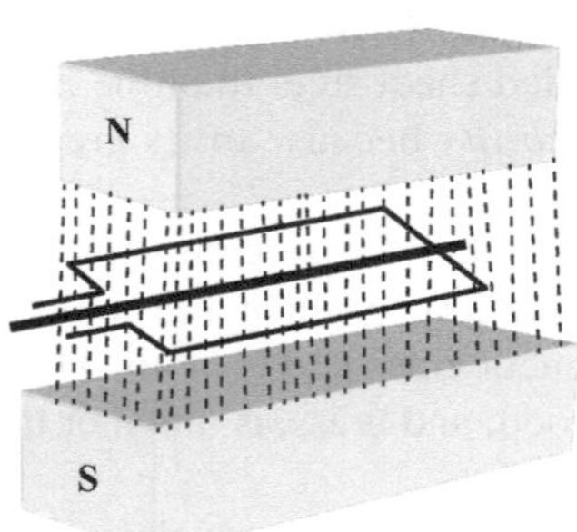

FIGURE 3.37 Loop rotating in a magnetic field produces an A-C voltage.

number of "cuttings" and the resulting emf are increased by (1) increasing the number of lines of force by using more magnets or stronger electromagnets, (2) using more conductors or loops, and (3) rotating the loops faster. [Note: both A-C and D-C generators are covered later.]

How an A-C generator operates to produce an A-C voltage and current is a fundamental concept today, taught in elementary and middle school science classes. Of course, we accept technological advances as commonplace today—we surf the internet, text our friends, watch cable television, use our cell phones, and take outer space flight as a given, and consider producing the electricity that makes all these technologies possible as our right. These technologies are bottom shelf to us today—we have them available to us, so we simply use them—no big deal, right? Not worth thinking about. This point of view surely was not held initially, especially by those who broke ground in developing technology and electricity.

In the groundbreaking years of electric technology development, the geniuses of the science of electricity (including George Simon Ohm) performed their technological breakthroughs in faltering steps. We tend to forget that those first faltering steps of scientific achievement in electricity were performed with crude, and for the most part, homemade apparatus. (It sounds like the more contemporary young garage-and-basement inventors who came up with the first basic user-friendly microcomputer and software packages, and hopefully the renewable energy innovations of tomorrow.)

Indeed, the innovators of electricity had to fabricate nearly all the laboratory equipment used in their experiments. At the time, the only convenient source of electrical energy available to these early scientists was the voltaic cell, invented some years earlier. Because cells and batteries were the only sources of power available, some of the early electrical devices were designed to operate from **direct current**.

Thus, initially, direct current was used extensively. However, when the use of electricity became widespread, certain disadvantages in the use of direct current became apparent. In a direct-current system, the supply voltage must be generated at the level required by the load. To operate a 240-V lamp, for example, the generator must deliver 240 V. A 120-V lamp could not be operated from this generator by any convenient means. A resistor could be placed in series with the 120-V lamp to drop the extra 120 V, but the resistor would waste an amount of power equal to that consumed by the lamp.

Another disadvantage of direct-current systems is the large amount of power lost due to the transmission wires' resistance to carrying current from the generating station to the consumer. Operating the transmission line at remarkably high voltage and low current could significantly reduce this loss. However, this is not a practical solution in a D-C system since the load must also operate at high voltage. As a result of the difficulties encountered with direct current, all modern power distribution systems use **alternating current (A-C).**

Unlike D-C voltage, A-C voltage can be stepped up or down by a transformer device. As shown in Figure 3.39, an A-C voltage and current can be produced when a conductor loop rotates through a magnetic field and cuts lines of force to generate an induced A-C voltage across its terminals. This describes the basic principle of operation of an alternating current generator or **alternator**. An alternator converts mechanical energy into electrical energy. It does this by utilizing the principle of **electromagnetic induction**. The basic components of an alternator are an armature, about which many turns of the conductor are wound, which rotates in a magnetic field, and some means of delivering the resulting alternating current to an external circuit.

[Note: We cover generator construction in more detail; in this section, we concentrate on the theory of operation].

3.15.1 Cycle

An A-C voltage changes in magnitude and periodically reverses in polarity. The zero axis is a horizontal line across the center. The vertical variations in the voltage wave show the changes in magnitude. The voltages above the horizontal axis have positive (+) polarity, while voltages below the horizontal axis have negative (−) polarity.

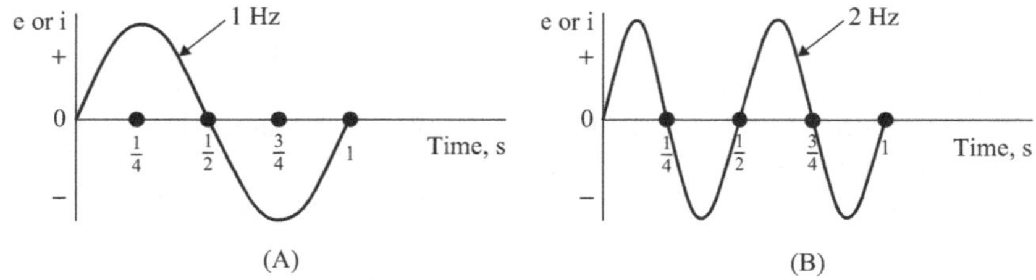

FIGURE 3.38 Comparison of frequencies.

3.15.2 Frequency, Period, and Wavelength

The *frequency* of an alternating voltage or current is the number of complete cycles occurring each second. It is indicated by the symbol f and is expressed in hertz (Hz). One cycle per second equals one hertz. Thus, 60 cycles per second (cps) equals 60 Hz. A frequency of 2 Hz [Figure 3.38(B)] is twice the frequency of 1 Hz [Figure 3.38(A)].

The period is the amount of time it takes to complete one cycle. It is indicated by the symbol T for time and expressed in seconds (s). Frequency and period are reciprocals.

$$f = \frac{1}{\mathrm{T}} \tag{3.26}$$

$$\mathrm{T} = \frac{1}{f} \tag{3.27}$$

Important point: The higher the frequency, the shorter the period.

The angle of 360° represents the time for one cycle, or the period T. So, we can show the sine wave's horizontal axis is in either electrical degrees or seconds (see Figure 3.39).

The *wavelength* is the length of one complete wave or cycle. It depends upon the frequency of the periodic variation and its transmission velocity. It is indicated by the symbol λ (Greek lowercase lambda). Expressed as a formula:

$$\lambda = \frac{\text{velocity}}{\text{frequency}} \tag{3.28}$$

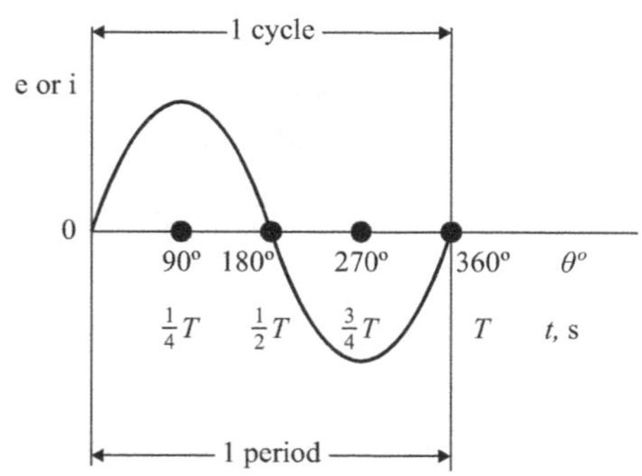

FIGURE 3.39 Relationship between electrical degrees and time.

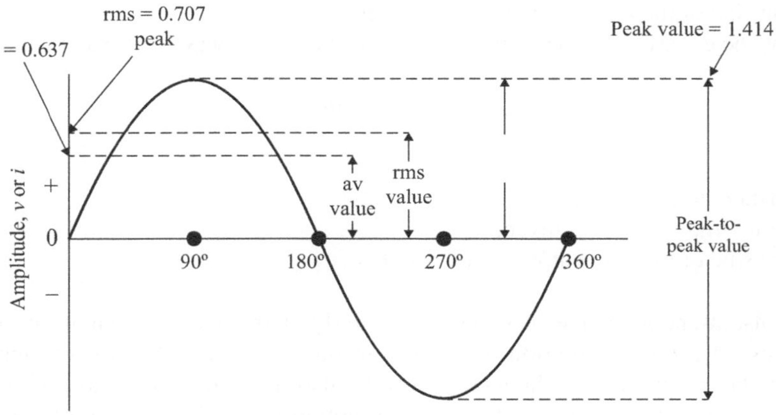

FIGURE 3.40 Amplitude values for A-C sine wave.

3.16 CHARACTERISTIC VALUES OF A-C VOLTAGE AND CURRENT

Because an A-C sine wave voltage or current has many instantaneous values throughout the cycle, it is convenient to specify magnitudes for comparing one wave with another. The peak, average, or root-mean-square (rms) value can be determined (see Figure 3.40). These values apply to current or voltage.

3.16.1 Peak Amplitude

One of a sine wave's most frequently measured characteristics is its amplitude. Unlike D-C measurement, the amount of alternating current or voltage in a circuit can be measured in various ways. One measurement method measures the maximum amplitude of either the positive or the negative alternation. The value of the current or voltage obtained is called the *peak voltage* or the *peak current*. An oscilloscope must be used to measure the peak value of current or voltage. The peak value is described in Figure 3.40.

3.16.2 Peak-to-Peak Amplitude

A second method of indicating the amplitude of a sine wave consists of determining the total voltage or current between the positive and negative peaks. This current or voltage value is the *peak-to-peak value* (see Figure 3.40). Because both alternations of a pure sine wave are identical, the peak-to-peak value is twice the peak value. Peak-to-peak voltage is usually measured with an oscilloscope, although some voltmeters calibrate a unique scale in peak-to-peak volts.

3.16.3 Instantaneous Amplitude

The *instantaneous value* of a sine wave of voltage for any angle of rotation is expressed by the formula:

$$e = E_m \times \sin\theta \tag{3.29}$$

where

e = the instantaneous voltage

E_m = the maximum or peak voltage

$\sin\theta$ = the sine of the angle at which e is desired.

Similarly, the equation for the instantaneous value of a sine wave of current would be:

$$i = I_m \times \sin\theta \tag{3.30}$$

where

i = the instantaneous current

I_m = the maximum or peak current

$\sin\theta$ = the sine of the angle at which i desired

Note: The instantaneous value of voltage constantly changes as the armature of an alternator moves through a complete rotation. Because current varies directly with voltage, according to Ohm's Law, the instantaneous changes in current also result in a sine wave whose positive and negative peaks and intermediate values can be plotted precisely as we plotted the voltage sine wave. However, instantaneous values do not help solve most A-C problems, so a practical value is used.

3.16.4 Effective or RMS Value

The *effective value* of an A-C voltage or current of a sine waveform is defined in terms of the equivalent heating effect of a direct current. The heating effect is independent of the direction of current flow.

Important point: Because all instantaneous values of induced voltage are somewhere between zero and E_M (maximum or peak voltage), the practical value of a sine wave voltage or current must be greater than zero and less than EM (the maximum or peak voltage).

The alternating current of the sine waveform, having a maximum value of 14.14 amps, produces the same amount of heat in a circuit having a resistance of 1 Ω as a direct current of 10 amps. Because this is true, we can calculate, having a maximum value of 14.14 amps, a constant value to convert any peak value to a corresponding effective value. X represents this constant in the simple equation below. Solve for X to three decimal places.

$$14.14\,X = 10$$

$$X = 0.707$$

The effective value is also called the *root-mean-square (rms)* value because it is the square root of the average of the squared values between zero and the maximum. The effective value of an A-C current is stated in terms of an equivalent D-C current. The phenomenon used as the standard comparison is the heating effect of the current.

Important point: Anytime an A-C voltage or current is stated without qualifications, it is assumed to be an effective value.

In many instances, converting from effective to peak or vice versa is necessary using a standard equation. Figure 3.40 shows that the peak value of a sine wave is 1.414 times the effective value; therefore, the equation we use is:

$$E_m = E \times 1.414 \tag{3.31}$$

where

E_m = maximum or peak voltage

E = effective or RMS voltage

and

$$I_m = I \times 1.414 \tag{3.32}$$

where

I_m = maximum or peak current

I = effective or RMS current

Occasionally, converting a peak current or voltage value to an effective value is necessary. This is accomplished by using the following equations:

$$E = E_m \times 0.707 \tag{3.33}$$

where

E = effective voltage

E_m = the maximum or peak voltage

$$I = I_m \times 0.707 \tag{3.34}$$

where

I = the effective current

I_m = the maximum or peak current

3.16.5 Average Value

Because the positive alternation is identical to the negative alternation, the *average value* of a complete cycle of a sine wave is zero. However, it is necessary to compute the average value of one alternation in certain circuits. Figure 3.40 shows that the average value of a sine wave is 0.637× the peak value and, therefore:

$$\text{Average value} = 0.637 \times \text{peak value} \tag{3.35}$$

cr

$$E_{avg} = E_m \times 0.637$$

where

E_{avg} = the average voltage of one alternation

E_m = the maximum or peak voltage

similarly

$$I_{avg} = I_m \times 0.637 \tag{3.36}$$

where

I_{avg} = the average current in one alternation

I_m = the maximum or peak current

TABLE 3.5
A-C Sine Wave Conversion Table

Multiply the Value	By	To Get the Value
Peak	2	Peak-to-peak
Peak-to-peak	0.5	Peak
Peak	0.637	Average
Average	1.637	Peak
Peak	0.707	RMS (effective)
RMS (effective)	1.414	Peak
Average	1.110	RMS (effective)
RMS (effective)	0.901	Average

Table 3.5 lists the various values of sine wave amplitude used to multiply in the conversion of A-C sine wave voltage and current.

3.17 RESISTANCE IN A-C CIRCUITS

If a sine wave of voltage is applied to a resistance, the resulting current will also be a sine wave. This follows Ohm's Law, which states that the current is directly proportional to the applied voltage. Figure 3.41 shows a sine wave of voltage and the resulting sine wave of current superimposed on the same time axis. Notice that as the voltage increases in a positive direction, the current increases along with it. When the voltage reverses direction, the current reverses direction. At all times, the voltage and current pass through the same relative parts of their respective cycles simultaneously. When two waves, such as those shown in Figure 3.41, are precisely in step, they are said to be **in phase**. To be in phase, the two waves must go through their maximum and minimum points simultaneously and in the same direction.

In some circuits, several sine waves can be in phase. Thus, it is possible to have two or more voltage drops in phase with each other and in phase with the circuit current.

Note: It is important to remember that Ohm's Law for DC circuits applies only to AC circuits with resistance.

Voltage waves are not always in phase. For example, Figure 3.44 shows a voltage E_1 considered to start at 0° (time 1). As voltage wave E_1 reaches its positive peak, a second voltage wave E_2 rises (time 2). Because these waves do not go through their maximum and minimum points at the same

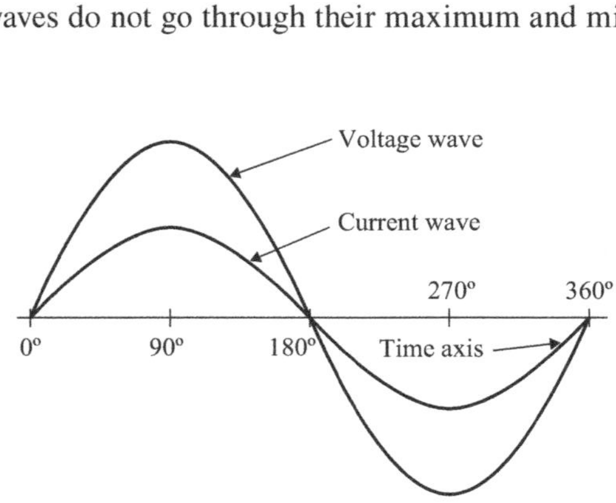

FIGURE 3.41 Voltage and current waves in phase.

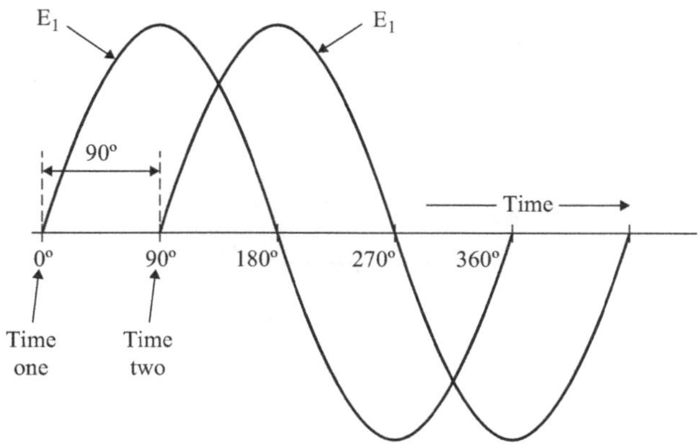

FIGURE 3.42 Voltage waves 90° out of phase.

instant of time, a **phase difference** exists between the two waves. The two waves are said to be out of phase. For the two waves in Figure 3.42, this phase difference is 90°.

3.18 PHASE RELATIONSHIP

In the preceding section, we discussed the essential concepts of **phase** and **phase difference**. Another vital phase concept is the phase angle. The angular difference at a given instant is the phase angle between two waveforms of the same frequency. For example, the phase angle between waves B and A (see Figure 3.43) is 90°. Take an instant of time at 90°. The horizontal axis is shown in angular units of time. Wave B starts at maximum value and reduces to zero at 90°, while Wave A begins at critical zero and increases to maximum at 90°. Wave B reaches its maximum value of 90° ahead of wave A, so wave B **leads** wave A by 90° (and wave A **lags** wave B by 90°). This 90-phase angle between Waves B and A is maintained throughout the complete and successive cycles. At any instant, wave B has the value that wave A will have 90° later. Wave B is a cosine wave displaced 90° from Wave A, a sine wave.

Important point: The amount by which one wave leads or lags another is measured in degrees.

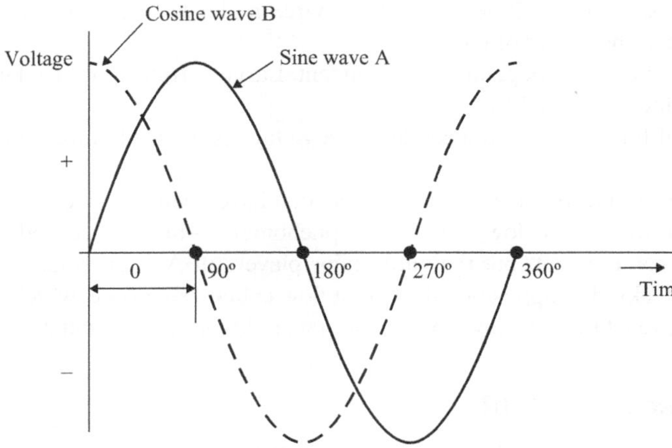

FIGURE 3.43 Wave B leads wave A by a phase angle of 90°.

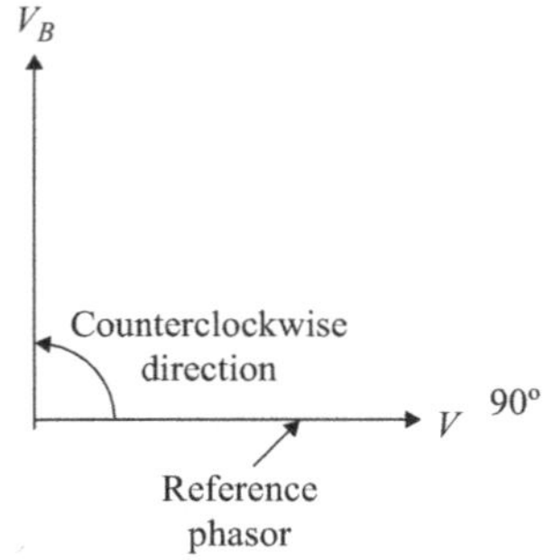

FIGURE 3.44 Phasor diagram.

To compare phase angles or phases of alternating voltages or currents, it is more convenient to use vector diagrams corresponding to the voltage and current waveforms. A *vector* is a straight line denoting a given quantity's magnitude and direction. Magnitude is denoted by the length of the line drawn to scale, and the direction is indicated by the arrow at one end of the line, together with the angle the vector makes with a horizontal reference vector.

Note: Since different directions represent time expressed as a phase relationship in electricity, an electrical vector is called a **phasor**. In an A-C circuit containing only resistance, the voltage and current co-occur or are in phase. To indicate this condition using phasors, all that is necessary is to draw the phasors for the voltage and current in the same direction. The length of the phasor indicates the value of each.

Figure 3.44 shows a vector, or phasor, diagram. Vector VB is vertical to show the phase angle of 90° with respect to vector VA, which is the reference. Since lead angles are shown in the counterclockwise direction from the reference vector, V_B leads V_A by 90°.

3.19 INDUCTANCE

To this point, we have learned the following key points about magnetic fields:

- A field of force exists around a wire carrying a current.
- This field has concentric circles around the wire, in planes perpendicular to the wire, and with the wire at the center of the circles.
- The strength of the field depends on the current. Large currents produce large fields; small currents produce small fields.
- When lines of force cut across a conductor, a voltage is induced in the conductor.

Moreover, to this point, we have studied circuits that have been resistive (i.e., resistors presented the only opposition to current flow). Two other phenomena—inductance and capacitance—exist in D-C circuits to some extent, but they are major players in A-C circuits. Both inductance and capacitance present a kind of opposition to current flow called **reactance**, which we will cover later. Before we examine reactance, however, we must first study inductance and capacitance.

3.19.1 Inductance: What Is It?

Inductance is the characteristic of an electrical circuit that makes itself evident by opposing the starting, stopping, or changing of current flow. A simple analogy can be used to explain inductance.

We are all familiar with how difficult it is to push a heavy load (a cart full of heavy materials). It takes more work to start the load than it does to keep it moving. This is because the load possesses the property of **inertia**. Inertia is the characteristic of mass that opposes a **change** in velocity. Therefore, inertia can hinder us in some ways and help us in others. Inductance exhibits the same effect on current in an electric circuit as inertia does on the velocity of a mechanical object. Inductance can be either beneficial or detrimental.

Important point: Inductance is the characteristic of an electrical conductor that opposes a **change** in current flow.

Because inductance is the property of an electric circuit that opposes any **change** in the current through that circuit if the current increases, a self-induced voltage opposes this change and delays the growth; on the other hand, if the current decreases, a self-induced voltage tends to aid (or prolong) the current flow, slowing the reduction. Thus, current can neither increase nor decrease as fast in an inductive circuit as in a purely resistive circuit.

This effect becomes significant in A-C circuits because it affects the **phase** relationships between voltage and current. Earlier, we learned that voltages (or currents) can be out of phase if induced in separate alternator armatures. In that case, the voltage and current generated by each armature were in phase. When inductance is a factor in a circuit, the voltage and current generated by the **same** armature are out of phase. We shall examine these phase relationships later. We aim to understand the nature and effects of inductance in an electric circuit.

The unit for measuring inductance, L, is the *Henry* (named for the American physicist, Joseph Henry), abbreviated h and generally written in lower case, henry. Figure 3.45 shows the schematic symbol for an inductor. An inductor has an inductance of 1 Henry if an emf of 1 V is induced in the inductor when the current through the inductor changes at the rate of 1 amp per second. The relation between the induced voltage, inductance, and the rate of change of current over time is stated mathematically as

$$E = L\frac{\Delta I}{\Delta t} \tag{3.37}$$

where

E = the induced emf in volts
L = the inductance in henrys
ΔI = is the change in amperes occurring in Δt seconds

Note: The symbol Δ (Delta) means "a change in …."

The henry is a large unit of inductance used with relatively large inductors. The unit employed with small inductors is the millihenry (mh). For still smaller inductors, the unit of inductance is the microhenry (μh).

As previously explained, current flow in a conductor always produces a magnetic field surrounding or linking with it. The magnetic field changes when the current changes, and an emf is induced in the conductor. This emf is called a *self-induced emf* because it is induced in the conductor carrying the current.

FIGURE 3.45 Schematic symbol for an inductor.

Note: Even a perfectly straight length of conductor has some inductance.

The direction of the induced emf has a definite relation to the direction in which the field that induces the emf varies. When a circuit's current increases, the circuit's linking flux increases. The flux through the conductor induces an emf that opposes the increase in current and flux. This emf is sometimes referred to as **counterelectromotive force** (cemf). The two terms are used synonymously throughout this manual. Likewise, when the current decreases, an emf is induced in the opposite direction and opposes the decrease in current.

Important point: The effects just described are summarized by **Lenz's Law**, which states that the induced emf in any circuit is always in a direction opposed to the impact that produced it.

Shaping a conductor so that the electromagnetic field around each portion of the conductor cuts across some other portion of the same conductor increases the inductance. This is shown in its simplest form in Figure 3.46(A). A loop of conductor is looped so that two portions of the conductor lie adjacent and parallel to one another. These portions are labeled Conductor 1 and Conductor 2. Upon closing the switch, electron flow through the conductor creates a standard concentric field around every conductor segment. The field is shown in a single plane (for simplicity) that is perpendicular to both conductors. Although the field originates simultaneously in both conductors, it is considered to originate in Conductor 1, and its effect on Conductor 2 will be noted. With increasing current,

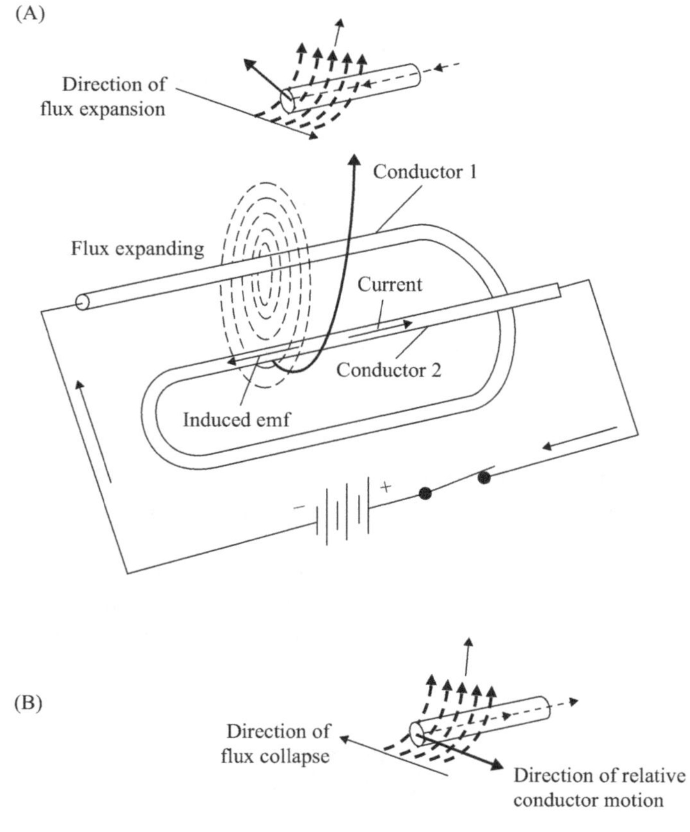

FIGURE 3.46 Self-inductance.

FIGURE 3.47 (A) Few turns, low inductance; (B) More turns, higher inductance.

the field expands outward, cutting across a portion of Conductor 2. The dashed arrow shows the resultant induced emf in Conductor 2. Note that it opposes the battery current and voltage, according to Lenz's Law.

In Figure 3.46(B), the same section of Conductor 2 is shown, but with the switch opened and the flux collapsing.

Important point: From Figure 3.46, the critical point to note is that the voltage of self-induction opposes both **changes** in current. It delays the initial buildup of current by opposing the battery voltage and the current breakdown by exerting an induced voltage in the same direction as the battery voltage.

Four major factors affect the self-inductance of a conductor or circuit.

1. *Number of turns:* Inductance depends on the number of wire turns. Wind more turns to increase inductance. Take turns off to decrease the inductance. Figure 3.47 compares the inductance of two coils made with different numbers of turns.
2. *Spacing between turns:* Inductance depends on the spacing between turns or the inductor's length. Figure 3.48 shows two inductors with the same number of turns. The first inductor's turns have a wide spacing. The second inductor's turns are close together. The second coil, though shorter, has a more considerable inductance value because of its close spacing between turns.
3. *Coil diameter:* Coil diameter, or cross-sectional area, is highlighted in Figure 3.49. The larger-diameter inductor has more inductance. Both coils shown have the same number of turns, and the spacing between turns is the same. The first inductor has a small diameter, and the second one has a larger diameter. The second inductor has more inductance than the first one.

FIGURE 3.48 (A) Wide spacing between turns, low inductance; (B) close spacing between turns, higher inductance.

FIGURE 3.49 (A) Small diameter, low inductance; (B) larger diameter, higher inductance.

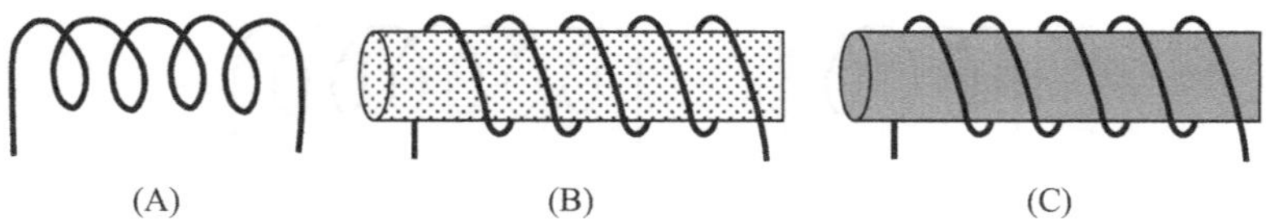

FIGURE 3.50 (A) Air core, low inductance; (B) powdered iron core, higher inductance; (C) soft iron core, highest inductance.

4. *Type of core material:* **Permeability**, as pointed out earlier, measures how easily a magnetic field goes through a material. Permeability also tells us how much stronger the magnetic field will be with the material inside the coil. Figure 3.50 shows three identical coils. One has an air core, a powdered iron core in the center, and a soft iron core. This figure illustrates the effects of core material on inductance. The inductance of a coil is affected by the current magnitude when the core is a magnetic material. When the core is air, the inductance is independent of the current.

Key point: A coil's inductance increases rapidly as the number of turns increases. It also increases as the coil is shorter, the cross-sectional area is larger, or the core's permeability is increased.

Suppose a battery is connected across a pure inductance. In that case, the current builds up to its final value at a rate determined by the battery voltage and the internal resistance. The current buildup is gradual because of the counter-emf (cemf) generated by the self-inductance of the coil. When the current starts to flow, the magnetic lines of force move out, cut the turns of wire on the inductor, and build up a cemf that opposes the emf of the battery. This opposition causes a delay in the time it takes for the current to build up to a steady value. When the battery is disconnected, the lines of force collapse, again cutting the turns of the inductor and building up an emf that tends to prolong the current flow.

Although the analogy is not exact, electrical inductance resembles mechanical inertia. A boat moves on the water's surface when a constant force is applied. At this instant, its rate of change in speed (acceleration) is the greatest, and all the applied force is used to overcome the boat's inertia. After a while, the speed of the boat increases (its acceleration decreases), and the applied force is used up to overcome the friction of the water against the hull. As the speed levels off and the acceleration becomes zero, the applied force equals the opposing friction force at this speed, and the inertia effect disappears. In the case of inductance, electrical inertia must be overcome.

Key point: When a switch is first closed to complete a circuit, inductance opposes the buildup of current in the circuit.

Note: In many electronic circuits, the time required for the growth or decay of current is essential. However, these applications are beyond the scope of this manual.

The time required for the current through an inductor to increase to 63.2% (63%) of the maximum current or decrease to 36.7% (37%) is known as the circuit's time constant. R in the L/R equation is always in ohms, and the time constant is on the same order of magnitude as L. Two practical relations used in calculating L/R time constants are as follows:

$$\frac{\text{L (in henrys)}}{\text{R (in ohms)}} = \text{t(in seconds)} \tag{3.38}$$

$$\frac{\text{L (microhenrys)}}{\text{R(in ohms)}} \tag{3.39}$$

Key point: The time constant of an L/R circuit is always expressed as a ratio between inductance (or L) and resistance (or R).

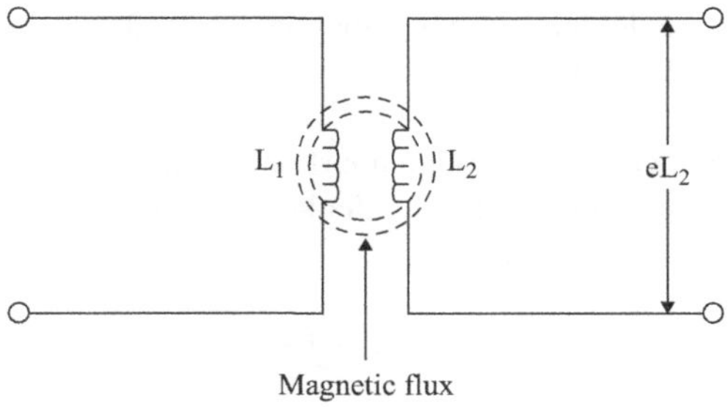

FIGURE 3.51 Mutual inductance between L_1 and L_2.

3.19.2 Mutual Inductance

When the current in a conductor or coil changes, the varying flux can cut across any nearby conductor or coil, thus inducing voltages in both. Therefore, a varying current in L_1 induces a voltage across L_1 and L_2 (see Figure 3.51; see Figure 3.52 for the schematic symbol for two coils with mutual inductance). When the induced voltage e_{L2} produces current in L_2, its varying magnetic field induces a voltage in L_1. Hence, the two coils, L_1 and L_2, have mutual inductance because the current change in one coil can induce a voltage in the other. The unit of mutual inductance is the henry, and the symbol is L_M. Two coils have L_M of 1H when a current change of 1 A/s in one coil induces 1 E in the other coil.

The mutual inductance of two adjacent coils depends on several factors.

- physical dimensions of the two coils
- number of turns in each coil
- distance between the two coils
- relative positions of the axes of the two coils
- the permeability of the cores

Important point: The amount of mutual inductance depends on the relative position of the two coils. If the coils are separated considerably, the flux standard for both is small, and the mutual inductance is low. Conversely, if the coils are close together so that nearly all the flow of one coil links the turns of the other, mutual inductance is high. The mutual inductance can be significantly increased by mounting the coils on a common iron core.

3.19.3 Calculation of Total Inductance

Note: In the study of advanced electrical theory, knowing the effect of mutual inductance in solving for total inductance in both series and parallel circuits is necessary. However, for our purposes, we

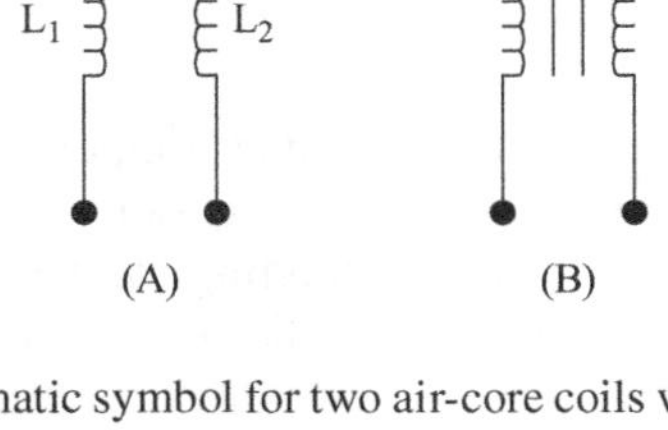

FIGURE 3.52 (A) depicts the schematic symbol for two air-core coils with mutual inductance. (B) Two coils (iron core) with mutual inductance.

do not attempt to make these calculations in this manual. Instead, the basic total inductance calculations are shown, and examples are solved.

If inductors in series are located far enough apart or well shielded to make the effects of mutual inductance negligible, the total inductance is calculated in the same manner as for resistances in series; we merely add them:

$$L_t = L_1 + L_2 + L_3 \ldots (\text{etc.}) \tag{3.40}$$

EXAMPLE 3.15

Problem:

If a series circuit contains three inductors whose values are 40 µh, 50 µh, and 20 µh, what is the total inductance?

SOLUTION:

$$\begin{aligned} L_t &= 40\ \mu\text{h} + 50\ \mu\text{h} + 20\ \mu \\ &= 110\ \mu\text{h} \end{aligned}$$

In a parallel circuit containing inductors (without mutual inductance), the total inductance is calculated in the same manner as for resistances in parallel:

$$\frac{1}{L_t} = \frac{1}{L_1} + \frac{1}{L_2} + \frac{1}{L_3} + \ldots (\text{etc.}) \tag{3.41}$$

EXAMPLE 3.16

Problem:

A circuit contains three shielded inductors in parallel. The values of the three inductances are 4 mh, 5 mh, and 10 mh. What is the total inductance?

SOLUTION:

$$\begin{aligned} \frac{1}{L_t} &= \frac{1}{4} + \frac{1}{5} + \frac{1}{10} \\ &= 0.25 + 0.2 + 0.1 \\ &= 0.55 \end{aligned}$$

$$\begin{aligned} L_t &= \frac{1}{0.55} \\ &= 1.8\ \text{mh} \end{aligned}$$

3.20 CAPACITANCE

No matter how complex the electrical circuit is, it is composed of no more than three basic electrical properties: resistance, inductance, and capacitance. Accordingly, understanding these three basic properties is necessary for understanding electrical equipment. Because resistance and inductance have been covered, the last of the basic three, capacitance, is covered in this section.

Earlier, we learned that inductance opposes any change in current. *Capacitance* is the property of an electric circuit that opposes any voltage change in a circuit. If the applied voltage increases, capacitance opposes the change and delays the voltage increase across the circuit. If the applied voltage is decreased, the capacitance maintains the higher original voltage across the circuit, thus delaying the decrease.

Capacitance is also defined as the property of a circuit that enables energy to be stored in an electric field. Natural capacitance exists in many electric circuits. However, in this manual, we are concerned only with the capacitance designed into the circuit using devices called **capacitors**.

Key point: The most noticeable effect of capacitance in a circuit is that voltage can neither increase nor decrease rapidly in a capacitive circuit as it can in a circuit that does not include capacitance.

3.20.1 The Capacitor

A *capacitor*, or condenser, is a manufactured electrical device that consists of two conducting plates of metal separated by an insulating material called a *dielectric* (see Figure 3.53). (Note: the prefix "di–" means "through" or "across").

There is a short current pulse when a capacitor is connected to a voltage source. A capacitor stores this electric charge in the dielectric (it can be charged and discharged, as we shall see later). To form a capacitor of any appreciable value, however, the area of the metal pieces must be pretty large, and the thickness of the dielectric must be relatively small.

Key point: A capacitor is essentially a device that stores electrical energy.

The capacitor is used in several ways in electrical circuits. It may block D-C portions of a circuit since it is effectively a barrier to direct current (but not to A-C current). It may be part of a tuned circuit—one such application is tuning a radio to a particular station. It may be used to filter A-C out of a D-C circuit. Most of these are advanced applications beyond this presentation's scope; however, a basic understanding of capacitance is necessary for the fundamentals of A-C theory.

Important point: A capacitor does not conduct a D-C current. The insulation between the capacitor plates blocks the flow of electrons. We learned earlier that there is a short current pulse when we first connect the capacitor to a voltage source. The capacitor charges the supply voltage, and the current stops.

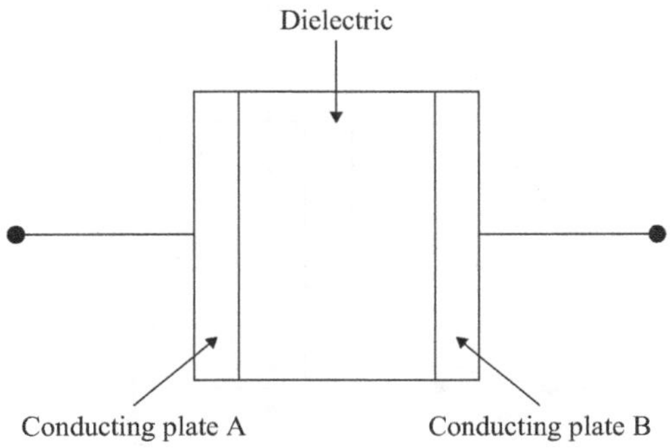

FIGURE 3.53 Capacitor.

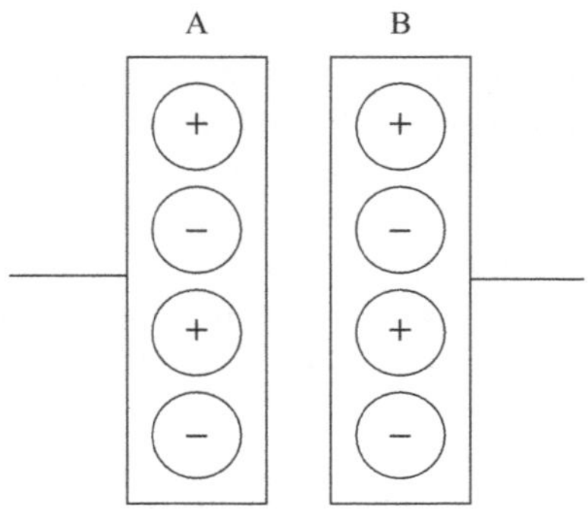

FIGURE 3.54 Two plates of a capacitor with a neutral charge.

The two plates of the capacitor shown in Figure 3.54 are electrically neutral since each plate contains as many protons (positive charge) as electrons (negative charge). Thus, the capacitor has **no charge**.

Now, a battery is connected across the plates [see Figure 3.55(A)]. When the switch is closed [see Figure 3.55(B)], the negative charge on plate A attracts the battery's positive terminal. This movement of charges will continue until the difference in charge between plates A and B equals the battery's electromotive force (voltage).

The capacitor is now **charged**. Because almost none of the charge can cross the space between plates, the capacitor will remain in this condition even if the battery is removed [see Figure 3.56(A)]. However, if a conductor is placed across the plates [see Figure 3.56(B)], the electrons find a path back to plate A, and the charges on each plate are again neutralized. The capacitor is now **discharged**.

Important point: In a capacitor, electrons cannot flow through the dielectric because it is an insulator. Because it takes a definite quantity of electrons to charge ("fill up") a capacitor, it is said to have **capacity**. This characteristic is referred to as *capacitance*.

3.20.2 Dielectric Materials

Similar to the permeability phenomenon in magnetic circuits, various materials differ in their ability to support electric flux (lines of force) or serve as dielectric material for capacitors. Materials are

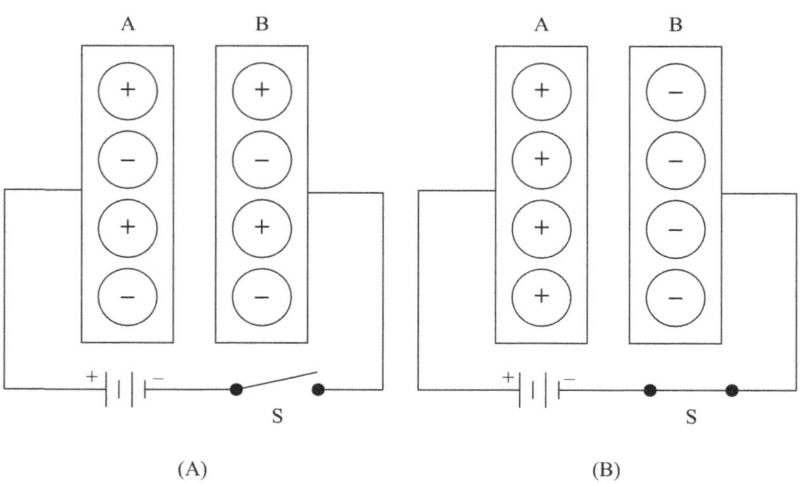

FIGURE 3.55 (A) Neutral capacitor; (B) charged capacitor.

TABLE 3.6
Dielectric Constants

Material	Constant
Vacuum	1.0000
Air	1.0006
Paraffin paper	3.5
Glass	5–10
Quartz	3.8
Mica	3–6
Rubber	2.5–35
Wood	2.5–8
Porcelain	5.1–5.9
Glycerin (15°C)	56
Petroleum	2
Pure water	81

Note: Table 3.6 clearly shows that "pure" water is the best dielectric. Keep in mind that the keyword is "pure." Water capacitors are used today in high-energy applications in which differences in potential are measured in thousands of volts.

rated in their ability to support electric flux using a *dielectric constant.* Other factors being equal, the higher the dielectric constant value, the better a dielectric material is. Dry air is the standard (the reference) by which other materials are rated.

Dielectric constants for some common materials are given in Table 3.6.

3.20.3 Unit of Capacitance

Capacitance equals the amount of charge stored in a capacitor divided by the voltage applied across the plates.

$$C = \frac{Q}{E} \tag{3.42}$$

where
C = capacitance, F (farads)
Q = charge, C (coulombs)
E = voltage, V

EXAMPLE 3.17

Problem:
What is the capacitance of two metal plates separated by one centimeter of air if 0.002 C of charge is stored when a potential of 300 V is applied to the capacitor?

SOLUTION:

$$\text{Given}: Q = 0.001\,\text{C}$$
$$F = 200\ \text{V}$$

$$C = \frac{Q}{E}$$

Converting to power of ten

$$C = \frac{10\times10^{-4}}{2\times10^{2}}$$

$$C = \frac{10\times10^{-4}}{2\times10^{2}}$$

$$C = 5\times10^{-6}$$

$$C = 0.000005\,F$$

Note: Although the capacitance value obtained in Example 3.17 appears small, many electronic circuits require capacitors of much smaller value. Consequently, the farad, named after the English physicist Michael Faraday (1791–1867), is a cumbersome unit far too large for most applications. The microfarad is a more convenient unit, 1 millionth of a farad (1×10^{-6} F). The symbols used to designate microfarads are μf. The following SI prefixes cover most electrical and electronic applications:

- 1 mF (millifarad, 1 thousandth (10^{-3}) of a farad) = 0.001 F = 1,000 μF = 1,000,000,000 pF
- 1 μF (microfarad, 1 millionth (10^{-6}) of a farad) = 0.000,001 F = 1,000 nF = 1,000,000 pF
- 1 nF (nanofarad, 1 billionth (10^{-9}) of a farad) = 0.000,000,001 F = 0.001 μF = 1,000 pF
- 1 pF (picofarad, 1 trillionth (10^{-12}) of a farad) = 0.000,000,000,001 F = 0.001 nF

Equation (3.42) can be rewritten as follows:

$$Q = CE \tag{3.43}$$

$$E = \frac{Q}{C} \tag{3.44}$$

Important point: Do not deduce the mistaken idea that capacitance depends on charge and voltage. Capacitance is determined entirely by physical factors, which are covered later.

The symbol used to designate a capacitor is (C). The unit of capacitance is the farad (F). The farad stores one coulomb of charge in the dielectric when the voltage applied across the capacitor terminals is one volt.

3.20.4 Factors Affecting the Value of Capacitance

The capacitance of a capacitor depends on three main factors: plate surface area, distance between plates, and the dielectric constant of the insulating material.

- *Plate surface area*: Capacitance varies directly with plate surface area. We can double the capacitance value by doubling the capacitor's plate surface area. Figure 3.56 shows a capacitor with a small surface area and another with a large surface area.

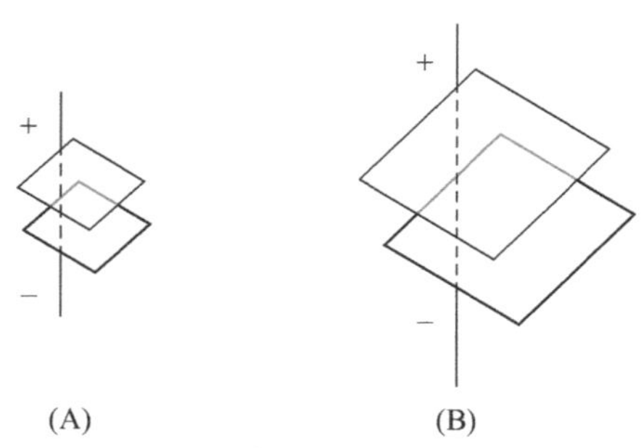

FIGURE 3.56 (A) Small plates, small capacitance; (B) larger plates, higher capacitance.

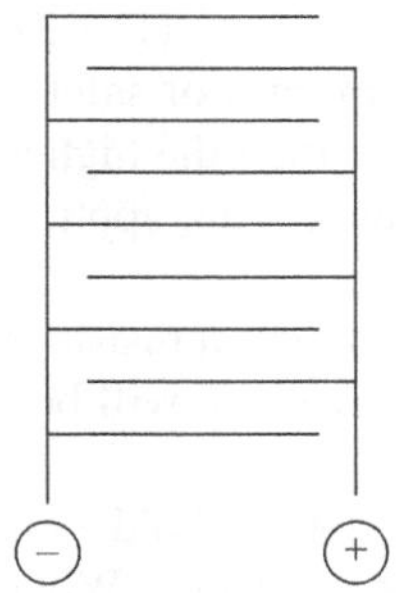

FIGURE 3.57 Shows several sets of plates connected to produce a capacitor with more surface area.

Adding more capacitor plates can increase the plate surface area. Figure 3.57 shows alternate plates connecting to opposite capacitor terminals.

- *Distance between plates*: Capacitance varies inversely with the distance between plate surfaces. The capacitance increases when the plates are closer together. Figure 3.58 shows capacitors with the same plate surface area but different spacing.
- *Dielectric constant of the insulating material*: An insulating material with a higher dielectric constant produces a higher capacitance rating. Figure 3.59 shows two capacitors. Both have the same plate surface area and spacing. Air is the dielectric in the first capacitor, and mica is the dielectric in the second. Mica's dielectric constant is 5.4 times greater than air's dielectric constant. The mica capacitor has 5.4 times more capacitance than the air-dielectric capacitor.

3.20.5 Voltage Rating of Capacitors

There is a limit to the voltage that may be applied across any capacitor. Excessive voltage can overcome the dielectric's resistance, forcing current through and possibly burning a hole. A short circuit exists in this event, and the capacitor must be discarded. The maximum voltage that may be applied to a capacitor is the **working voltage** and must never be exceeded.

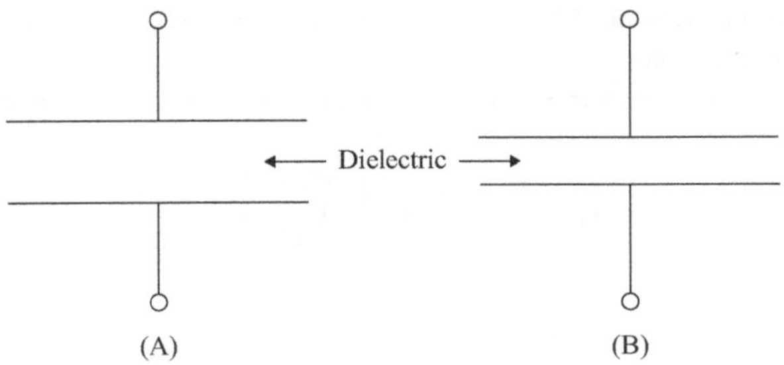

FIGURE 3.58 (A) Wide plate spacing, small capacitance; (B) narrow plate spacing, larger capacitance.

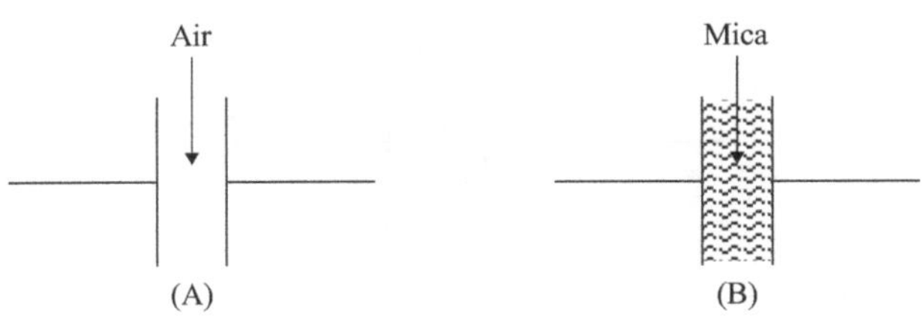

FIGURE 3.59 (A) Low capacitance; (B) higher capacitance.

The working voltage of a capacitor depends on (1) the type of material used as the dielectric and (2) the thickness of the dielectric. As a margin of safety, the capacitor should be selected so that its working voltage is at least 50% greater than the highest voltage to be applied. For instance, if a capacitor is anticipated to experience a maximum applied voltage of 200 V, its rated working voltage should be at least 300 V.

According to Ohm's Law, the voltage across a resistance is equal to the current through it times the value of the resistance. This means a voltage will be developed across a resistance **only when current flows through it**.

As previously stated, a capacitor can store or hold a charge of electrons. When uncharged, both plates contain the same number of free electrons. When charged, one plate contains more free electrons than the other. The difference in the number of electrons is a measure of the charge on the capacitor. The accumulation of this charge builds up a voltage across the terminals of the capacitor, and the charge continues to increase until this voltage equals the applied voltage. The greater the voltage, the greater the charge on the capacitor. Unless a discharge path is provided, a capacitor keeps its charge indefinitely. Any practical capacitor, however, has some leakage through the dielectric so that the voltage will gradually leak off.

The time constant T for an RC circuit is

$$T = RC \tag{3.45}$$

The time constant of an RC circuit is usually very short because the circuit's capacitance may be only a few microfarads or even picofarads.

Key point: An RC time constant expresses a capacitor's charge and discharge times.

3.20.6 Capacitors in Series and Parallel

Like resistors or inductors, capacitors may be connected in series, parallel, or series-parallel combinations. Unlike resistors or inductors, however, total capacitance in series, in parallel, or a series-parallel combination is found differently. Simply put, the rules are not the same for calculating total capacitance. This difference is explained as follows:

Parallel capacitance is calculated like series resistance, and series capacitance is calculated like parallel resistance. For example:

When capacitors are connected in **series** (see Figure 3.60), the total capacitance C_T is

$$\text{Series}: \frac{1}{C_T} = \frac{1}{C_1} + \frac{1}{C_2} + \frac{1}{C_3} + \ldots + \frac{1}{C_n} \tag{3.46}$$

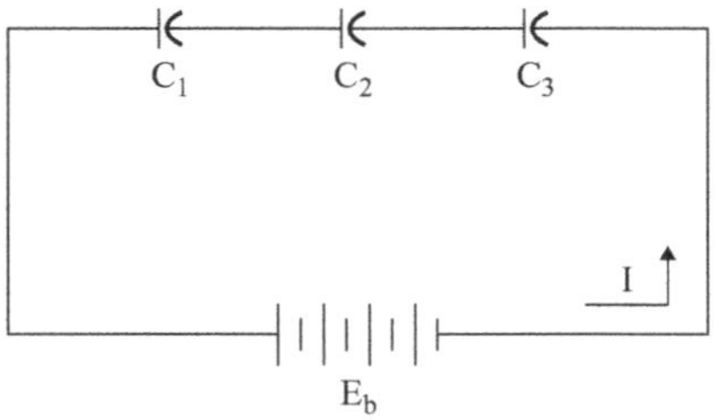

FIGURE 3.60 Series capacitive circuit.

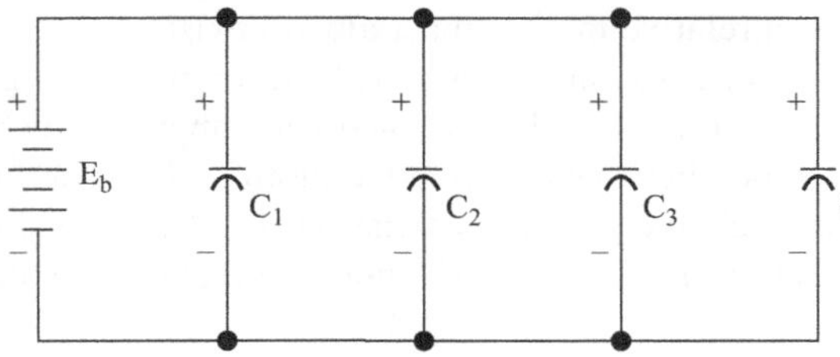

FIGURE 3.61 Parallel capacitive circuit.

EXAMPLE 3.18

Problem:
Find the total capacitance of a 3-μF, a 5-μF, and a 15-μF capacitor in series.

SOLUTION:

Write a equation for three capacitors in series.

$$\frac{1}{C_T} = \frac{1}{C_1} + \frac{1}{C_2} + \frac{1}{C_3}$$

$$= \frac{1}{3} + \frac{1}{5} + \frac{1}{15} = \frac{9}{15} = \frac{3}{5} - \frac{5}{3} = 1.7\ \mu\text{F}$$

When capacitors are connected in **parallel** (see Figure 3.61), the total capacitance C_T is the sum of the individual capacitances.

$$\text{Parallel}: C_T = C_1 + C_2 + C_3 + \ldots + C_n \tag{3.47}$$

EXAMPLE 3.19

Problem:
Determine the total capacitance in a parallel capacitive circuit:

$$\text{Given}: C_1 = 2\ \mu\text{F}$$
$$C_2 = 3\ \mu\text{F}$$
$$C_3 = 0.25\ \mu\text{F}$$

SOLUTION:

Write an equation for three capacitors in parallel:

$$C_T = C_1 + C_2 + C_3$$
$$= 2 + 3 + 0.25$$
$$= 5.25\ \mu\text{F}$$

3.21 GENERATORS

3.21.1 D-C Generators

A *D-C generator* is a rotating machine that converts mechanical energy into electrical energy. This conversion is accomplished by rotating an armature, which carries conductors, in a magnetic field, thus inducing an emf in the conductors. As stated previously, for an EMF to be induced in the conductors, a relative motion must always exist between the conductors and the magnetic field so that the conductors cut through the field. In most D-C generators, the armature is the rotating member, and the field is the stationary member. A mechanical force is applied to the shaft of the rotating member to cause relative motion. Thus, when mechanical energy is put into the machine as a mechanical force or twist on the shaft, causing the shaft to turn at a certain speed, electrical energy in the form of voltage and current is delivered to the external load circuit.

Important point: Mechanical power must be constantly applied to the shaft as long as the generator supplies electrical energy to the external load circuit.

To gain a basic understanding of the operation of a D-C generator, consider the following explanation.

A simple D-C generator with an armature with a single turn of wire is shown in Figure 3.62(A) and (B).

Note: The armature coils used in large D-C machines are usually wound in their final shape before being put on the armature. The sides of the preformed coil are placed in the slots of the laminated armature core—this armature coil cuts across the magnetic field to produce voltage. If a complete path is present, the current will move through the circuit in the direction the arrows show [see Figure 3.62(A)]. In this coil position, commutator segment 1 is in contact with brush 1, while commutator segment 2 is in contact with brush 2. As the armature rotates a half turn clockwise, the contacts between the commutator segments and the brushes are reversed [see Figure 3.62(B)]. Currently, segment 1 is in contact with brush 2, and segment 2 is in contact with brush 1. Because of this commutator action, that side of the armature coil, in contact with either of the brushes, always cuts across the magnetic field in the same direction. Thus, brushes 1 and 2 have constant polarity, and a **pulsating D-C current** is delivered to the external load circuit.

Note: In D-C generators, the voltage induced in individual conductors is A-C. The commutator, which rotates in contact with carbon brushes, converts it to D-C (rectified) so that the current generated is in one direction or direct current.

There are several different types of D-C generators. They take their names from the kind of field excitation used (i.e., they are classified according to how the field windings are connected to the armature circuit). For example, when the generator's field is excited (or supplied) from a separate

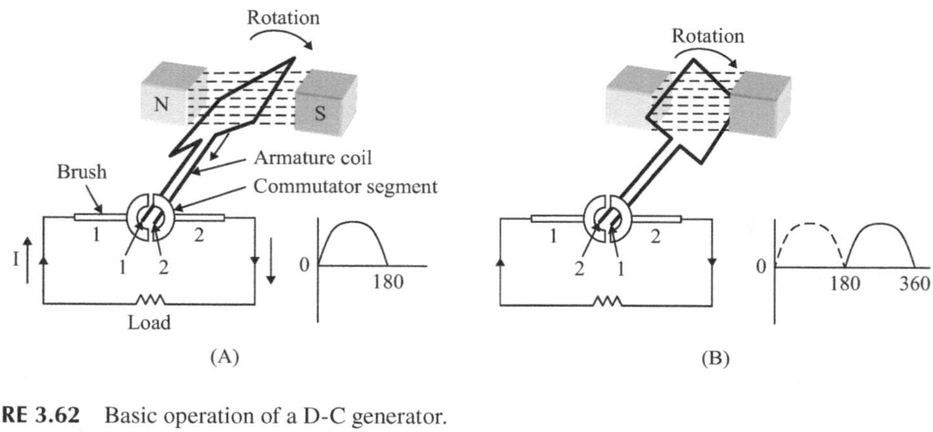

FIGURE 3.62 Basic operation of a D-C generator.

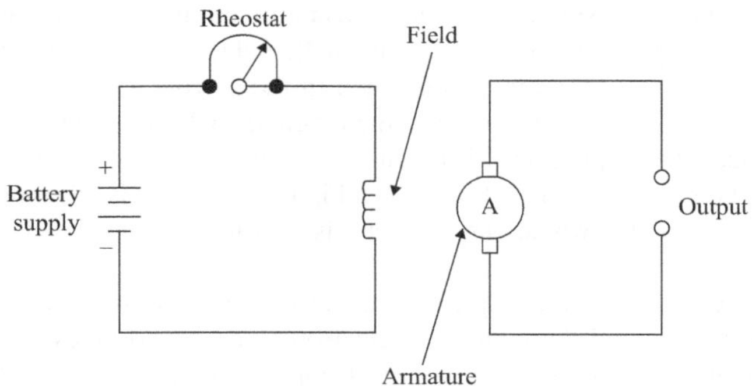

FIGURE 3.63 Separately excited D-C generator.

D-C source (such as a battery) other than its armature, it is called a **separately excited D-C generator** (see Figure 3.63).

Note: As central generating stations increased in size along with their number and power distribution distances, D-C generating systems were replaced by A-C generating systems to reduce power transmission costs. This was due to the high-power losses in long D-C transmission lines.

3.21.2 A-C Generators

Most electric power utilized today is generated by *alternating-current generators* (also called **alternators**). They are made in many different sizes, depending on their intended use. Regardless of size, however, all generators operate on the same basic principle: a magnetic field cutting through conductors or conductors passing through a magnetic field. They are (1) a group of conductors in which the output voltage is generated and (2) a second group of conductors through which direct current is passed to obtain an electromagnetic field of fixed polarity. The conductors in which the electromagnetic field originates are always called field windings.

In addition to the armature and field, there must be motion between them. To provide this, AC generators are built in two major assemblies: the **stator** and the **rotor**. The rotor rotates inside the stator.

The revolving-field A-C generator (see Figure 3.64) is the most widely used type. In this type of generator, direct current from a separate source is passed through windings on the rotor using slip rings and brushes. [*Note:* Slip rings and brushes are adequate for the D-C field supply because the

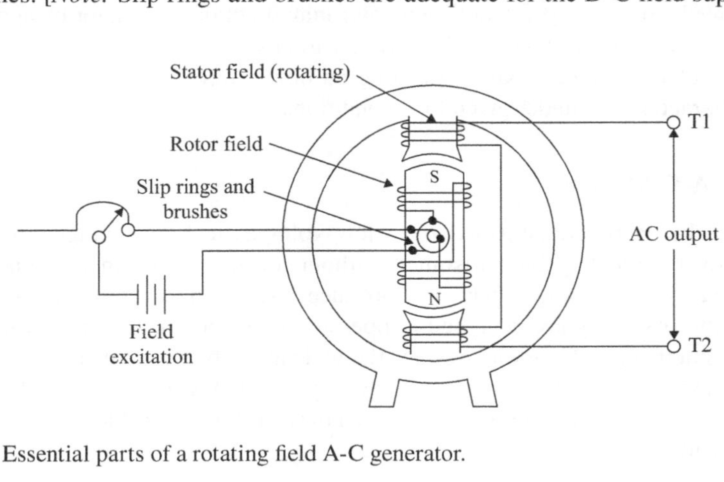

FIGURE 3.64 Essential parts of a rotating field A-C generator.

power level in the field is much smaller than in the armature circuit]. This maintains a rotating electromagnetic field of fixed polarity. The rotating magnetic field, following the rotor, extends outward and cuts through the armature windings embedded in the surrounding stator. As the rotor turns, A-C voltages are induced in the windings since the magnetic fields of one polarity and the other cut through them. Because the output power is taken from stationary windings, the output may be connected through fixed output terminals T1 and T2 in Figure 3.64. This is advantageous because there are no sliding contacts, and the whole output circuit is continuously insulated.

Important point: In A-C generators, frequency and electromagnetic wave cycles per second depend on how fast the rotor turns and the number of electromagnetic field poles. The voltage generated depends on the speed of the rotor, the number of coils in the armature, and the magnetic field strength.

3.22 ELECTRIC MOTORS

Electric motors perform almost endless tasks when operating industrial and domestic equipment. An *electric motor* is a machine that changes electrical energy into mechanical energy to do the work. [*Note:* Recall that a generator does just the opposite; that is, a generator changes mechanical energy to electrical energy.]

Previously, we pointed out that a magnetic field is produced around the wire when a current passes through it. If this magnetic field passes through a stationary magnetic field, the fields either repel or attract, depending on their relative polarity. If both are positive or negative, they repel. If they are of opposite polarity, they attract.

Applying this basic information to motor design, an electromagnetic coil causes the armature to rotate on a shaft. The armature and shaft assembly are called the rotor. The rotor is assembled between the poles of a permanent magnet, and each end of the rotor coil (armature) is connected to a commutator mounted on the shaft. A commutator comprises copper segments insulated from the shaft and each other by an insulating material. As the electromagnet poles in the rotating armature pass the stationary permanent magnet poles, they are repelled, continuing the motion. As the opposite poles are near each other, they attract, continuing the motion.

3.22.1 D-C Motors

The construction of a D-C motor is essentially the same as that of a D-C generator. However, it is essential to remember that the D-C generator converts mechanical energy into electrical energy, and then back into mechanical energy. A D-C generator may function as a motor by applying a suitable D-C voltage source across the standard electrical terminals.

Various types of D-C motors exist depending on how the field coils are connected. Each has advantageous characteristics under given load conditions.

3.22.2 Basic A-C Motors

A-C voltage can be easily transformed from low voltages to high voltages or vice versa and can be moved over a much greater distance without too much loss in efficiency. Most of the power-generating systems today, therefore, produce alternating current. Thus, it follows that most electrical motors today are designed to operate on alternating current. However, there are other advantages in using A-C motors besides the wide availability of AC power. In general, A-C motors are less expensive than D-C motors. Most types of A-C motors do not employ brushes and commutators. This eliminates many maintenance and wear problems and also eliminates dangerous sparking.

A-C motors are manufactured in many different sizes, shapes, and ratings for use in an even greater number of jobs. They are designed for use with either polyphase or single-phase power systems.

Important note: A three-phase (3–θ) system is a combination of three single-phase (1–θ) systems. In a 3—balanced system, the power comes from an AC generator that produces three separate but equal voltages, each of which is out of phase with the other voltages by 120. Although 1-circuits are widely used in electrical systems, most generation and distribution of AC current is 3–θ.

3.23 A-C VERSUS D-C

In the 1880s and early 1890s, there was a personal war called the "War of Currents" between two American geniuses: Thomas Edison, the American inventor and businessman who established the first investor-owned electric utility in 1882, basing its infrastructure on D-C power, and George Westinghouse, the American entrepreneur and engineer who introduced the rival A-C distribution system. Westinghouse came out on top of the argument for one crucial reason: A-C power can be transformed; it can be stepped up and stepped down in the voltage delivered to customers. Edison's side of the argument was doomed because D-C power is straight-line power; it cannot be stepped up and down without huge voltage losses along the transmission lines. A-C power can be sent and transmitted in a steady distribution to a point to counter voltage drops due to transmission line resistance. A-C power can then be stepped up to continue its transmission and then stepped down or further up to suit the user's needs. This steady supply of adjustable electricity is possible because of transformers.

A *Transformer* is an electric control device (with no moving parts) that raises or lowers voltage or current in an electric distribution system. The basic transformer consists of two electrically insulated coils from each other and wound upon a common core (see Figure 3.65). Magnetic coupling is used to transfer electric energy from one coil to another. The coils that receive energy from AC sources are called primary. The coil that delivers energy to A-C loads is called the **secondary**. The core of transformers used at low frequencies is generally made of magnetic material, usually laminated sheet steel. Cores of transformers used at higher frequencies are made of powdered iron and ceramics or nonmagnetic materials. Some coils are wound on nonmagnetic hollow forms such as cardboard or plastic, so the core material is air.

In operation, an alternating current will flow when an A-C voltage is applied to the primary coil of a transformer. This current produces a field of force that changes as the current changes. The magnetic core carries the changing magnetic field to the secondary coil, where it cuts across the turns of that coil. In this way, an A-C voltage in one coil is transferred to another, even though there is no electrical connection between them. The number of lines of force available in the primary is determined by the primary voltage and the number of turns on the primary, each turn producing a

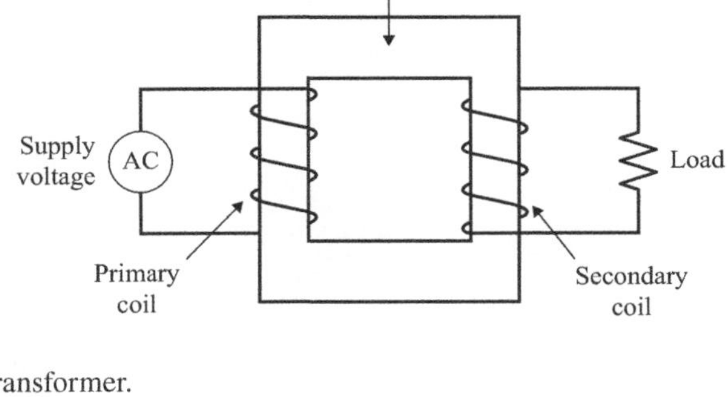

FIGURE 3.65 Basic transformer.

given number of lines. If there are many turns on the secondary, each line of force will cut many turns of wire and induce a high voltage. If the secondary contains only a few turns, there will be few cuttings and low induced voltage. The secondary voltage, then, depends on the number of secondary turns compared to the number of primary turns. If the secondary voltage has twice as many turns as the primary voltage, the secondary voltage will be twice as large as the primary voltage. If the secondary has half as many turns as the primary, the secondary voltage will be one-half as large as the primary voltage.

Important point: The voltage on a transformer's coils is directly proportional to the number of turns on the coils.

A voltage ratio 1:4 (read as "1 to 4") means that for each volt on the primary, there are 4 V on the secondary. This is called a **step-up** transformer. A step-up transformer receives a low voltage on the primary and a high voltage from the secondary. A voltage ratio of 4:1 (read as "4 to 1") means that for 4 V on the primary, there is only 1 V on the secondary. This is called a **step-down** transformer. A step-down transformer receives a high voltage on the primary and delivers a low voltage from the secondary.

3.24 THE BOTTOM LINE

The focus of this chapter is to present fundamental aspects of electricity. This information is crucial because storing energy is essential, especially for providing electricity on demand. Stored electricity is critical, mainly from variable renewable sources like wind, solar, and hydroelectric power. AI is essential in energy storage because it provides benefits such as the potential to cut energy waste and provide lower energy costs globally. AI also increases the use of renewable energy sources. AI is important in the planning process, in operations, and in control. AI can predict and mitigate grid disruptions. AI enhances resilience and ensures consistent power supplies. Currently, AI neural network technology is a commonly used method to predict power load. The absolute bottom in using AI in energy storage is system security. This is accomplished using deep learning to automatically identify network attack features, detect malware and intrusion, and provide network security protection for the power systems.

NOTE

1 Much of the information in this section is adapted from F.R. Spellman & J. Drinan, Electricity (2001), Boca Raton, FL: CRC Press.

4 Battery Technology

Artificial intelligence (AI) systems use weather and satellite data, numerical weather prediction models, and statistical analysis to make energy production and consumption predictions. Many of these AI systems are backed up by battery-supplied electricity. Energy storage systems are not electricity sources; they do not generate electricity from fuel or natural resources. Instead, they store electricity that has already been created from an electricity generator or the electric power grid, which makes energy storage systems *secondary* sources of electricity. Note that energy storage systems use more electricity for charging than they provide when supplying electricity to the grid. Secondary sources of electricity, such as batteries, are included because they offer the capacity to meet the load, even though energy storage systems *do not* generate electricity directly.

4.1 BATTERIES: ELECTROCHEMICAL STORAGE DEVICES

If someone asked what the standard means of storing energy is, those unaware would scratch their heads, throw their hands in the air, and eventually say gasoline, diesel, kerosene, white gas, coal, and nuclear. However, others would answer that the battery is the primary storehouse of energy. Storage batteries indeed store energy, contrary to those who think batteries create energy, which they do not—store, yes, but battery devices create no energy. Electrochemical batteries store energy in a form that can be easily converted into electrical energy.

Electrical Energy/Electricity. What is it? Employees in industrial complexes have little difficulty in recognizing electrical equipment. Electrical equipment is everywhere and is easy to spot. For example, many typical industrial sites are outfitted with electrical equipment to:

- Generate electricity (a generator—or emergency generator)
- Store electricity (batteries)
- Change electricity from one form to another (transformers)
- Transport or transmit and distribute electricity throughout the plant site (wiring distribution systems)
- Measure electricity (meters)
- Convert electricity into other forms of energy (rotating shafts – mechanical energy, heat energy, light energy, chemical energy, or radio energy)
- Protects other electrical equipment (fuses, circuit breakers, or relays)
- Operates and controls other electrical equipment (motor controllers)
- Convert some condition or occurrence into an electric signal (sensors)
- Converts some measured variable to a representative electrical signal (transducers or transmitters).

Recognizing electrical equipment is easy because we use so much of it. Suppose we ask typical workers where such equipment is in their plant site. They know because they operate these devices or their ancillaries. If we ask these same operators, they could tell us what a particular electrical device does. If we were to ask if their plant electrical equipment were essential to plant operations, the chorus would resound, "Absolutely."

However, another question does not always result in such a resounding note of assurance. If we asked these same workers to explain in fundamental terms how electricity works to make their worksite equipment operate, the answers we would receive would be varied, messy, disjointed—and

DOI: 10.1201/9781003637851-5

not all that accurate. Even on a more fundamental level, how many workers would be able to answer the question: What is electricity?

Very few. Why do so many workers in many industries know so little about electricity? Part of the answer is that many industrial workers are expected to know so much (and they do) but are given minimal opportunity to be trained appropriately. We all know that experience is excellent training. For example, let us look at what an operator assigned to change the bearings on a 5-hp three-phase motor would need to know to accomplish this task. (Remember, it is common for workers in many fields to maintain and operate industrial equipment.)

Note to Reader: Let us examine batteries and their function using a simple electrical circuit for illustrative purposes, battery terminology, and essential operation. This discussion is relevant, particularly in the current shift towards electric vehicles (EVs). Moreover, batteries are one of the primary sources of stored energy, and their operation must be understood.

Note: Battery-supplied direct current (DC) electricity (batteries do not store A-C) is widely used in countless applications. Applications include providing electrical energy in workplace vehicles and emergency diesel generators, material handling equipment (forklifts), portable electric/electronic equipment, backup emergency power for light-packs, hazard warning signal lights and flashlights, and standby power supplies or uninterruptible power supplies (UPS) for computer systems. In some instances, they are used as the only power source, while in others (as mentioned above), they are used as a secondary or standby power supply.

DID YOU KNOW?

In the early days of electrical experimentation and research, a standard method of detecting electric current was using a frog's leg galvanoscope. The experimenter kept a large number of live frogs at hand, ready to have the legs prepared for the galvanoscope. Frogs were common in other experiments; their legs were sensitive to electric current. A frog battery is an electrochemical battery consisting of several dead frogs, which form the battery cells connected in a series arrangement. This is a biobattery. Biobatteries are totally non-polluting, renewable, and environmentally friendly. These batteries are still in development state.

4.1.1 Battery Terminology

- A voltaic cell is a combination of materials used to convert chemical energy into electrical energy.
- A battery is a group of two or more connected voltaic cells.
- An electrode is a metallic compound, or metal, with an abundance of electrons (negative electrode) or positive charges (positive electrode).
- An electrolyte is a solution capable of conducting electric current.
- Specific gravity is the ratio comparing any liquid's weight to an equal volume of water.
- An ampere-hour (Ah) is defined as a current of one-ampere flowing for 1 hour.

4.1.2 The Voltaic Cell

The simplest cell (a device that transforms chemical energy into electrical energy) is a voltaic (or galvanic) cell (see Figure 4.1). It consists of a piece of carbon (C) and a piece of zinc (Zn) suspended in a jar containing a solution of water (H_2O) and sulfuric acid (H_2SO_4).

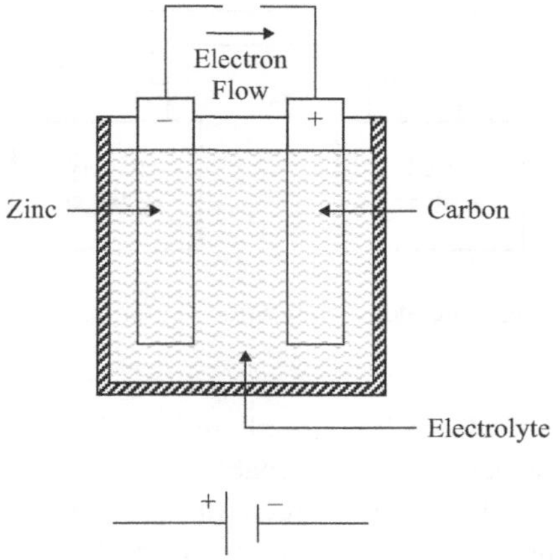

FIGURE 4.1 Simple voltaic cell. The schematic symbol for a battery.

Note: A simple cell consists of two strips, or electrodes, placed in a container that holds the electrolyte. A battery is formed when two or more cells are connected.

The electrodes are the conductors by which the current leaves or returns to the electrolyte. The simple cell described above contains carbon and zinc strips in the electrolyte. Zinc includes an abundance of negatively charged atoms, while carbon has an abundance of positively charged atoms. When the plates of these materials are immersed in an electrolyte, chemical action between the two begins. In the dry cell (see Figure 4.2), the electrodes are the carbon rod in the center and the zinc container in which the cell is assembled.

The electrolyte is the solution that acts upon the electrodes placed in it. It may be salt, an acid, or an alkaline solution. The electrolyte is liquid in the simple voltaic cell and the automobile storage battery, while in the dry cell (see Figure 4.2), it is a moist paste.

4.1.3 Primary and Secondary Cells

Primary cells usually cannot be recharged or returned to good condition after their voltage drops too low. Dry cells in flashlights and transistor radios are examples of primary cells. Some primary cells have been developed to a state where they can be recharged. A *secondary cell* is one

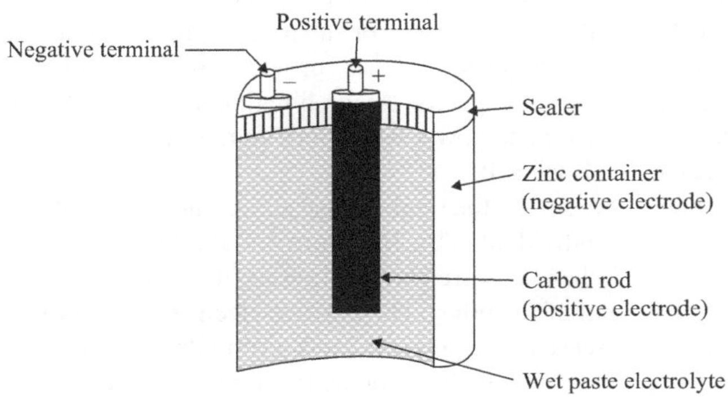

FIGURE 4.2 Dry cell (cross-sectional view).

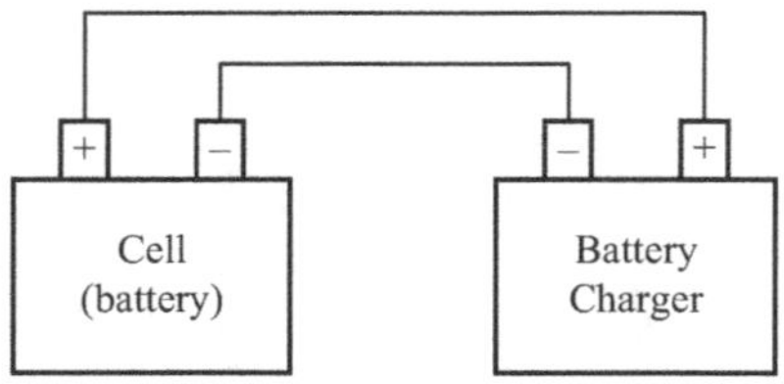

FIGURE 4.3 Hookup for charging a secondary cell with a battery charger.

in which the electrodes and the electrolyte are altered by the chemical action that occurs when the cell delivers current. These cells are rechargeable. The chemicals that provide electric energy are restored to their original condition during recharging. Recharging is accomplished by forcing an electric current through them in the opposite direction of discharge. Connecting, as shown in Figure 4.3, recharges a cell. Some battery chargers have a voltmeter and an ammeter that indicate the charging voltage and current. The automobile storage battery is the most common example of a secondary cell.

Batteries consist of two or more cells placed in a standard container. Depending on the battery's voltage and current requirements, the cells are connected in series, parallel, or some combination of series and parallel.

The chemical reaction within a battery provides voltage. This occurs when a conductor is connected externally to the electrodes of a cell, causing electrons to flow under the influence of a difference in potential across the electrodes from the zinc (negative plate) through the external conductor to the carbon (positive), returning within the solution to the zinc. After a short period, the zinc will begin to waste away because of the acid.

The voltage across the electrodes depends upon the materials from which the electrodes are made and the composition of the solution. The difference in potential between the carbon and zinc electrodes in a dilute solution of sulfuric acid and water is about 1.5 V.

The current that a primary cell may deliver depends upon the resistance of the entire circuit, including that of the cell itself. The internal resistance of the primary cell relies on the size of the electrodes, the distance between them in the solution, and the resistance of the solution. The larger the electrodes and the closer they are in solution (without touching), the lower the internal resistance of the primary cell and the more current it can supply to the load.

Note: When current flows through a cell, the zinc gradually dissolves in the solution, neutralizing the acid.

Again in many operations, battery-powered devices may require more electrical energy than one cell can provide. Various devices may require either a higher voltage or more current, and in some cases, both. Under such conditions, combining or interconnecting enough cells to meet the higher requirements is necessary. Cells connected in series provide a higher voltage, while cells connected in parallel provide a higher current capacity. A combination series-parallel network of cells must be interconnected to provide adequate power when both voltage and current requirements are more significant than the capacity of one cell.

When cells are connected in series, the total voltage across the battery of cells is equal to the sum of the voltage of each of the individual cells. In Figure 4.4, the four 1.5-V cells in series provide a total battery voltage of 6 V. When cells are placed in series, the positive terminal of one cell is connected to the negative terminal of the other. The positive electrode of the first cell and the negative electrode of the last cell then serve as the power takeoff terminals of the battery. The current flowing through such a battery of series cells is the same as from one cell because the same current flows through all the series cells.

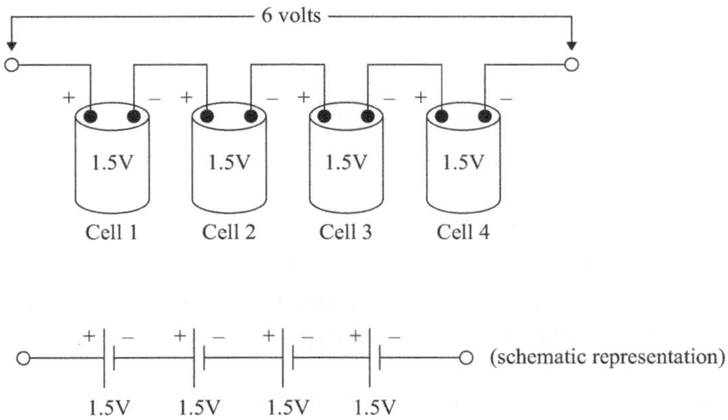

FIGURE 4.4 Cells in series.

A battery has cells connected in parallel to obtain a more significant current, as shown in Figure 4.5. In this parallel connection, all the positive electrodes are connected to one line, and all negative electrodes are connected to the other. Any point on the positive side can serve as the battery's positive terminal, and any point on the negative side can be the negative terminal. The total voltage output of a battery of three parallel cells is the same as that for a single cell (Figure 4.6), but the available current is three times that of one cell; the current capacity has been increased. Identical cells supply equal parts of the current to the load in parallel. For example, three parallel cells produce a load current of 210 mA, and each cell contributes 70 mA.

Figure 4.6 depicts a schematic of a series-parallel battery network supplying power to a load requiring both a voltage and current greater than one cell can provide. To provide the necessary increased voltage, groups of three 1.5-V cells are connected in series. Four series groups are connected in parallel to provide the required increased amperage.

Several different types of batteries have been developed in the past 25 years. This text briefly discusses five types: the dry cell, lead-acid battery, alkaline cell, nickel-cadmium battery, and mercury cell.

The *dry cell*, or carbon-zinc cell, is called because its electrolyte is not liquid (however, it is a moist paste). The dry cell battery is one of the oldest and most widely used commercial types of dry cell. The carbon, in the form of a rod placed in the cell's center, is the positive terminal. The cell case is made of zinc, the negative terminal (see Figure 4.2). Between the carbon electrode and the zinc case is the electrolyte, a moist chemical paste-like mixture. The cell is sealed to prevent the liquid in the paste from evaporating. The voltage of a cell of this type is about 1.5 V.

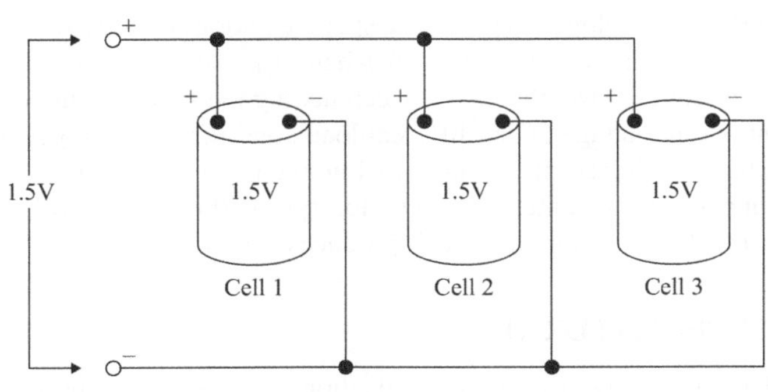

FIGURE 4.5 Cells in parallel.

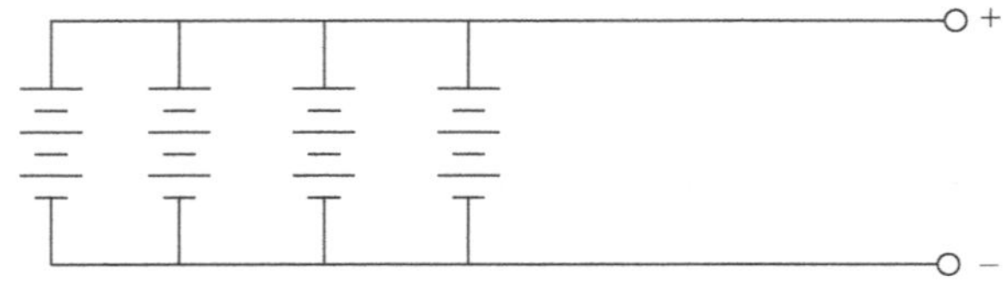

FIGURE 4.6 Series-parallel connected cells.

The *lead-acid battery* is a secondary cell, commonly termed a *storage battery,* which stores chemical energy until it is released as electrical energy.

Note: The lead-acid battery differs from the primary cell-type battery because it may be recharged, whereas most primary cells are not recharged. In addition, the term "storage battery" is deceiving because this battery does not store electrical energy but is a source of chemical energy that produces electrical energy.

As the name implies, the lead-acid battery consists of several lead-acid cells immersed in a dilute sulfuric acid solution. Each cell has two groups of lead plates; one set is the positive terminal, and the other is the negative terminal. Active materials within the battery (lead plates and sulfuric acid electrolyte) react chemically to produce a flow of DC whenever current-consuming devices are connected to the battery terminal posts. This current is made by a chemical reaction between the active material of the plates (electrodes) and the electrolyte (sulfuric acid). This type of cell produces slightly more than 2 V. Most automobile batteries contain six cells connected in series so that the output voltage from the battery is slightly more than 12 V. Besides being rechargeable, the primary advantage of the lead-acid storage battery over the dry cell battery is its capability to provide current for a significantly longer duration than an average dry cell.

Safety Note: Whenever a lead-acid storage battery charges, the chemical action produces dangerous hydrogen gas; thus, the charging operation should only occur in a well-ventilated area.

An *alkaline cell* is a secondary cell named after its alkaline electrolyte, potassium hydroxide. Another type of battery, sometimes called an *alkaline battery*, has a negative zinc electrode and a positive manganese dioxide electrode. It generates 1.5 V.

The *nickel-cadmium cell,* or Ni-Cad cell, is the only dry battery with a proper storage battery with a reversible chemical reaction, allowing recharging many times. In the secondary nickel-cadmium dry cell, the electrolyte is potassium hydroxide, the negative electrode is nickel hydroxide, and the positive electrode is cadmium oxide. The operating voltage is 1.25 V. Because of its rugged characteristics (it stands up well to shock, vibration, and temperature changes) and availability in various shapes and sizes, it is ideally suited for powering portable communication equipment.

The *mercury cell* was developed because of space exploration activities, the development of small transceivers, and miniaturized equipment where a power source of miniaturized size was needed. In addition to reduced size, the mercury cell has a good shelf life and is very rugged; it also produces a constant output voltage under different load conditions. Two types of mercury cells are available. One is a flat cell shaped like a button, while the other is a cylindrical cell that looks like a standard flashlight cell. The advantage of the button-type cell is that several can be stacked inside one container to form a battery. A mercury cell produces 1.35 V.

4.2 BATTERY-SUPPLIED ELECTRICITY

Battery-supplied DC electricity has many applications and is widely used in household, commercial, and industrial operations. Applications include providing electrical energy in industrial vehicles and emergency diesel generators, material handling equipment (forklifts), portable electric/electronic equipment, backup emergency power for light-packs, hazard warning signal lights and flashlights, and standby power supplies or UPS for computer systems. In some instances, they are used as the only power source, while in others (as mentioned above), they are used as a secondary or standby power supply. In renewable energy applications, batteries are used to store electrical energy. A battery pack stores electricity produced by a solar electric system. Today, they are commonly used in specialized applications such as plug-in hybrid EV energy storage systems. Batteries are often used in wind-hybrid systems to store excess wind energy and provide supplementary energy when the wind cannot generate sufficient power to meet the electric load. Batteries are also used in roadside solar-charging systems; that is, solar energy is used to charge a battery pack that supplies electrical power to some electrical signaling device, navigation buoy, or other remote application (e.g., emergency telephone) where it would be impractical and costly to run miles of electrical supply cable.

In the past 30 years, several different types of batteries have been developed. This text briefly discusses the dry cell battery and those currently used to store electrical energy in a reversible chemical reaction applied to renewable energy production. Renewable energy sources (solar, wind, or hydro) produce energy, and the battery stores it when there is or is not renewable energy production. The types of batteries used in this application and discussed here include lead-acid, alkaline, nickel-cadmium, mercury, nickel-metal hydride, lithium-ion, and lithium-ion polymer batteries. Remember that a battery does not create energy; instead, it stores energy. For most renewable energy applications, the preferred battery type is deep-cycle. A deep-cycle battery is designed to deliver a constant voltage as the battery discharges. A car starting battery, in contrast, is designed to provide sporadic current spikes. Battery-driven vehicles, such as forklifts, golf carts, and floor sweepers, commonly use deep-cycle batteries. Deep-cycle batteries require less charging current than regular batteries.

- *Dry Cell*: The dry cell, or carbon-zinc cell, is so called because its electrolyte is not liquid (however, it is a moist paste). The dry cell battery is one of the oldest and most widely used commercial types of dry cell. The carbon, in the form of a rod placed in the cell's center, is the positive terminal. The cell case is made of zinc, the negative terminal (see Figure 4.2). Between the carbon electrode and the zinc case is the electrolyte, a moist, chemical, paste-like mixture. The cell is sealed to prevent the liquid in the paste from evaporating. The voltage of a cell of this type is about 1.5 V.
- *Lead–Acid Battery*: **A** *lead-acid battery* is a secondary cell—commonly termed storage or rechargeable—that stores chemical energy until it is released as electrical energy. The lead-acid battery differs from the primary cell-type battery mainly because it may be recharged, whereas most primary cells are not recharged. As the name implies, the lead-acid battery consists of several lead-acid cells immersed in a dilute sulfuric acid solution. Each cell has two groups of lead plates; one set is the positive terminal, and the other is the negative terminal. Active materials within the battery (lead plates and sulfuric acid electrolyte) react chemically to produce a flow of DC whenever current-consuming devices are connected to the battery terminal posts. This current is made by a chemical reaction between the active material of the plates (electrodes) and the electrolyte (sulfuric acid). This type of cell produces slightly more than 2 V. Most automobile batteries contain six cells connected in series so that the output voltage from the battery is slightly more than 12 V. Besides being rechargeable, the main advantage of the lead-acid storage battery over the dry cell battery is that the storage battery can supply current for a much longer time than the average dry cell. Lead-acid batteries can be designed to be high power and are inexpensive, safe, and reliable. Recycling infrastructure is in place for them. However, low specific energy, poor cold temperature performance, and short calendar and cycle life are still impediments to their use. Advanced high-power, deep-cycle lead-acid batteries are being developed for hybrid electrical vehicle (HEV) applications. However, lead-acid batteries are used for residential solar electric systems because of their low maintenance requirements and cost.

Safety Note: Whenever a lead-acid storage battery charges, the chemical action produces dangerous hydrogen gas; thus, the charging operation should only occur in a well-ventilated area.

- *Alkaline Cell*: The *alkaline cell* is a secondary cell named after its alkaline electrolyte, potassium hydroxide. Another type of battery, sometimes called the "alkaline battery," has a negative electrode of zinc and a positive electrode of manganese dioxide. It generates 1.5 V.
- *Nickel–Cadmium Cell*: The *nickel-cadmium cell*, or Ni-Cad cell, is the only dry battery that is an actual storage battery with a reversible chemical reaction, allowing recharging many times. In the secondary nickel-cadmium dry cell, the electrolyte is potassium hydroxide, the negative electrode is nickel hydroxide, and the positive electrode is cadmium oxide. The operating voltage is 1.25 V. Because of its rugged characteristics (it stands up well to shock, vibration, and temperature changes) and availability in various shapes and sizes, it is ideally suited for powering portable communication equipment. Ni–Cad batteries are costly. Moreover, although nickel-cadmium batteries, used in many electronic consumer products, have higher specific energy and better life cycle than lead-acid batteries, they have low efficiency (65–80%), do not deliver sufficient power, and are not considered for hybrid electric vehicle (HEV) applications. Cadmium is a toxic heavy metal that is costly to dispose of, making it less desirable for hybrid applications (NREL, 2009).
- *Mercury Cell*: The *mercury cell* was developed due to space exploration activities, the development of small transceivers, and miniaturized equipment where a power source of miniaturized size was needed. In addition to reduced size, the mercury cell has a good shelf life and is very rugged; it also produces a constant output voltage under different load conditions. There are two different types of mercury cells. One is a flat cell shaped like a button, while the other is a cylindrical cell that looks like a standard flashlight cell. The advantage of the button-type cell is that several can be stacked inside one container to form a battery. A cell generates 1.35 V.
- *Nickel–Metal Hydride*: Nickel-metal hydride batteries, used routinely in computers and medical equipment, offer reasonable specific energy and specific power capabilities. Their components are recyclable, but a recycling structure is not yet in place. Nickel-metal hydride batteries have a much longer life cycle than lead acid batteries and are safe and abuse-tolerant. These batteries have been used successfully in producing EVs and recently in the low-volume production of HEVs. The main challenges with nickel-metal hydride batteries are their high cost, high rate of self-discharge, very high gas consumption, and heat generation at high temperatures. There is a need to control hydrogen losses and their low cell efficiency (as low as 50%, typically 60–65%) (NREL, 2009).

The chemical reaction within a battery provides voltage. This occurs when a conductor is connected externally to the electrodes of a cell, causing electrons to flow under the influence of a difference in potential across the electrodes from the zinc (pessimistic) through the external conductor to the carbon (positive), returning within the solution to the zinc. After a short period, the zinc will begin to waste away because of the acid.

Again, the voltage across the electrodes depends upon the materials from which the electrodes are made and the composition of the solution. The difference in potential between the carbon and zinc electrodes in a dilute solution of sulfuric acid and water is about 1.5 V.

The current that a primary cell may deliver depends upon the resistance of the entire circuit, including that of the cell itself. The internal resistance of the primary cell relies on the size of the electrodes, the distance between them in the solution, and the resistance of the solution. The larger the electrodes and the closer they are in solution (without touching), the lower the internal resistance of the primary cell and the more current it can supply to the load.

Note: When current flows through a cell, the zinc gradually dissolves in the solution, and the acid is neutralized.

Again, in many operations, battery-powered devices may require more electrical energy than one cell can provide. Various devices may require either a higher voltage or more current, and in some cases, both. Under such conditions, combining or interconnecting enough cells to meet the higher requirements is necessary. Cells connected in series provide a higher voltage, while cells connected in parallel provide a higher current capacity. A combination series-parallel network of cells must be interconnected to provide adequate power when both voltage and current requirements are more significant than the capacity of one cell.

When cells are connected in **series,** the total voltage across the battery of cells is equal to the sum of the voltage of each of the individual cells. Four 1.5-V cells in series provide a total battery voltage of 6 V. When cells are placed in series, the positive terminal of one cell is connected to the negative terminal of the other. The positive electrode of the first cell and the negative electrode of the last cell then serve as the power takeoff terminals of the battery. The current flowing through such a battery of series cells is the same as from one cell because the same current flows through all the series cells.

A battery connects cells in parallel to obtain a more significant current. In this parallel connection, all the positive electrodes are connected to one line, and all the negative electrodes are connected to the other. Any point on the positive side can serve as the battery's positive terminal, and any point on the negative side can be the negative terminal.

The total voltage output of a battery of three parallel cells is the same as that for a single cell, but the available current is three times that of one cell; the current capacity has been increased.

Identical cells supply equal parts of the current to the load in parallel. For example, three different parallel cells produce a current load of 210 mA, and each cell contributes 70 mA.

A **series-parallel** battery network supplies power to a load requiring both voltage and current greater than one cell can provide. For example, groups of three 1.5-V cells are connected in series to provide the necessary increased voltage, and four series groups are connected in parallel to provide the required increased amperage.

Note that batteries are generally classified by their various characteristics. Parameters such as internal resistance, specific gravity, capacity, shelf life, "c" rate, mid-point voltage (MPV), gravimetric energy density, volumetric energy density, constant-voltage charges, constant-current charge, and specific power are used to describe and classify batteries by operation and type.

Regarding **internal resistance**, it is essential to remember that a battery is a DC voltage generator. As such, the battery has internal resistance, or equivalent series resistance (ESR). In a chemical cell, the resistance of the electrolyte between the electrodes is responsible for most of the cell's internal resistance. Because any current in the battery must flow through internal resistance, this resistance is in series with the generated voltage. With no current, the voltage drop across the resistance is zero, so the full generated voltage develops across the output terminals. This is the open-circuit voltage, or no-load voltage. If a load resistance is connected across the battery, the load resistance is in series with internal resistance. When the current flows in this circuit, the internal voltage drop decreases the battery's terminal voltage.

The ratio of a specific volume of liquid to the weight of the same volume of water is called the **specific gravity** of the liquid. Pure sulfuric acid has a specific gravity of 1.835 since it weighs 1.835 times as much as water per unit volume. The specific gravity of a mixture of sulfuric acid and water varies with the strength of the solution from 1.000 to 1.830.

The specific gravity of the electrolyte solution in a lead-acid cell ranges from 1.210 to 1.300 for new, fully charged batteries. Higher specific gravity reduces internal resistance and increases the current load. As the cell discharges, water dilutes the acid, reducing the specific gravity to about 1.150, indicating full discharge.

The specific gravity of the electrolyte is measured with a **hydrometer**, which has a compressible rubber bulb at the top, a glass barrel, and a rubber hose at the bottom of the barrel. In taking readings with a hydrometer, the decimal point is usually omitted. For example, a specific gravity of 1.260 is read simply as "twelve-sixty." A hydrometer reading of 1210–1300 indicates full charge; about 1250 is half-charge; and 1150–1200 is complete discharge.

The **capacity** of a battery is measured in Ah. The Ah capacity is equal to the product of the current in amperes and the time in hours during which the battery supplies this current and is defined as the amount of current that a battery can deliver for 1 hour before the battery voltage reaches the end-of-life point. In other words, 1 Ah is the equivalent of drawing 1 amp steadily for 1 hour or 2 amps for half an hour. A typical 12-V system may have 800 Ah of battery capacity. The battery can draw 100 amps for 8 hours if fully discharged and starting from a fully charged state. This equals 1200 watts for 8 hours (power in watts = amps × volts). The Ah capacity varies inversely with the discharge current. The size of a cell is determined by its Ah capacity.

The capacity of a storage battery determines how long it will operate at a given discharge rate and depends upon many factors; the most important of these are as follows:

- The area of the plates in contact with the electrolyte
- The quantity and specific gravity of the electrolyte
- The type of separators
- The general condition of the battery (degree of sulfating, plates bucked, separators warped, sediment in the bottom of cells, etc.)
- The final limiting voltage

A cell's shelf life is when it can be stored without losing more than approximately 10% of its original capacity. The loss of capacity of a stored cell is due primarily to the drying out of its electrolyte in a wet cell and to chemical actions that change the materials within the cell. A cell's shelf life can be extended in a cool, dry place.

The **"c" rate** is a current numerically equal to the Ah-rating of the cell. Charge and discharge currents are typically expressed in fractions of multiples of the c rate.

The **MPV** is the cell's nominal voltage, measured when the battery has discharged 50% of its energy.

The **gravimetric energy density** of a battery measures how much energy it contains compared to its weight.

The **volumetric energy density** of a battery measures how much energy it contains compared to its volume.

A **constant-voltage charger** is a circuit that recharges a battery by sourcing only enough current to force the battery voltage to a fixed value.

A **constant-current charger** is a circuit that charges a battery by supplying a fixed current, regulating battery voltage.

In batteries, **specific power** usually refers to the power-to-weight ratio, measured in kilowatts per kilogram (kW/kg).

4.3 THE BOTTOM LINE

AI can assist in anticipating and mitigating grid disruptions, optimizing grid operations, and addressing challenges in energy research. Moreover, AI is being applied to batteries and energy storage in various ways.

REFERENCE

NREL (2009). *Battery Requirements for Plug-in Hybrid Electric Vehicles—Analysis and Rationale.* Paper Presented at National Renewable Energy Laboratories Conference in Anaheim, CA.

Part II

Energy Storage Technologies

5 Lead-Acid Storage Batteries

Artificial intelligence (AI) plays a significant role in the dynamic landscape of energy storage technology. AI integration in energy storage technology becomes necessary to meet the evolving challenges and technological demands of the 21st century.

5.1 FIRST RECHARGEABLE BATTERY

In 1859, French physicist Gaston Plante invented the lead-acid battery, also known as an accumulator. A lead-acid battery is a secondary cell, commonly termed a storage or rechargeable device, that stores chemical energy until it is converted into electrical energy.

The lead-acid battery differs mainly from the primary cell type because it may be recharged. Whereas most primary cells are not recharged.

As the name implies, the lead-acid battery consists of several lead-acid cells immersed in a diluted sulfuric acid electrolyte solution. Each cell contains two sets of lead components (plates); one set is the positive terminal, and the other is the negative terminal. Active materials within the battery (lead plates and sulfuric acid electrolyte) react chemically to produce a flow of direct current whenever current-consuming devices are connected to the battery terminal posts. This current is made by a chemical reaction between the active material of the plates (electrodes) and the electrolyte (sulfuric acid solution).This type of cell produces slightly more than 2 V. Most automobile batteries contain six cells connected in series so that the output voltage from the battery is slightly more than 12 V.

Besides being rechargeable, the primary advantage of the lead-acid storage battery over the dry cell battery is its ability to store electrical energy more efficiently and supply current for a much longer than the average dry cell. Lead-acid batteries can be designed to be high power and are inexpensive, safe, and dependable. Recycling infrastructure is in place for them. However, low specific energy, poor cold temperature performance, and short calendar and cycle life are still impediments to their use. Advanced cost-effective, secure, and dependable high-power, deep-cycle lead-acid batteries are being developed for Hybrid Electrical Vehicle (HEV) applications. Lead-acid batteries are also selected for residential solar systems because of their low maintenance requirements and efficiency.

5.2 LEAD-ACID BATTERY CHARACTERISTICS

When compared to modern rechargeable batteries, lead-acid batteries have low energy density. However, they are able to supply high surge currents and it is these features, along with their low-cost that make them attractive for use in motor vehicles to provide the high current required by starter motors. The specific capacity of a battery depends on its specified discharge rate. If discharged at a faster rate the delivery capacity is less.

5.2.1 Peukert's Law

A power relationship between the discharge current and delivery capacity over some specified range of discharge currents is described by Peukert's law, which expresses approximately the change in capacity of rechargeable lead-acid batteries at different rates of discharge. Applying this law, if Peukert's constant k (the exponent) were equal to unity, the delivered capacity would be independent of the current. An actual battery has an exponent greater than unity in operation, and its capacity

DOI: 10.1201/9781003637851-7

decreases as the discharge rate increases. The ***k*** for a lead-acid battery is typically between 1.1 and 1.3. The Peukert constant (which is dimensionless) usually worsens as the battery ages. Consider the battery's self-discharge current at low rates. Batteries generally offer less capacity than expected under high current flow with a fixed exponent. Equation 5.1 ignores the temperature's impact on battery capacity.

For a one-ampere discharge rate, Peukert's law is often stated as

$$C_p = I^k t \tag{5.1}$$

where:

C_p = the capacity at a one-ampere discharge rate, must be expressed in ampere hours
I = the current drawn from a load in amperes
t = represents the actual time to discharge the battery, must be expressed in hours
k = the dimensionless Peukert constant

Note that Peukert's law is a valuable tool for estimation. However, it has limitations. The equation does not include the effects of temperature on batteries, consider battery age, or account for self-discharge rates at low discharge rates. Additionally, the Peukert exponent increases with battery age.

5.3 LEAD-ACID (PbA) BATTERIES

When the lead-acid (PbA) battery is in the charged state, the positive electrode is lead dioxide (PbO_2) and the negative electrode is sponge lead (metallic lead [Pb]); when discharged in the sulfuric acid electrolyte, both electrodes convert to lead sulfate ($PbSO_4$), this is why many in the battery field say that a fully discharged lead-acid battery is called *sulfated*. The storage of electricity occurs when the electrodes transition between chemical states. PbA batteries have a lower energy density (25–100 kWh/m^3) compared to lithium-ion batteries (150–500 kWh/m^3), but they offer excellent low-temperature stability (Kebede et al., 2022). PbA batteries have several advantages, including user-familiarity with low-cost, globally abundant raw materials that are accessible, safe because of their aqueous electrolyte, and a 99% recycling rate, which minimizes the health and environmental risks due to their exclusion from the throwaway waste problem. The PbA battery has a strong history of market influence in automotive starting, lighting, and ignition but also is used in material handling equipment such as forklifts and critical data center backup (Pilot, 2019). Note that data centers are one of the most energy-intensive, consuming up to 50 times the energy per floor space of a typical commercial office building—note that the data centers also use copious amounts of freshwater for cooling (Spellman, 2025).

To support automotive starting, lighting, and ignition market needs, PbA batteries have transitioned from the conventional flooded to recombinant (value-regulated lead-acid) designs, and from prismatic (objects with two parallel bases) to tubular. Battery engineering remains crucial, with current trends aiming to extend battery lifespan, optimize energy efficiency, and reduce costs. This requires innovations such as advanced bipolar electrode designs (with floating electrodes) and balance of plant optimizations.

PbA batteries are now used in grid energy storage. The electrical energy grid is rarely considered but is crucial for progress. PbA batteries are utilized in grid energy storage due to the intermittent nature of renewable energy sources like solar and wind. As the grid's renewable content increases to a level characteristic of deep decarbonization, durations more significant than 10 hours will be required for long-duration energy storage (LDES). The truth is, LDES entities or markets require exceptionally low-cost technology solutions, and the only potentially viable storage chemistries are those derived from super-low-cost and abundant raw materials, such as lead and sulfuric acid. The problem is that PbA battery designs that are appropriate for today's automotive starting, lighting, and ignition and backup power applications cannot meet LDES performance requirements. Today's stationary markets need batteries with deep charge/discharge cycles and high capacity for cost efficiency, but current models lack round-trip efficiency. The challenge is improving these batteries.

5.4 APPLICATION OF AI TO BATTERIES

Machine learning (ML) uses various algorithms (for clustering, neural networks, penalized regression, and Gaussian process regression) to compress better the time needed to characterize battery performance, lifetime, and safety. ML provides a gateway to monitoring and achieving manufacturing quality. Crowd-sourced data is used—Internet, social media, and smartphone apps—for batteries, which aid in providing and sorting through real-world/complex data. AI is used in battery research to make them better performers with longer staying power and for safety reasons.

5.5 THE BOTTOM LINE

Research into finding the "perfect battery" is ongoing, and there is no need to express how important this research is. The production of battery power is growing and growing and growing some more. AI enables inventors, designers, producers, and operators to obtain the maximum benefit possible with energy storage in batteries commonly called accumulators.

REFERENCES

Kebede, A., Kalogiannis, T., Mierio, J., and Berecibar, M. (2022). A Comprehensive Review of Stationary Energy Storage Devices for Large-Scale Renewable Energy Sources Grid Integration. *Renew Sust Energ Rev*. vol. 159, no. 112213.

Pilot, C. (2019). The Rechargeable Battery Market and Main Trends 2018–2030. Advanced Automotive Battery Conference, May.

Spellman, F.R. (2025). *The Science of AI in Environmental Engineering*. Boca Raton, FL: CRC Press.

6 Lithium-Ion Batteries

Whenever someone uses an electric toothbrush or looks at a battery-powered watch, or operates a laptop, a smartphone, a tablet, or they power their homes by renewable energy, or they use a portable power pack, or a personal mobility device, or a digital personal assistant being used, or they use e-bikes, wheelchairs, or some mobile vehicle, and they have a pacemaker, and finally, they may have some energy storage capability; batteries may power all of these. Here is the point: whenever anyone uses these devices, all is well if the device works as designed. However, if someone goes to use any of these, they do not work.

The first action is removing and replacing the batteries with new ones. Then they evaluate to see if the fault is battery failure, and if it is, the problem is solved. If you request a battery replacement for the device, you need to know what type of batteries it uses and how many are required. Are they able to use lithium batteries?[1]

6.1 INTRODUCTION

Lithium-ion batteries (LIBs) are rapidly entering the laptop and cell phone markets because of their high specific energy. They also have high specific power, high energy efficiency, high-temperature performance, and low self-discharge. Components of LIBs could also be recycled. These characteristics make LIBs suitable for Hybrid Electrical Vehicle (HEV) applications. However, further development is needed to make them commercially viable for HEVs, including improved calendar and cycle life, higher cell and battery safety, abuse tolerance, and acceptable cost (NREL, 2009). A few essential properties of lithium are given in Table 6.1. International Electrotechnical Commission (IEC) battery nomenclature for lithium batteries is listed in Table 6.2.

TABLE 6.1
Properties of Lithium

Property	Value
Atomic mass	6.941 gm
Density	0.535 gm/cm^3
Atomic number	3
Atomic symbol	Li
Appearance	Silvery white
Atomic radiance	152 pm
Melting point	180.5°C (356.9°F)
Boiling point	1,342°C (2,448°F)
Critical point	3,220K, 67 MPa
Heat of fusion	3.00 kJ/mol
Heat of vaporization	136 kJ/mol
Molar heat capacity	24.860 j/(mol·K)
Oxidation state	+1
Electronegativity[a]	0.98 (Pauling scale)
Electronic configuration	[He]2s^1

[a] Electronegativity, symbolized by Y, is the tendency for an atom of a given chemical element to attract shared electrons while forming a chemical bond. As shown in Table 2.1, lithium's electronegativity is 0.98 on the Pauling scale. This value, 0.98, is based on calculation because electronegativity cannot be measured. It must be calculated.

DOI: 10.1201/9781003637851-8

TABLE 6.2
IEC Nomenclature and Definitions for Lithium Batteries (IEC, 2019)

Letter Code	Negative Electrode	Electrolyte	Positive-Electrode	Nominal Voltage
B	Lithium	Organic electrolyte	Carbon monofluoride	3.0
C	Lithium	Organic electrolyte	Manganese dioxide	3.0
E	Lithium	Non-aqueous inorganic electrolyte	Thionyl chloride	3.6
F	Lithium	Organic electrolyte	Iron disulfide	1.5
G	Lithium	Organic electrolyte	Copper(II) oxide	1.5

Note that in addition to the active material at the electrodes, LIBs also contain an electrolyte that consists of organic liquid or polymer compounds and a Li-containing salt.

DID YOU KNOW?

Most people have never heard about a paper (cellulose-based spacer) battery. This is surprising to a degree because paper batteries can be used as both batteries and capacitors. They are flexible, ultra-thin, long-lasting, non-toxic, produce steady power, and are biodegradable—some would say all of this is great. Well, with the good, always find something that is bad. The paper battery is prone to tearing, and nanotubes made from carbon are expensive because of procedures like electrolysis and laser ablation. From a safety point of view, they should not be inhaled to damage the lungs.

DID YOU KNOW?

Most people know about batteries and how they are used in various applications. This is usually the average person's only inclination or knowledge about them. Moreover, battery shape is only a concern for the user when installing or replacing the batteries. These same users see the size and shape of the batteries they need but are unfamiliar with battery shape codes. Shape codes are the following:

R Round (coin, button, or cylindrical)
P Prismatic
F Flat (layer built)
S Square (or rectangular or prismatic)

6.1.1 Atomic Structure of Lithium

A neutral lithium atom has three protons and three electrons (number 3 of 118 elements on the periodic table; Figure 6.1 shows lithium's chemical symbol). On the periodic table, an element's period, or row, indicates how many energy levels are occupied by the electrons in its atoms. Lithium is in period 2, indicating the element's three electrons are found on two energy levels. Two electrons are held at the first energy level. The last electron is found on the second, outermost energy level. Note that electrons found on the outermost energy level of an atom are called *valence* electrons. Valence electrons are necessary because they are the electrons that engage in chemical bonding with other atoms.

FIGURE 6.1 Lithium chemical symbol indicating atomic number 3 and approximate atomic weight of 7.

DID YOU KNOW?

Lithium (Li) is very reactive and if unable to sustain normal room temperature it may catch fire. Kerosine oil is used to ignite fire but is a safe haven for storage of reactive chemicals such as Li.

6.2 LITHIUM-ION POLYMER BATTERIES

Lithium-ion polymer batteries with high specific energy (i.e., high energy per unit mass), initially developed for cell phone applications, also have the potential to provide high particular power for HEV applications. The other key characteristics of lithium polymer are safety and good cycle life. The battery could be commercially viable if the cost is lowered and higher specific power batteries are developed (NREL, 2009).

DID YOU KNOW?

Lithium (Li) is very reactive and may catch fire if it cannot sustain normal room temperature. Kerosine oil ignites a fire but is a haven for storing reactive chemicals such as Li.

6.3 PART OF OUR DAILY LIFE

As pointed out above, LIBs have become the standard in everyday applications and are not given much thought when installing or replacing them. LIBs have been used since the 1990s in consumer electronics, electric vehicles, and stationary energy storage applications. Their energy density has driven much of the shift in electrification over the past decades. Their large-scale deployment has been supported by the ability to mass produce cells and make closed-case batteries in several sizes and shapes that are favorable to arrangement in different high-energy configurations. The mass production of large LIB cells for electric vehicles has also benefited grid deployment.

6.4 LITHIUM COMPOUNDS

Many different lithium salts can be used as medication, including lithium carbonate, lithium acetate, lithium sulfate, lithium citrate, lithium orotate, and lithium gluconate. Note that lithium in pure form has limited but essential uses, but lithium compounds such as lithium carbonate, lithium stearate, and lithium hydroxide are widely used in everyday life and are briefly described below for our purposes.

6.4.1 Lithium Carbonate

Lithium Carbonate (Li_2CO_3), only found in the anhydrous state, is the lithium white salt of carbonate and is an inorganic compound widely used in metal oxide processing. Lithium carbonate is extracted primarily from spodumene in pegmatite deposits and lithium salts from underground brine reservoirs. Lithium carbonate is added to glass or ceramic to make the materials stronger. Pyrex™ glass cookware is a good example of where lithium carbonate is used. It is also used in mirrors and lenses for telescopes. Lithium carbonate is also used to extract aluminum metal from aluminum ores. Lithium carbonate is used mainly in LIBs, where its compounds are used as the cathode and the electrolyte. As early as the 1840s, lithium carbonate was used to treat stones in the bladder, and by the 1850s, lithium carbonate salts were used to treat gravel, urinary calculi, rheumatism, mania, depression, and headache. By the 1940s, lithium carbonate became a standard treatment for bipolar disorder.

6.4.2 Lithium Stearate

Lithium stearate ($LiC_{18}H_{35}O_3$) is a white solid powder mixed with petroleum to make a heavy, lubricated grease used in many industrial machines and soaps. The grease does not break down at high temperatures or get hard at low temperatures. It also does not react with oxygen in air or water, so its chemical composition has a long shelf life.

6.4.3 Lithium Hydroxide

Lithium hydroxide (LiOH) is an organic compound that can be anhydrous or hydrated and is useful in absorbing carbon dioxide.

6.5 THE BOTTOM LINE

Because demand is soaring lithium is often referred to as 'the new oil' and also as 'the new white gold' because it is seen as a valuable alkali metal that can store vast amounts of energy, all of it squeezed into a tiny area, space, and/or containers such as a storage battery. LIBs are currently the most common form of energy storage. The demand for lithium is soaring, competition to find and produce lithium is significant, and, in some cases, is filled with drama. LIBs remain the current batteries of choice because they can store much energy in a compact space. With proper use, these batteries last much longer than other alternatives. They can also recharge more quickly compared to traditional battery systems. AI is being used in various aspects of LIBs. AI assists with accelerated material discovery, replacing traditional manual, serial, and human-intensive work. AI also helps in battery design, manufacturing, materials, and device characterization, life cycle, and safety assessments. AI uses machine learning algorithms to analyze mountains of datasets to predict the properties and performance of potential new battery materials.

NOTE

1 From Spellman's, F.R. (2024). *The Science of Lithium*. Boca Raton, FL: CRC Press.

REFERENCE

NREL (2009). *Battery Requirements for Plug-in Hybrid Electric Vehicles—Analysis and Rationale*. Paper Presented at National Renewable Energy Laboratories Conference in Anaheim, CA.

7 Flow Batteries

Flow batteries store their electric charge within their electrolytes rather than their electrodes.

7.1 FLOW BATTERY OR REDOX FLOW BATTERY

The two names, flow batteries (FBs) and redox FBs (RFBs), are interchangeable in most cases. This innovative technology offers a bidirectional energy storage system that uses redox-active energy carriers dissolved in liquid electrolytes. Thaller of the National Aeronautics and Space Administration first proposed the principle of the FB systems in 1974 (Bartolozzi, 1989). In recent years, there has been significant progress in improving their performance and reducing their costs. RFBs, especially vanadium FBs (VFBs) and zinc-bromine RFBs, are relatively mature technologies actively deployed in various applications.

RFBs work by pumping negative and positive electrolytes through energized electrodes in electrochemical reactors (stacks), allowing energy to be stored and released as needed. With the promise of cheaper, more reliable energy storage, FBs are poised to transform how we power our homes and businesses and usher in a new era of sustainable energy.

Various designs are used for RFBs, including traditional, hybrid, and redox-targeting RFBs. In a conventional dual-FB system (see Figure 7.1) with dissolved active species, two electrolyte tanks containing dissolved active species are separated by a membrane. During the charging and discharging, the active species undergo redox reactions. All the reactants and products are in the solution phase, and no metals are plated on the electrodes during charging. So, the redox flow cell stores energy in the solutions so that the size of the electrolyte tanks controls the system's capacity, while the size of the cell stacks determines the system's power. This indicates that the redox flow cell is more like a rechargeable fuel cell than a battery.

A hybrid FB (HFB) system employs solid anolyte-active and dissolved catholyte-active species, providing extra capacity and higher energy density. In contrast, a redox shuttle design stores solid

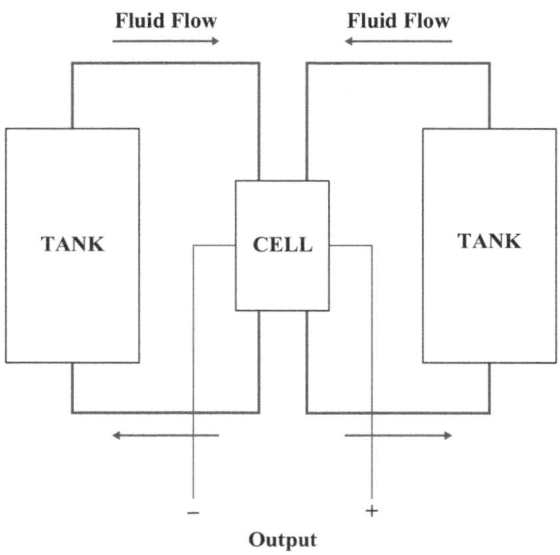

FIGURE 7.1 Simplified diagram of a flow battery.

DOI: 10.1201/9781003637851-9

active material in multiple tanks and a separate tank with a redox shuttle to transport the active species between the solid active tanks.

The significance of RFBs is commonly noted for their variable duration capabilities, utilizing a material supply chain separate from lithium batteries and having the flexibility to scale power and energy separately. This independence of power and energy primarily applies to traditional RFBs and redox-targeting RFBs; note that this is not the case with hybrid RFBs, as those contain a solid anode.

Sometimes, RFBs are categorized as either organic or inorganic redox couples. Success has come to aqueous inorganic vanadium RFBs (VFBs) because the system is *symmetric*, where the same species can be used as a positive charge storer (catholyte) and a negative charge storer for an (anolyte). These two types of electrolytes are used in electrochemical processes. *Anolyte* (the oxidizing agent) refers to the electrolyte solution, the positively charged electrode at the anode. The *catholyte* (the reducing agent) refers to the electrolyte solution at the cathode, the negatively charged electrode. These solutions are crucial in FBs and have critical applications in fuel cells and electrolysis. Note that the composition and properties of these electrolytes can vary depending on the specific application.

A significant aspect of the symmetric design is that it is beneficial because species crossover is no longer a major issue, and electrolyte rebalancing effectively allows decades of reliability. The drawback with chemistry is its volatile accessibility and cost, which often impede development.

DID YOU KNOW?

Because anolyte is able to generate free radicals and reactive oxygen species, these highly reactive species play a crucial role in various applications, including water treatment, disinfection, and sterilization. Because of its strong antimicrobial properties, makes it an effective solution for killing bacteria, viruses, and other pathogens. The catholyte is alkaline in nature and contains reducing agents; it is excellent for neutralizing acids and balancing pH levels.

7.2 FLOW BATTERIES AND THE AI CONNECTION

The addition of new sources of renewable energy like wind and solar power to the electric grid will require specially designed large batteries that can charge when the sun is shining and provide energy at night. The FB fits these requirements. As mentioned, FBs contain two tanks of electrically active chemicals that exchange charges and can have large volumes that hold a lot of energy. The key to providing reliable FBs involves finding target molecules that can store a lot of energy and remain stable for long periods. The active molecules in a FB are anolyte redoxmers (i.e., organic active molecules storing electrochemical energy). Each redoxmer has three beneficial properties. The first two, reduction potential and solvation-free energy (i.e., the energy required to dissolve a solute in solvent), relate to how much energy the molecule can store. The third, fluorescence, is a self-reporting marker indicating the battery's overall health.

In researching methods to optimize the operation of FBs, machine learning and artificial intelligence (AI) techniques are called *active learning*, in which a model can train itself to identify increasingly plausible methodologies. Active learning is a special case of machine learning in which a learning algorithm can interactively query some information source (aka the teacher or oracle) or a human user to label new data points with the desired outputs. If the human user is employed as the information source, the person must possess knowledge/expertise in the problem domain, including the ability to consult/research authoritative sources when necessary (Das et al., 2016; Rubens et al., 2016; Settles, 2020).

One thing is sure: Unlabeled data is abundant, but manual labeling is expensive. In such a scenario, the white knight is the learning algorithm that can actively query the user/teacher for labels.

Again, this type of iterative supervised learning (i.e., repetition until the goal is reached) is called active learning. Note that innovation via research is ongoing to avoid overwhelming the algorithm with uninformative examples. In light of these issues, recent developments are dedicated to multi-label active learning (Yang et al., 2009), hybrid active learning (Lughofer, 2012), and active learning in a single-pass (online) context (Lughofer, 2012), combining concepts from the field of machine learning (e.g., conflict and ignorance) with adaptive, incremental learning policies in the field of online machine learning. The significant advantage of active learning is that it allows for faster development of a machine learning algorithm when comparative updates require a sophisticated super or quantum computer (exploits quantum mechanical properties; Novikov, 2021).

DID YOU KNOW?

The potato/lemon battery is a simple electrochemical cell created by high-voltage-producing lemon or low-voltage-producing potato.

7.3 THE BOTTOM LINE

Flow batteries assisted with AI applications are long-life, stored energy sources. They have a good heat transfer effect and high safety, and the consistency between single cells is better than that of other batteries. In addition, deep charge and deep discharge have no adverse effect on the batteries' performance.

The operation of VFBs is more stringent than that of lithium batteries. Flow batteries are more expensive than iron-chromium FBs. Divalent vanadium is easily oxidized, resulting in system capacity degradation. Despite these drawbacks, today's AI-assisted technology in FBs has increased their popularity and usefulness.

REFERENCES

Bartolozzi, M. (1989). Development of Redox Flow Batteries. A Historical Biography. *J Power Sourc.* vol. 27, no. 3, pp. 219–234.

Das, S., Wong, W.-K., Dietterich, T., Fern, A., and Emmott, A. (2016). Incorporating Expert Feedback into Active Anomaly Discovery. In Bonchi, F., Domingo-Ferrer, J., Baeza-Yates, R., Zhou, A-H., Xindong (eds.). *IEEE 16th International Conference on Data Mining.* Kunming, China.

Lughofer, E. (2012). Hybrid Active Learning Reduces Operators' Annotation Effort in Classification Systems. *Pattern Recogn.* vol. 45, no. 2, pp. 884–896.

Novikov, I. (2021). The MLIP Package: Moment Tensor Potentials With MPI and Active Learning. *Mach Learn: Sci Technol.* vol. 2, no. 2, pp. 3, 4.

Rubens, N., Mehdi, F., Sugiyama, M., and Kaplan, D. (2016). Active Learning in Recommender Systems. In Ricci, Francesco, Rokach, L., and Shapira, B. (eds.). *Recommender System Handbook.* New York: Springer Publishing Company.

Settles, B. (2020). *Active Learning Literature Survey.* Accessed 3/8/25 @ http://pages.cs.wisc.edu/~bsettles/pub/settles.activelearning.pdf

Yang, B., Sun, J.-T., Wang, T., and Chen, Z. (2009). *Effective Multi-Label Active Learning for Text Classification.* Accessed 3/8/25 @ https://www.microsoft.com/en-us/research/wp-content/uploads/2009/01/sigkdd09-yang.pdf

8 Zinc Batteries

Lithium-ion batteries have long been the prevailing norm, the standard for energy storage. However, zinc-based batteries are emerging as a more sustainable, cost-effective, and high-performing alternative.[1]

8.1 ZINC

Zinc is a periodic table chemical element, with the chemical symbol Zn and atomic number 30, group number 30 (zinc group), block d, and an atomic weight of 65.38 Ar (see Figure 8.1). It is classified as a transition metal and is part of group 12 (zinc group). At room temperature, zinc is solid and is chemically like magnesium.

8.2 BUTTON CELLS

Since the early 1800s, zinc (Zn) has been used as the negative electrode (anode) when Alessandro Volta formed early voltaic piles from alternating copper and Zn stacks. Zinc has been a good choice because of its low-cost, higher energy density, safety, and worldwide availability, making Zn-based batteries attractive for development for more than 220 years. This Zn-based battery was widespread until 1980, when the popularly used Zn-carbon battery began to have limited use because of the introduction of the alkaline Zn-MnO_2 batteries (Kordesch and Taucher-Mautner, 2009; Lim et al., 2021; Marsal et al., 1960; Wruck et al., 1991). Commonly recognized as AA, AAA, C, D, and 9 V batteries, these batteries have become some of the most commercially successful. Zn-Ni batteries were developed in the 1920s and were used as rechargeable batteries capable of ~1000 deep discharge cycles, potentially suitable for application in electric vehicles (Lim et al., 2021; McLarnon and Cairns, 1991; Walsh, 1980). It was in 1933 that zinc-air batteries, commonly recognized as "button cells" today, became widely used in hearing aids and some film cameras. These batteries have been around for some time and have set the foundation and inspiration for several of the most commercially advanced batteries used for grid-scale storage.

8.3 ZINC BATTERIES AND OTHERS

Zinc batteries have a negative electrode made of zinc. The positive electrode for zinc batteries depends on the type and includes manganese dioxide, oxygen, mercuric oxide, silver oxide, or nickel oxyhydroxide. The electrolyte varies from ammonium chloride, zinc chloride, or Alkali metal

FIGURE 8.1 Zinc chemical symbol indicating atomic number 30 and approximate atomic weight of 65.38.

DOI: 10.1201/9781003637851-10

hydroxide, and the nominal voltage for zinc batteries ranges from a low of 1.3 V to a maximum open circuit voltage (V) as high as 1.75 V.

So, how do zinc batteries compare to other types of batteries?

There are a few significant differences between alkaline and zinc batteries, including the shelf life of alkaline batteries, which is longer than that of zinc batteries. Alkaline batteries also provide more power than zinc batteries. One plus side of zinc over alkaline batteries is that zinc batteries are cheap. In addition, zinc batteries also have high specific energy and high safety. In contrast, alkaline batteries are ideal for high-drain devices, while zinc batteries are best for low-drain devices.

DID YOU KNOW?

Alkaline electrodes have limited rechargeability and stability due to dendrite formation, passivation, and corrosion during periods of inactivity.

Okay, how about zinc batteries and lithium-ion batteries?

Lithium-ion batteries are the popular storage battery choice, with more than 90% of the market. Zinc battery electrolytes are more common and cheaper than the materials used in lithium-ion batteries. The problem with lithium-ion batteries is that due to their popularity, their components are constrained and becoming difficult to procure. On the other hand, zinc-ion batteries offer a cheaper and ultimately safer energy storage option.

8.3.1 Zinc-Air Batteries

Zinc-air batteries are metal-air electrochemical cells powered by the oxidation of zinc with oxygen from the air. During discharge, a mass of zinc particles forms a porous anode saturated with an electrolyte. Hydroxyl ions are formed when oxygen from the air reacts at the cathode. The hydroxyl ions migrate into the zinc paste and form zincate ($Zn(OH)^{2-}{}_4$). This triggers electrons to advance to the cathode—recall that the movement of electrons in a conductor is electricity.

Note that zinc-air batteries need oxygen in the air and <u>cannot</u> be sealed in a battery holder.

DID YOU KNOW?

Battery technology might not be new—the keywords are, might be. Consider the Baghdad Battery, for example. In 1936, a set of three artifacts—a ceramic pot, a copper tube, and an iron rod—were found together. Their purpose and origin remain a mystery. Some have suggested that the object functioned as a galvanic cell, possibly used for electroplating or some kind of electrotherapy. Is the Bagdad Battery just that, a battery? We do not know for sure.

8.4 CURRENT USE

Current commercial usage of alkaline $ZnMnO_2$ and Zn-air batteries is limited to powering smaller portable consumer electronics. Recent demonstrations, deployment, and applications of grid-scale Zn-MnO_2 include assurance (backup power), grid stabilization, and renewable solar integration for residential and commercial microgrid applications. USDOE (2023) points out that Urban Electric Power's alkaline battery backup system installation for the San Diego Supercomputer Center demonstrates a more extensive deployment of zinc battery technology.

Zinc-bromine rechargeable batteries (ZBRBs) are finding traction for microgrids, peak shifting (e.g., behind-the-meter [BTM] product that bypasses the electric meter and is not measured by the utility company) applications, and renewables integration. These ZBRBs are one of the most potent contenders for next-generation energy storage due to the potentially lower material cost, deep discharge capability, non-flammable electrolytes, relatively long lifetime, and good reversibility. Note, however, that these batteries are still a work in progress to improve efficiency and stability for long-life operation. ZBRB, for commercial purposes, comprises both static and flow battery configurations. Like the Zn-air batteries, the ZBRB uses an aqueous Zn-halide electrolyte and relies on a Zn metal anode's reversible plating (reduction) and stripping (oxidation). The following reactions present the overall discharge electrochemistry for both systems (Gao et al., 2020).

$$\textbf{Cathode}: \mathrm{Br} + 2\mathrm{e}^{-} [\mathrm{E}^{0} = +1.09\,\mathrm{V}]$$

$$\textbf{Anode}: \mathrm{Zn} \quad \mathrm{Zn}^{2+} + 2\mathrm{e}^{-} [\mathrm{E}^{0} = +0.76\,\mathrm{V}]$$

$$\textbf{Overall}: 2\mathrm{Zn} + \mathrm{Br}_2 \quad 2\mathrm{Zn}^{2+} + 2\mathrm{Br}\ [\mathrm{E}^{0}{}_{\mathrm{cell}} = +1.85\,\mathrm{V}]$$

Due to their hazardous nature, Br_2 batteries are assembled in the discharged state. When charged, Zn metal deposits on the anode while Br_2 forms on the cathode, complexing with Br to form soluble Br^{3-} species. This easily reversible reaction leads to high cycle life (full depth of discharge) with daily cycles for 10 years (flow battery) and 20 for static, sealed cells.

Note that there are other promising adaptations of Zn-based batteries, many of which may be considered Zn-ion batteries. Some of the current variations in development use slightly acidic or neutral pH electrolytes and rely on protons or Zn ions to balance charge during electrochemical cycling.

DID YOU KNOW?

Research continues on battery development, emphasizing power availability, shelf life, and non-toxic energy storage devices. One emerging battery is the sugar biobattery. The sugar battery generates an electric current by oxidizing the glucose unit of maltodextrin (i.e., white powder made from rice, potato, corn wheat, or tapioca).

8.5 THE BOTTOM LINE

Zn-based batteries are cheaper than alkaline ones and have a long shelf life. They can be stored for several years and kept at full charge. In addition, Zn-based batteries are not toxic. Zinc-based batteries have good power for their size. On the downside, Zn-based batteries have a high risk of leakage and have lower energy density.

Artificial intelligence (AI) is applied to zinc batteries in various ways. A good example is using machine learning (ML) to find a dendrite-free aqueous zinc-ion layer for the battery. AI is also used to address complex challenges in high-throughput organic battery development. The real bottom line is that AI plays a significant role in advancing zinc battery technology.

NOTE

1 From Spellman, F.R. (2025) *The Science of Energy*. Boca Raton, FL: CRC Press.

REFERENCES

Gao, et al. (2020). A High-Performance Aqueous Zinc-Bromine Static Battery. *Science*. vol. 23.

Kordesch, K. and Taucher-Mautner, W. (2009). Primary Batteries—Aqueous Systemsl Leclanché and Zinc-Carbon. In *Encyclopedia of Electrochemical Power Sources*. pp. 43–54.

Lim, M.B., Lambert, T.N., and Chalamala, B.R. (2021). Rechargeable Alkaline Zinc-Manganese Oxide Batteries for Grid Storage: Mechanisms, Challenges, and Developments. *Mater. Sci. R*. vol. 143.

Marsal, P.A., Kordesch, K., and Urry, L.F. (1960). *Dry Cell*. U.S. Patent 2,960,558.

McLarnon, F.R. and Cairns, E.J. (1991). The Secondary Alkaline Zinc Electrode. *J Electrochem Soc*. vol. 138, no. 2, pp. 645–656.

USDOE (2023). *Technology Strategy Assessment: Zinc Batteries*. Washington, DC: United States Department of Energy.

Walsh, W.J. (1980). Advanced Batteries for Electric Vehicles Look to the Future. *Phys Today*. vol. 33, no. 6.

Wruck, W.J., Bullock, B.K.R., and Kao, W.H.. (1991). *J Electrochem Soc*. vol. 138, no. 12, pp. 3560–3567.

9 Sodium Batteries

Think about battery-operated portable light devices. This type of light is powered by energy stored within the battery in chemical form, which is converted into electrical energy to power the bulb. The bulb uses electrical energy and is converted into radiant energy. Note that for large-scale energy storage, sodium (Na) is attractive due to its global abundance and distribution, making it widely available.[1]

9.1 SODIUM

Sodium is a chemical element with the symbol Na and atomic number 11 (see Figure 9.1). It is a soft, silvery-white, highly reactive metal. Sodium is an alkali metal in group 1 of the periodic table.

9.2 SIXTH MOST ABUNDANT ELEMENT

Sodium (Na) batteries are candidates for large-scale energy storage because sodium is the sixth most abundant element in the Earth's crust and the fourth most abundant element in the ocean. It is also globally plentiful and inexpensive. Significant research and development of Na batteries have occurred for more than 50 years. In the 1960s, molten Na batteries began with the sodium-sulfur (NaS) battery as a potential high-temperature power source for vehicle electrification (Kummer and Weber, 1968). In the 1970s, the NaS battery was followed by the sodium-metal halide battery (NaMH-sodium-nickel chloride), also known as the ZEBRA battery (i.e., Zero Emission Battery Research Activities), also with transportation applications in mind (Werth, 1975). At the same time, sodium-ion batteries (NaIBs) were developed along with lithium-ion batteries (LIBs) in the 1980s. The problem was the limitations of charge/discharge rate, cyclability, energy density, and stable voltage profiles, which historically made NaIBs less competitive than their lithium-based counterparts (Hwang et al., 2017). Solid-state sodium batteries (SSSBs) have begun to emerge as candidate commercial products. This solid-state battery is currently one of the most promising options for next-generation battery technology, promising improved operational safety and higher energy and power density compared to state-of-the-art lithium-ion batteries (LIBs) using organic liquid electrolytes. This is the case even though SSSBs have not been established at the present time in terms of their applicability to large-scale, long-duration storage.

DID YOU KNOW?

Another type of energy storage system is the **Carnot battery**, which stores electricity in thermal energy storage. During this, electricity is turned into heat and kept in storage. During discharge, the stored heat is converted back into electricity (Dumont et al., 2020; IEA, 2023).

FIGURE 9.1 Sodium chemical symbol indicating atomic number 11 and approximate atomic weight of 22.98.

DOI: 10.1201/9781003637851-11

9.3 SODIUM BATTERY CHEMISTRY

Both NaS and NaMH batteries employ a molten Na anode and ceramic sodium-ion conduction solid-state separator, most commonly beta-alumina solid electrolyte (BASE) (or β''-alumina), but the molten cathode chemistries differ (Spoerke et al., 2020, 2021). Both chemistries typically operate at elevated temperatures near 300°C to ensure the active material's molten state and the BASE's high conductivity. Each class of molten Na batteries is described below (USDOE, 2023).

9.3.1 Sodium-Sulfur (NaS) Batteries

Sodium-sulfur batteries (see Figure 9.2) are molten-salt batteries that use liquid sodium and liquid sulfur electrodes. During electrochemical cycling, traditional NaS batteries oxidize (discharge) and oxidize (charge) molten sulfur (S) at the cathode. To balance these reactions, oxidized Na+ shuttles between the electrodes through an ion-conducting ceramic separator and participates in the reversible formation of sodium polysulfides (Spoerke et al., 2020).

$$xS + 2Na \leftrightarrow Na_2S_x\,(3 \le x \le 5)\,E_{cell} \sim 2.08\,V\,at\,350°C$$

9.3.2 Sodium Metal Halide (NaMH) Molten Salt Batteries

Sodium metal halide (NaMH) molten-salt batteries, such as sodium-nickel chloride (Na-$NiCl_2$ or ZEBRA), are battery class that uses molten salts as an electrolyte and offers high energy and power densities. Like the NaS battery, these batteries rely on the oxidation and reduction of Na at the anode and utilize an ion-conducting ceramic separator; however, they rely on the reduction and oxidation of a nickel chloride/nickel-based cathode ($NiCl_2Ni$). Typically, the Ni cathode takes the form of powders suspended in a supporting metal halide molten salt "catholyte" (located near the cathode in an electrolytic cell), usually sodium tetrachloroaluminate ($NaAlCl_4$). These batteries are also referred to as salt or molten salt batteries. The electrochemical reaction of the Na-$NiCl_2$ is given by Spoerke et al. (2020):

$$NiCl_2(s) + 2Na(l) \leftrightarrow 2\,NaCl + Ni(s)E_{cell} \sim 2.58\,V\,at\,300°C$$

Over the past few years, variations of these NaMH molten salt batteries have been developed under DOE-funded research, looking to replace the costly Ni metal in NaMH batteries with low-cost and domestically abundant iron, zinc, or aluminum cathodes and lower the operation temperature

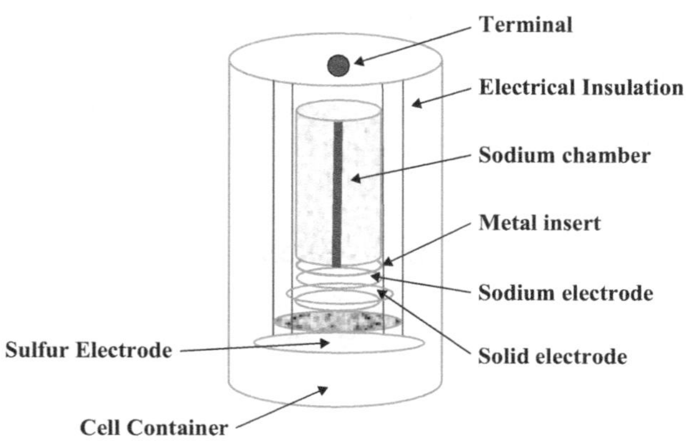

FIGURE 9.2 Sodium-sulfur battery.

by using iodide or other halide salts (Lu et al., 2013; Small et al., 2017; Zhan et al., 2020). Emerging chemistries have also led to lower temperature (<200°C) systems. This innovation is expected to lower costs and increase battery life, and the iodide chemistry exhibits voltages as high as 3.6 V with selected molten salt compositions (Gross et al., 2020, 2021).

9.3.3 Sodium-Ion Batteries (NaIBs)

Sodium-ion batteries (NaIBs) have features comparable to LIBs in that, in some cases, the working principle and cell construction are similar. For one thing, sodium belongs to the same group in the periodic table as lithium, meaning they both have comparable chemical properties. However, this comparable chemical property makes NaIBs significantly different from molten Na batteries (Abraham, 2020; Hwang et al., 2017). These batteries operate near ambient temperature and can employ the two-dimensional transition metal layered oxide (TMLO) cathode. In addition, TMLOs are non-selective and include an insulating porous polymer separator and a non-graphitic carbon or a titanate anode. Moreover, TMLOs use an organic or aqueous liquid electrolyte (Abraham, 2020; Hwang et al., 2017; Jin et al., 2020; Xiao et al., 2018). During the charging of the anode and the discharge of the cathode, Na ions are alternately intercalated. Probably the best way to use an example in a generalized way is to show the TMLO-based Na-ion chemistry as being analogous to traditional lithium-ion chemistry by providing the following reaction for review (Abraham, 2020):

$$Na_xC_6 + Na_{1-x}MO_2 \leftrightarrow NaMO_2 + C_6$$

Depending on chemistry and the state of charge, the voltage of these batteries varies from −2 V to nearly 4 V (Hwang et al., 2017). However, their cycle life, energy density, and power (their performance) still suffer somewhat compared to lithium equivalents. NaIBs do not rely on cobalt or nickel; instead, they use abundant Na. Also, they do not require expensive copper negative electrode current collectors, as with lithium-ion systems.

Prussian blue (PB) and its equivalents (analogs, PBAs) are cathode materials for practical applications in sodium-ion batteries. These PBAs with a nominal composition of $Na_xM[R(CN)_6]$ (M = Ni, Cu, Co, Fe; R = Fe, Mn, or Cr; x varies with the state of charge) are being utilized as sodium-ion alternatives to conventional metal oxide cathode material (and possibly some anode material) (Hurlbutt et al., 2018; Pasta et al., 2013; Wessells et al., 2011). Time and again, these PBAs are ferric ferrocyanide salts with a cubic crystal structure containing large channels to accommodate the rapid movement of Na+ ions in and out of the material with minimal volume change. Consequently, these materials have unusually high-rate capability (enabling high power) and cycling stability (up to 100,000 cycles is feasible in commercial systems). Note that these batteries significantly differ from traditional TMLO or phosphate polyanion chemistry.

The bottom line: From a safety perspective, PBA batteries use non-flammable aqueous electrolytes, which help provide a simplified, non-hazardous battery management system.

DID YOU KNOW?

Although PB dye was initially commercialized in the dye industry, it has more recently been used in diverse energy technologies, including electrocatalysis, thermal power generation, and energy storage.

9.3.4 Solid-State Sodium Batteries (SSSBs)

Earlier, it was pointed out that SSSBs have begun to emerge as candidate commercial products. Moreover, this solid-state battery is currently one of the most promising options for next-generation battery technology, promising improved operational safety and higher energy and power density compared to state-of-the-art LIBs using organic liquid electrolytes. Recall that this is the case even though SSSBs have not been established regarding their applicability to large-scale, long-duration storage. Unlike molten Na of NalBs, relatively less mature SSSBs do not use a (significant) liquid electrolyte to facilitate ion transport through the batteries. However, it is essential to note that they borrow many of the cathode chemistries (e.g., TMLO, PBAs) from NalBs, and they often rely on solid-state electrolytes (SSEs) like those used in molten NaS and NiMH batteries. These systems would ideally use metallic Na for the highest energy density, but Na composites, alloys, and other materials continue to be developed. SSSBs are attractive emerging batteries because they are intended to replace batteries with large volumes of flammable, leakable electrolytes. Significant challenges still exist, such as dynamic, solid-state interfaces, material stability, and the efficacy of solid electrolytes at ambient atmospheres and temperatures, for these technologies to advance commercially. Other issues need to be addressed; for example, if SSSBs are to be considered potential replacements for NalBs and LIBs for solid-state lithium batteries, they will also have to be addressed for the Na-based systems. In many ways, SSSB development is a parallel effort to current, aggressive lithium solid-state battery development (USDOE, 2023).

9.4 THE BOTTOM LINE

Sodium batteries are rechargeable batteries that utilize sodium ions as the charge carriers between positive and negative electrodes. They are similar to LIBs in that they are designed to store and release electrical energy by moving ions back and forth between the electrodes during charging and discharging cycles. These sodium batteries have advantages in that sodium is more abundant and widely available than lithium. Sodium is available at a lower cost and can lead to more affordable battery production. Existing manufacturing processes and infrastructure for LIBs have the same chemistry as sodium batteries; these lithium-like processes facilitate the production of sodium batteries. Sodium batteries can be used in grid-scale energy storage applications, which help to integrate renewable energy sources. Then, sodium batteries have environmental advantages in that they reduce environmental impact compared to lithium batteries during disposal or recycling. These are all the positives of sodium over lithium, but like anything else in the real world, there are two sides to each coin, and this is the case in using sodium instead of lithium batteries. For example, sodium battery use has not been around as long as lithium technology. Some sodium-based batteries may exhibit a low life cycle, which could lead to short battery lifespans. With sodium-based batteries, it is a challenge to achieve consistent performance.

Regarding artificial intelligence (AI) and sodium-based batteries, note that AI is used in Na batteries to optimize material composition and develop high-efficiency batteries. AI used with sodium-based batteries promotes environmentally friendly ecosystem solutions. Machine learning (ML) is being used to make sodium-based batteries more efficient, cost-effective, and safer.

NOTE

1 From Spellman, F.R. (2025). *The Science of Energy*. Boca Raton, FL: CRC Press.

REFERENCES

Abraham, K.M. (2020). How Comparable Are Sodium-Ion Batteries to Their Lithium-Ion Counterparts? *ACS Energy Lett.* vol. 5, pp. 3544–3547.

Dumont, O., Frate, G.F., Pillai, A., Lecompte, S., De Paepe, M., and Lemont, V. (2020). Carnot Battery Technology: A State-of-the-Art Review. *J Energy Storage.* vol. 32, p. 101756.

Gross, M.M., Percival, L.J., Small, L.J., Lamb, J., Peretti, A.S., and Spoerke, E.D. (2020). Low-Temperature Molten Sodium Batteries. *ACS Appl Energy Mater.* vol. 3, no. 11, pp. 11456–11462.

Gross, M.M., Percival, L.J., Small, L.J., Lee, R.Y., Peretti, A.S., and Spoerke, E.D. (2021). Molten Halide Catholyte Chemistry Enables a High-Voltage, Low-Temperature Molten Sodium Battery. *Cell Rep Phys Sci.* vol. 2, no. 7.

Hurlbutt, K., Wheeler, S., Capone, I., and Pasta, M. (2018). Prussian Blue Analogs as Battery Materials. *Joule.* vol. 2, no. 10, pp. 1950–1960.

Hwang, J.Y., Myung, S.T., and Sun, Y.K. (2017). Sodium-Ion Batteries: Present and Future. *Chem. Soc. Rev.* vol. 46, pp. 3529–3614.

IEA (2023). *Task 36 Carnot Batteries Final Report.* Paris: International Energy Agency.

Jin, T., Li, K., Zhu, P.F., Wang, P., Liu, and Jiao, L. (2020). Polyanion-Type Cathode Materials for Sodium-Ion Batteries. *Chem Soc Rev.* vol. 49, no. 8, pp. 2342–2377.

Kummer, J.T., and Weber, N. (1968). *Secondary Battery Employing Molten Alkali Metal Reactant.* United States: SAE International Warrendale, PA.

Lu, X., Li, G., Kim, J.Y., Lemmon, J.P., Sprinkle, V.L., and Yang, Z. (2013). A Novel Low-Cost Sodium-Zinc Chloride Battery. *Energy Environ Sci.* vol. 6, pp. 1837–1843.

Pasta, S., et al (2013). Full Open-Framework Batteries for Stationary Energy Storage. *Nature Commun.* vol. 5.

Small, L.J., et al (2017). Next-Generation Molten NaI Batteries for Grid-Scale Energy Storage. *J Power Sources.* vol. 360, no. 31, p. 6.

Spoerke, E.D., Gross, M.M., Percival, S.J., and Small, L.J. (2020). Molter Sodium Batteries. In Alston, M. and Lambert, T. (eds.). *Energy-Sustainable Advanced Materials.* New York, NY: Springer.

Spoerke, E.D., Gross, M.M., Small, L.J., and Percival, S.J. (2021). Sodium-Based Technologies for Grid Scale Energy Storage. In *DOE Energy Storage Handbook.* Washington, DC: U.S. Department of Energy.

USDOE (2023). *Technology Strategy Assessment: Sodium Batteries.* Washington, DC: United States Department of Energy.

Werth, J.J. (1975). *Alkali Metal-Metal Chloride Battery.* Patent Appl. US3877984: Washington, DC: United States of America.

Wessells, C., Huggins, R.A., and Cui, R. (2011). Copper Hexacyanoferrate Battery Electrodes With Long Cycle Life and High Power. *Nature Commun.* vol. 2.

Xiao, B., Rojo, T., and Li, X. (2018). Hard Carbon as Sodium-Ion Battery Anodes: Progress and Challenges. *ChemSusChem.* vol. 12, no. 1, pp. 133–144.

Zhan, X., et al. (2020). Na-FeCl2 Batteries: A Low-Cost, Durable Na-$FeCl_2$ Battery With Ultrahigh Rate Capability. *Adv Energy Mater.* vol. 10, no. 10.

10 Flywheel Energy

$E = ½ • I • w^2$

10.1 FLYWHEEL ENERGY STORAGE (FES)

One of the mechanical energy storage systems is the flywheel that, for short periods, can provide high power repeatedly. This system works by accelerating a flywheel (rotor) to an extremely high speed and maintaining energy in the system as rotational kinetic energy. The principle of energy conservation comes into play when energy is extracted from the system, reducing the flywheel's rotational speed. To add energy to the flywheel system, the speed of the flywheel must be increased. Safety is an issue with flywheels; they can disintegrate while in operation, causing substantial damage in areas where they are located (USDOE, 2023).

10.1.1 Technical Description

Flywheels are nothing new; they have been used for thousands of years. However, very few know about or understand flywheels and their operation. Recently, the most common flywheel application has been internal combustion energy. Probably a more straightforward example of the application of a flywheel in operation is the potter's wheel. A flywheel is a simple form of mechanical (kinetic) energy storage. This energy storage is the form of the angular momentum of a spinning mass called a rotor. Again, the work to spin the mass is stored as kinetic energy. The flywheel system transfers kinetic energy into AC power through power conversion systems and controls. The stored energy is proportional to the flywheel's mass and the square of the rotational speed; this is shown in Equation (10.1).

$$E = ½ \bullet I \bullet w^2 \tag{10.1}$$

where:

E = energy stored in a flywheel (its capacity) in the form of rotational energy
I = momentum of inertia of the rotation component of the flywheel
w = angular velocity of the rotating component
Measured in rads per sec (rads/s)

Advances in power electronics, artificial intelligence (AI), magnetic bearings, flywheel materials, and innovative component integration have resulted in direct current (D-C) flywheel energy storage systems that can substitute or supplement batteries in uninterruptible power supply (UPS) systems. Regarding first-cost expense, flywheel systems are more expensive than batteries but have longer life and more straightforward maintenance. These systems offer a competitive advantage, making them attractive as a battery alternative (USDOE, 2023).

Note that batteries used for UPS applications are typically sized for about 15 minutes of full-load power. However, experience has shown that most power disturbances last 5 seconds or less. So, because many UPSs rely on fuel-fired generators that come up to full power within 10 seconds, the typical D-C flywheel system, designed to provide 15 seconds of full load power, could be substituted for batteries in UPS systems with fuel-fired generators. Or else, D-C flywheel systems could be used in combination with batteries. Remember that frequent discharging and recharging are much more harmful to battery life than to flywheel life.

DOI: 10.1201/9781003637851-12

10.1.2 Flywheel Types

In general, flywheels can be low-speed or high-speed. The high-speed types operate at revolutions per minute (rpm) measured in thousands, while the slow-speed type operates at rpm measured in tens of thousands. Note that doubling the rpm quadruples the stored energy; all else equal, increasing rpm significantly increases a flywheel's energy density. Flywheel design is based on speed; operating at higher rpm necessitates fundamental differences in design approach. While high-speed flywheels are typically made from carbon or carbon and fiberglass composite materials that will withstand the higher stresses associated with higher rpm, low-speed flywheels are usually made from steel. Higher rpm also increases concern with friction losses from bearings and air drag. High-speed flywheels universally employ magnetic bearings and vacuum enclosures to reduce or eliminate the two sources of friction. Magnetic bearings allow the flywheel to levitate to eliminate frictional losses associated with conventional bearings. Low-speed flywheels typically use conventional bearings, but most flywheels use a combination of the two bearing types. Note that vacuums are also employed in some low-speed flywheels.

10.1.3 Flywheel Applications

The limiting factor in employing flywheel energy storage systems is their initial cost. However, D-C flywheel energy storage systems are generally more reliable than batteries. So, again, batteries will usually have a lower first cost than flywheels; nevertheless, they suffer a significantly shorter equipment life and higher annual operation and maintenance expenses. Accordingly, flywheels will look especially attractive in operating environments that are detrimental to battery life, such as the following:

- Main power supply problems necessitating frequent cycling.
- Unconditioned spaces lead to high temperatures.

10.1.4 Flywheel Limitations

Certain situations involving flywheels need to be avoided. Let's first look at UPS. UPS batteries are sized to provide backup power for a period measured in minutes. Note that the period ranges from about 5 minutes up to close to 1 hour but is commonly about 15 minutes. A period of 15 minutes or thereabouts is generally presumed adequate to allow an orderly equipment shutdown. On the other hand, flywheels provide backup power for periods measured in approximately 15 seconds of backup power. This relatively short period is usually enough time to handle most power disruptions that last 5 seconds or less, and it still has time to cover slightly more extended outages until a backup generator can come up to speed and can usually provide full power in 10 seconds. However, remember that a flywheel alone will not offer backup power for a period long enough to allow an orderly process shutdown in most cases. Accordingly, the following limitations should be considered.

- Flywheels should not be used alone without some backup.
- If the backup generator cannot come up to speed and full load in less than 10 seconds, flywheels alone should not be used.

10.2 THE BOTTOM LINE

A D-C flywheel system can handle power disturbances, saving batteries for extended outages and significantly increasing battery life. A flywheel could also be used alone for applications where long-term backup capability is not required or economically justified. Operational changes to the grid, caused by the restructuring of the electric utility industry and advancements in electricity storage technology, have created an opportunity for storage systems to provide unique services to the evolving grid. Note that applications requiring millions of charge-discharge cycles in batteries and a calendar life of decades will benefit from flywheels, as battery cycle life remains significantly lower. Regarding the impact of AI in flywheel storage systems, it works to optimize and enhance flywheel energy systems, improving reliability and efficiency in operation. It is economically competitive with other energy storage systems. Although the initial cost will usually be higher, flywheels offer a much longer life, reduced maintenance, a smaller footprint, and better reliability than batteries.

REFERENCE

USDOE (2023). *Technology Strategy Assessment: Flywheel Energy System*. Washington, DC: United States Department of Energy.

11 Compressed-Air Energy
Electricity to Compression to Storage to Expansion to Electricity

11.1 OFF-PEAK ELECTRICITY

Another mechanical energy storage option using cheap off-peak electricity is compressed-air energy storage (CAES). An attractive feature of this technology is its relatively simple process. It can store electric energy as potential (compressed air) and be deployed near central power plants and distribution centers. In the process, a compressor is powered by available electricity to charge the air compressor, which is then stored in a chamber until the energy is needed. In response to demand, the stored energy can be discharged by expanding the stored air with a turboexpander generator (see Figure 11.1) (USDOE, 2023).

The question is: what are the attributes of CAES? CAES's characteristics make it attractive because it includes a wide range of energy storage capacities, from a few megawatts to several gigawatts. Other attributes include it being an environmentally friendly process (this is the case when no fossil fuel is used for combustion), long life and durability, low self-discharge (because of loss of pressure and temperature), and energy storage cost. The main challenges of technology include high upfront capital costs, the need for heat during the expansion step, lower round-trip efficiency (RTE), locating and permitting challenges, difficulty in identifying and preparing natural caverns for storage, low depth for discharge, and longer response times.

DID YOU KNOW?

Energy storage capacity in megawatts is determined by the amount of air and thermal energy that can be stored. Higher pressure, volume, and temperature of the thermal energy storage (TES) media equate to more energy storage capacity. A high energy-to-power ratio enables longer charge and discharge periods. The point is that the incremental cost of a more extensive underground reservoir is lower than that of multiple storage tanks.

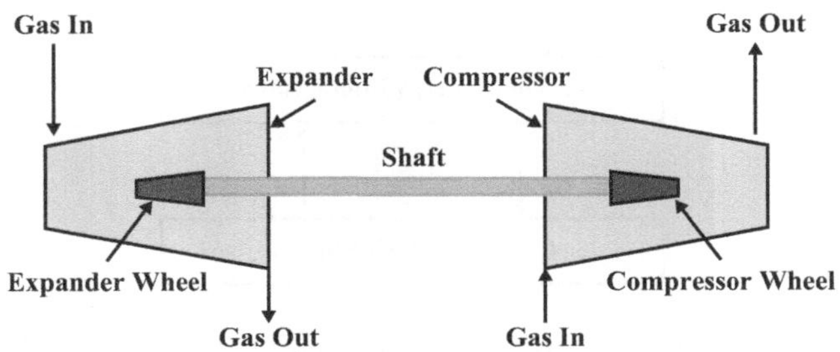

FIGURE 11.1 Diagram of turboexpander, centrifugal or axial-flow turbine, driving a compressor.

DOI: 10.1201/9781003637851-13

11.2 NOTHING NEW ON THE BLOCK

When teaching energy storage systems to college students, I always begin my lecture on CAES technology by stating that compressed air has been used for mechanical processes around the globe since 1870; relatively speaking, it is nothing new on the block. Compressed-air energy has been used to move clock arms every minute. In 1986, compressed air was used to power homes and industry in Paris. In Huntorf, Germany, in 1978, the first utility-scale diabatic CAES project was developed for grid applications.

DID YOU KNOW?

The diabatic process is a change of state in which the system exchanges energy with its surroundings by changing its temperature. Simply put, no heat is transferred into or out of the system.

11.3 THE PROCESS

The CAES process is not rocket science; it's simpler and involves compressing and storing high-pressure air (see Figure 11.2). The air can be stored in another TES medium, rejected, or used in other integrated applications, thereby improving the RTE. RTE is one of those quintessential metrics and is usually defined as the output electrical energy discharged after storage as a percentage of the incoming energy. Note that the management of thermal energy is a key element in the design of the process, each with its pluses and minuses. CAES processes can be classified as (1) diabatic, where the heat during compression is either rejected or recovered, and fuel is burned during the expansion process, with an RTE of 46–54%; (2) adiabatic, where the heated and compressed air is either stored in the reservoir during charging and is available at discharge, with an RTE upper bound of 70%; or (3) isothermal, where the air is compressed, stored, and expanded at close to constant temperature. The temperature is controlled to a set temperature using electric heat. The isothermal process is thermodynamically more efficient, potentially reaching 80% (Bazdar et al., 2022), with the various innovative processes being studied, but remember that many of these processes are still considered developmental. Note that these processes have

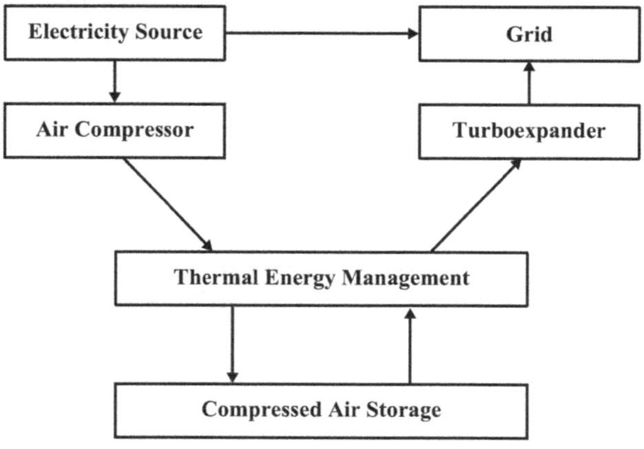

FIGURE 11.2 Simplified diagram of a CAES process. (Adapted from USDOE (2023). *Technology Strategy Assessment: Compressed-Air Energy Storage Technology*. Washington, DC: United States Department of Energy.)

multiple variations depending on the temperature and pressure, the use of TES, the reservoir type, and other integration options.

The longest-running CAES systems in Huntorf, Germany, with a 321-MW plant and 532,00 m^3 of underground storage (Borri et al., 2022), and McIntosh, Alabama, with 270,000 m^3 of underground storage (Bazdar et al., 2022; Olabi et al., 2021), can be classified as diabatic processes. They use underground salt caverns to store the compressed air at pressures in the metric 4–7-bar range—1 bar = 100,000 Pa.

Note that variations of the basic process result from configurations designed to match the location. Bazdar et al. (2022) point out that depleted gas wells, salt mines, porous rocks, and caverns are well-suited for CAES (80% of the United States may be geologically suited for CAES). These available storage volumes can be underground at a constant volume or variable pressure (isochoric, meaning $\Delta V = 0$) or underwater tanks with steady pressure and variable volume (isobaric, meaning pressure remains constant). The storage volumes need to match the following (USDOE, 2023):

- The scale of application, whether an individual factory, grid, etc.
- Storage duration needs
- Power and energy needs
- The mode of thermal management
- The availability of fuel
- Other considerations

The plant must balance the needs of energy storage (megawatt-hours, MWh), power (megawatts, MW), initial and operating costs, and plant life. The last two factors and RTE increase the price per kilowatt-hour of stored energy.

DID YOU KNOW?

Higher power charge and discharge rates in megawatts require large compressors and expanders. The volumetric capacity of these units is correlated to the square of the cylinder diameter. Large turbines are available with capacities of hundreds of megawatts (Maxie, 2025), with fan diameters of 3 m or more (Blain, 2025).

11.3.1 Key Components

This discussion is essential in key technology areas, such as compressed-air storage volume, TES, and management strategies, and integration of process steps with on-site energy production and consumption.

11.3.1.1 Storage Resources

CAES is stored in large volumes within natural aquifers, caverns, abandoned mines, and oil or gas fields. The value of these underground reservoirs depends on their proximity to power sources, suitability (i.e., >10% porosity, >500 mean drawdown—mD; Bowersox et al., 2022), acquisition cost, and the permitting time needed. Idle pipeline networks represent another excellent resource for CAES systems. Leveraging these storage volumes can save significant capital costs and enable the commercial success of such ventures. Inserting the location, volume, porosity, and permeability of potential storage locations into a database will aid developers in seeking to match their preferred locations with reservoirs nearby (Bowersox et al., 2022).

Finding an underground cavern, abandoned mine, or other potential location is one thing. Still, finding it is easy; ensuring that the space chosen is satisfactory for CAES storage is another. In addition, conducting a subsurface evaluation of porous rock for storage is essential. Several storage characteristics must be examined and analyzed to make an accurate assessment of storage capacity, including geophysical density measurements for the precise evaluation of storage capacity and rapid pressure testing for confirming reservoir deliverability; fast, low-cost rapid, low-cost tubular lining for storage; effective isolation process to isolate hydrocarbon layers from aquifer layers; monitoring and surveillance procedures to confirm well integrity and ensure that elimination of fugitive emissions; and workflows for rapidly assessing the feasibility of idle oil and gas well sites for storage (USDOE, 2023).

Another aspect of compressed gas energy storage systems in aboveground storage is the possible footprint, which has real estate implications. One way to reduce the footprint is to install vertical tanks with high length-to-diameter (L/D) ratios. High L/D allows thinner walls for a given pressure rating. One key application for CAES storage is to use idled natural gas pipelines; these have high L/D and pre-permitted access and use. Moreover, these repurposed pipelines can significantly reduce the cost of a plant.

11.3.1.2 Thermal Energy Storage

A CAES system must integrate the thermal energy between the compression and expansion steps. This means employing an efficient heat exchange system and a TES medium to charge and discharge this asynchronous operation. The medium's temperature range, the option of using and selecting a heat exchange fluid, the medium's containment and energy loss rate, and so on guide the selection options. Other innovations include the design of low-cost storage techniques (e.g., molten silicon, concrete, alumina spheres) that provide high capacity at a minimum cost and improved water-based storage with insulated tanks that enable longer-duration heat storage (USDOE, 2023).

11.3.1.3 Process Integration

Integration of multiple units and processing enables chemical plants to increase resource utilization. Adiabatic and isothermal CAES systems attempt to store and exchange thermal energy between charge and discharge. A significant fraction of that energy cannot be recovered economically due to low temperature (or quality) and must be wasted. Depending on the temperature, this energy can potentially be used to generate steam, hot water, or space heating for the CAES facilities or nearby buildings and processes. Likewise, the air at the turbine exit is at sufficient pressure. It may be usable in other low-pressure systems in nearby facilities to supply a compressed-air network in the plant or nearby industry, furnaces, and forges. Reducing wasted energy can improve the efficiency of the process and reduce plant operating costs.

11.4 THE BOTTOM LINE

CAES systems have advantages of low-cost, long service life, low maintenance, large-scale storage, use of renewable energy sources, and environmental friendliness. On the other hand, these systems have the disadvantages of low energy density, slow response time, limited underground storage sites, and significant upfront costs. Artificial intelligence (AI) can be said to be transforming CAES system technology. AI uses machine learning (ML) and advanced control algorithms to enhance CAES design, efficiency, and reliability. As AI continues to evolve, it promises to open even more avenues for innovation and growth in CAES system operations.

REFERENCES

Bazdar, E., Sameli, M., Nasiri, F., and Haghighat, F. (2022). Compressed Air Energy Storage in Integrated Energy Systems: A Review. *Renewable, and Sustainable Energy Reviews*. vol. 167.

Blain, L. (2025). New Atlas. Accessed @ https://newatlas.com/aircraft/roll-royce-ultrafan-
Borri, E., Tafone, G., Comodi, G., Romagnoli, A., and Cabeza, L.F. (2022). Compressed Air Energy Storage—An Overview of Research Trends and Gaps through a Bibliometric Analysis. *Energies*. vol. 15, p. 7692.
Bowersox, J.B., Hickman, J.B., and Skeese, D.A. (2022). *Assessing Compressed Air Energy Storage (CAES) Potential in Kentucky to Augment Energy Production from Renewable Resources*. Lexington, KY: Kentucky GEO Survey.
Maxie, K. (2025). Engineering.com: Gas Turbine. Accessed 3/25/25 @ https://www.engineering.com/story/testing-the-worlds-mostpowerful-gas-turbine.
Olabi, A.G., Wilberforce, T., Ramadan, M., and Abdelkareem, M.A. (2021). Compressed Air Energy Storage System: Components and Operating. *Journal of Energy Storage*. vol. 34.
USDOE (2023). *Technology Strategy Assessment: Compressed-Air Energy Storage Technology*. Washington, DC: United States Department of Energy.

12 Thermal Energy
Water, Molten Salts, Concrete, Aquifers, Boreholes = Phase-Change Materials

12.1 NOTHING NEW UNDER THE SUN

12.1.1 A Historical Perspective[1]

According to today's calendar, it is June 26, 15,543 BC. The place is a large natural cave set deep under a solid outcropping that formed a significant mountain meadow before the last glacial ice sheet gouged, gorged, ground, and pulverized it down to its present size and shape.

The colossal sheet of ice is in retreat. At its full width, depth, and length, it extended several hundred miles beyond the cave site to a V-shaped valley, where glacial melt fed a youthful, raging river that ran through the valley's bottomland.

A small, steady stream of meltwater courses in a straight line past one side of the cave, down toward that valley, where it will join and feed that river.

On the other side is a sloping field of young grass, brushes, and flowers everywhere. Up close, we see the stark remnants of the terminal moraine that formed this abrupt slope with its fresh cover of grass and blossoms. A closer look reveals a dark heap at the base of the hill—a pile of trash: skin, sinew, bone, decaying corpses, burnt remnants of past hunts, and feasts. We know exceptionally well the refuse, filth, and discards that people leave behind as their foulest signature; somebody, many bodies, live close by. Where? Of course—in the cave.

Let's look inside.

We wander up to the massive hole in the rock that forms the mouth of the cave. We tread carefully—we do not want to disturb (startle, frighten, anger) the occupants. Remember, we're talking about cavepeople here—no common language. There is no common culture. Could we have anything in common with such primitive people?

But all is quiet, and with no overt threat present, our curiosity overcomes our caution, and we walk into the opening chamber.

And something reaches out and grabs us—not a cave dweller but a stench—a horrible stench too horrible to describe. With our fingers clamped tightly to our noses, crushing to the bone (cartilage), we move on. Despite the reek, we are too interested in this strange environment to retreat.

We can see well in this chamber because daylight pours through the entrance. We take a few steps and look around, religiously not breathing through our noses.

The walls of the cave are covered with black soot. A pit near the cave wall to the right, under attack by millions of flies and other insects, provides much of the stench—the latrine. Like the dump outside, a heap of detritus offers the rest of the reek.

It dawns on us that this cave has been abandoned. We do not doubt why. The most significant byproduct manufactured by humankind has taken over—the cave is a garbage dump—mankind's garbage (we are assuming cave dwellers of this period were pre-human or human).

Perhaps cave paintings and remnants of the ceremony exist in other chambers deeper in the cave. But we don't have the tools to explore them with us today, so we retreated, grateful for a breath of fresh air.

DOI: 10.1201/9781003637851-14

Outside, a few hundred feet from the smelly cave and within sight of the garbage heap, we contemplate what we've seen. Seventeen thousand years from now, archaeologists will find this cave and explore it thoroughly, learning information that will give us insights into the world of the people whose former home we have just visited. However, the remains the archaeologists will find will be altered by 17,000 years of history. The picture they see will be incomplete, scattered by the natural interferences' life causes, giving mystery to our ancestors' short and brutal lives.

But here, right now, we see similarities. We foul our environment in the same ways and more. But the cave dweller had a massive advantage over modern people in that respect. When his living quarters became too foul for comfort, he could pick up whatever he considered of value that he and his clan could carry and move on. A fresh site was always just around the next bend in the river. The pollution he created was ultimately (eventually) naturally biodegradable—in a few years, this same foul cave could house humans again.

Although we have similarities with those far-off ancestors, one stark difference is plain: modern man cannot destroy and pollute his environment with impunity. We can no longer pick up the stakes and move on. What we do to our environment has ramifications on a scale that we cannot ignore—or avoid.

Also, note that one stark similarity is somewhat familiar to us today—well, to a degree. Cave occupants used fire (heat) to cook food and, ignoring the smoke and soot, just as importantly, to keep warm. For self-protection, cave dwellers used fire to ward off Smilodon (aka saber-toothed tiger) and other dangerous critters that roamed outside their cave. Certainly, cave dwellers took in as much thermal heat as possible from sunshine on sunny days. And those cave dwellers near the oceans, lakes, rivers, or ponds might have felt the heat from the sand and pebbles near the water. Today, anyone who has ever hot footed it barefoot across the beach on a sunny day walks away with a greater understanding of how much heat sand can retain.

Sand can act as a battery because it can store thermal energy.

Is the thermal energy stored in sand during a sunny day a viable thermal energy source? Is this energy storage medium an answer to energy storage in the future? Will we ignore this thermal energy source or move on to other approaches, such as thermal energy storage (TES) that uses molten salt or even superheated rocks? Predicting the future is difficult, at best.

However, we will consider storing the energy in a sand battery because we need long-term energy storage (LTES). And this is not a far stretch because sand, like air and water, is everywhere.

12.2 THE 411 ON THERMAL ENERGY STORAGE

Those next to a rock-lined roaring campfire, in or outside the cave, are aware of or become accidentally introduced to the heat retained in the rocks. The heat in the stones is mechanical thermal heat, known as *thermal storage*; more specifically, it is known as *sensible heat storage*, which also applies to heated concrete, sand, or liquid media such as molten salt. Energy stored using phase-change materials (e.g., metals, organics, salts) is *latent heat storage*. The energy stored in chemical bonds is known as *thermochemical storage*.

We usually consider storing heat energy when discussing TES. The truth is that TES also includes cold storage. TES technologies heat or cool a storage medium and, when needed, deliver the stored thermal energy to meet heating or cooling needs. TES systems are used in commercial buildings, industrial processes, and district energy installations to deliver stored thermal energy during peak demand, reducing energy use. TES systems are often integrated with electric or absorption chillers to reduce peak electricity costs and, in the case of new construction, to reduce capital costs by optimizing chiller size. Resizing equipment is key to improving overall efficiency for heating or cooling plants. TES systems can support renewable or fossil power generation sites, including combined heat and power (CHP) installations.

12.3 FROM ICE BOX TO OTHER USES

TES was first used in ice boxes in the 19th century to prevent peanut butter from melting. Modern usage of TES began with building heating and cooling and concentrating solar thermal technologies for power generation in the 1900s and again in the late 1970s (Dincer and Rosen, 2011). TES, compared to LTES technologies, differs from them in that TES is at a lower cost, has long operation lives, high energy density, synchronous power generation capability with the inertia that inherently stabilizes the grid, and the ability to output both heat and electricity (Enescu et al., 2020; Ma et al., 2021; Viswanathan et al., 2022).

Note that TES technologies can be combined with most renewable energy systems, including wind, photovoltaic, and concentrated solar thermal energy, and can be used for heat-to-heat, heat-to-electricity, electricity-to-heat, and bidirectional electricity (i.e., electricity-to-electricity) applications (Alva et al., 2017; Enescu et al., 2020; Kocak et al., 2020).

Figure 12.1 shows a bidirectional electricity TES (electrothermal energy storage, ETES) design emerging as a prime technology of LDES at a grid scale. One of the major plus sides of ETES technology is that it can be utilized without any geological limitations, as is the case with water for pumped storage hydro and underground natural caverns for compressed-air energy storage. It can be deployed anywhere in the United States and worldwide for extensive uses. Moreover, ETES technology can be in retired fossil fuel thermal power plants to reuse decommissioned assets, protect job security in associated communities, and provide grid-resilient and spinning (i.e., high-inertia) power.

Regarding heat input and output, there are many ways to integrate TES within heat-to-electricity, heat-to-heat, and electricity-to-heat applications, such as those used in concentrated solar power (CSP), buildings, district heating, and industry process heat applications. We can also classify these categories for low- and high-temperature applications. High-temperature thermal energy storage (HTTES) heat-to-electricity TES applications are currently associated with CSP systems for power generation. TES and CSP have proven their value to the electric grid, especially within the Southwestern United States. This region is rich in solar resources and has provided value for the electric grid. Significant potential exists for decarbonizing energy-intensive industrial processes, heat applications, electricity-to-heat applications, and heat-to-heat (HTTES) applications such as iron ore processing, iron smelting, cement production, glass manufacturing, mineral processing, and chemical production (USDOE, 2023). Note that some industrial processes require process heat at a temperature >1,400°C, so HTTES can be utilized to reduce fuel consumption in those processes through fuel, oxidizer, and process material pre-heating. TES for augmenting industrial process heat applications makes a much more attractive economic case because the energy penalty due to thermal-to-electric conversion is eliminated.

Low-temperature TES can be utilized for building and district heating and cooling, and some process heat applications in electricity-to-heat and heat-to-heat configurations. The optimal strategy for integrating TES has yet to be determined; however, note that lower-temperature TES (LTTES) can be added to electric input heat pump equipment, directly interacting with the refrigerant in the condenser, evaporator, or secondary heat-transfer fluid. It can also be integrated into the building envelope or the heating, ventilation, and air conditioning (HVAC) duct.

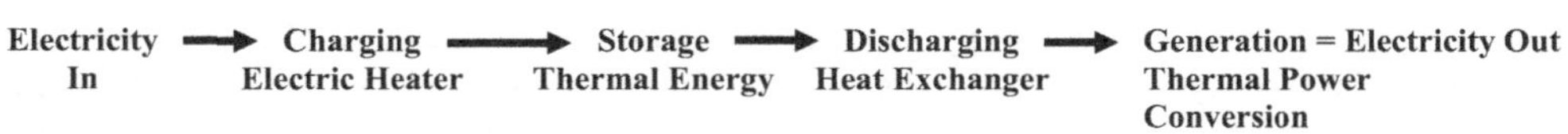

FIGURE 12.1 Stand-alone ETES application of electric-thermal energy storage independent from concentrated solar power. (Adaptation from USDOE (2023). *Technology Strategy Assessment: Thermal Energy Storage Technology*. Washington, DC: United States Department of Energy.)

12.4 CHARACTERIZATION OF TES TYPES AND MEDIA

A TES system's characteristics include storage media, storage containment, and its ability to support power generation or heat sources for efficient energy charging and discharging (i.e., heat exchange/transfer). TES covers various energy formats using multiple storage media and energy conversion methods. TES technologies are grouped into three categories (see Figure 12.2). For CHP, the common types of TES are sensible heat and latent heat.

When discussing heat related to energy, it is essential to point out that there might be some confusion about whether *latent* or *sensible heat* is a unique or special form of energy. It is not. Instead, latent and sensible heat describes heat exchanges under conditions specified in terms of their effect on a material for a thermodynamic system. Note that sensible heat and latent heat are observed in many processes while transporting energy in nature. The difference between the two temperatures, especially the atmospheric water vapor phase changes, is mostly vaporization and condensation. In contrast, sensible heat directly affects the temperature of the atmosphere.

Sensible heat storage is the most commercially deployed TES type for power generation and heating. In sensible heat, energy is stored by raising the temperature of a medium. The physical properties of the storage material, including density, volume, specific heat, and temperature change, are proportional to the amount of energy stored (Alva et al., 2018). For storing energy between 290 and 600°C, molten nitrate salt or solar salt, 60% $NaNO_3$ and 40% KNO_3 by weight are commonly used as the thermal storage medium in commercial TES systems (Bauer et al., 2021). Since it was first demonstrated in the 1990s, molten salt has been used as a storage medium in commercial CSP power plants (Reilly and Kolb, 2001). Solid media such as particles, concrete, and graphite have been developed or are under development and can be utilized as solid media in TES at a very high temperature (>1,000°C; Ho and Gerstle, 2021; Ma et al., 2020; Trevisan et al., 2022). Solid particles or rocks used as TES media are obtained from nature, where they are abundant, low-cost, and environmentally compatible. When ceramic or sand-type solid particles are used as thermal storage media, they overcome the corrosion issues and the low-temperature freezing concerns of molten salt and have high-temperature stability. Even though some of the operating challenges experienced in the molten nitrate salt are mitigated using a solid storage medium, new operation challenges arise when using solid media. This takes in erosion when applying sand-like particles and heat-transfer fluid issues, such as contamination and low energy density, in the case of stationary porous solid media. Geological structures such as aquifers and consolidated or unconsolidated rocks in the subsurface are natural solid media storage.

DID YOU KNOW?

An important compound of Earth's surface energy budget is "Sensible heat flux," which means the conductive heat flux from the Earth's surface to the atmosphere (Stull, 2000).

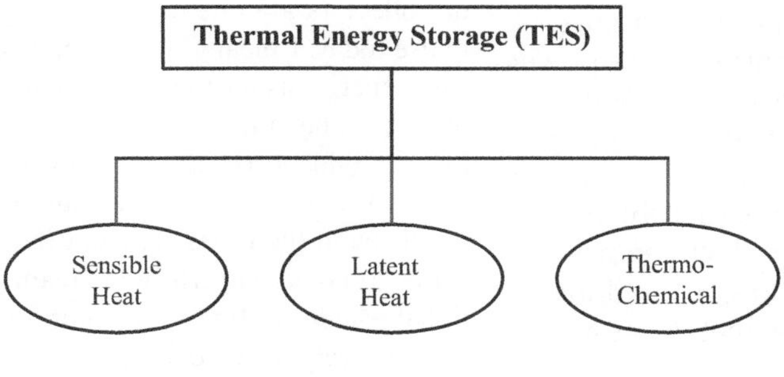

FIGURE 12.2 TES are grouped into three categories.

Latent heat energy storage depends on a phase-change material (e.g., salt or metal alloy) that can be used in TES (Bruno et al., 2021). Most latent heat technologies use frozen water as the phase-change material, although others have been employed, like eutectic salts. A relatively large amount of energy is required to change the phase of the phase-change materials compared with raising their temperature, and latent heat storage takes advantage of this characteristic. Using this characteristic increases energy density, with both latent and sensible heat contributing to the stored energy capacity. Note that phase-change materials can be utilized over various conditions, from low-temperature TES, such as building heating or cooling applications, to high-temperature TES for power generation or industrial process heat. HTTES phase-change materials have not been commercially deployed at a sizeable power scale simply because (to a point) they are less mature compared with sensible heat. There is nothing new about phase-change materials; they, like water/ice, have existed for centuries. New technologies, such as bio-based, paraffin wax, and salt hydrate materials, are being studied and developed for greater energy density and cyclability for promising building heating and cooling applications. Phase-change materials, including clothing and telephone pads, are used in many commercial applications requiring energy storage and/or stable temperatures.

Increasing penetrations of intermittent renewable energy sources (e.g., solar and wind energy) have increased the need for energy storage technologies to accommodate daily periods of overgeneration and peak loads. Promising thermochemical energy storage (TCES) can be employed to meet this need but has not reached industrial applications. This energy storage technology is used primarily at high temperatures (>700°C) because it allows heat storage through chemical reactions—breaking/reforming bonds.

So, what are the potential benefits of using TCES? There are several potential benefits of using TCES, including enabling more efficient high-temperature power cycles. Compared to sensible heat, TCES has the advantage of indefinite storage time, no risk of freezing, lower cost, increased energy density, and the ability to store the sun's energy at high densities for grid-scale concentrated solar power (CSP) electricity generation.

Note to Reader: Although concentrated solar power (CSP) was mentioned earlier, it is a term and technology some readers may not be familiar with. The following Sidebar 12.1 is provided herein to provide foundational material for understanding.

SIDEBAR 12.1 CONCENTRATING SOLAR POWER (CSP)[2]

Concentrating solar power (CSP) offers a utility-scale, reliable, firm, dispatchable renewable energy option that can help meet a nation's demand for electricity. EERE (2008a) points out that nine trough plants producing more than 400 MW of electricity have operated reliably in the California Mojave Desert since the 1980s. CSP plants produce power by using mirrors to focus sunlight on a working fluid. A thermal receiver is like a boiler tube. The receiver absorbs and converts sunlight into heat. Ultimately, this high-temperature fluid is used to spin a turbine or power an engine that drives an electricity generator.

Concentrating solar power systems can be classified by how they collect solar energy: linear concentrators, dish/engines, or power tower systems (NREL, 2009). All three are typically engineered with tracking devices that follow the sun seasonally and throughout the day to maximize the system's electrical output (Chiras, 2002).

Linear concentrating solar power (CSP) concentrators or collectors are one of the three types of CSP systems in use today. Linear CSP collectors capture or collect the sun's energy using long rectangular, curved (U-shaped) mirrors. The mirrors are tilted toward the sun, focusing sunlight on tubes (or receivers) that run the length of the mirrors. The reflected sunlight heats a fluid flowing through the tubes. The hot fluid is then used to boil water to create superheated steam that spins a conventional steam turbine generator to produce electricity. Alternatively, steam can be generated directly in the solar field, eliminating the need for costly heat exchangers.

Linear concentrating collector fields consist of many collectors in parallel rows that are typically aligned in a north-south orientation to maximize annual and summertime energy collection. With a single-axis sun-tracking system, this configuration enables the mirrors to track the sun from east to west during the day, ensuring that the sun reflects continuously onto the receiver tubes (EERE, 2008b).

There are two major types of linear concentrator systems: parabolic trough systems, where receiver tubes are positioned along the focal line of each parabolic mirror, and linear Fresnel reflector systems, where one receiver tube is positioned above several mirrors to allow the mirrors to create greater mobility in tracking the sun.

EERE (2008b) points out that the predominant CSP systems currently operating in the United States are linear concentrators using parabolic trough collectors. All rays of light that enter parallel to the axis of a parabolic-shaped mirror will be reflected to one point, the focus. Hence, it is possible to concentrate virtually all the radiation incidents upon the mirror in a relatively small area in the focus (Perez-Blanco, 2009). Parabolic trough systems form a long-curved array of mirrors (usually coated silver or polished aluminum). Trough systems have been developed in the United States, Spain, and Japan (Chiras, 2002). The receiver tube (Dewar Tube) runs its length and is positioned along the focal line of each parabola-shaped reflector (Duffie and Beckman, 1991; Patel, 1999). The tube is fixed to the mirror structure, and the heated either a heat-transfer fluid (usually oil) or water/stream flow through and out of the field of solar mirrors to where it is used to create steam (or, in the case of a water/steam receiver, it is sent directly to the turbine (i.e., the turbine is the prime mover of the electrical generator that produces an electrical current flow in an accompanying electrical distribution system)). Temperatures in these systems range from 150 to 750°F (80–400°C).

Currently, the largest individual trough systems generate 80 MW of electricity. However, individual systems being developed will generate 250 MW. In addition, individual systems can be collocated in power parks. This capacity would be constrained only by transmission capacity and the availability of contiguous land area.

Trough designs can incorporate thermal storage. Thermal storage systems are discussed later, but for informative purposes, we can see that in such a system, the collector field is oversized to heat a storage system during the day that can be used in the evening or during cloudy weather to generate additional steam to produce electricity. Parabolic trough plants can be designed as hybrids, meaning that they use fossil fuel to supplement the solar output during periods of low solar radiation. A natural gas-fired heater or gas-steam boiler/reheater is used in this design. Troughs may be integrated into existing or new combined-cycle natural gas and coal-fired plants (EERE, 2008b).

The linear Fresnel reflector system is a second linear concentrator technology, a solar power collector. Flat or slightly curved mirrors mounted on trackers on the ground are configured to reflect sunlight onto a water-filled receiver tube fixed in space above these mirrors. A small parabolic mirror is sometimes added atop the receiver to focus the sunlight further. The key advantage of Fresnel reflector systems is their simplicity compared to other systems.

Dish/engine systems use a mirrored dish-shaped parabolic mirror similar to a giant satellite dish. The dish-shaped surface directs and concentrates sunlight onto a thermal receiver (the interface between the dish and the engine/generator), which absorbs and collects the heat and transfers it to the engine generator. Solar dish/engine systems convert the energy from the sun into electricity at a very high efficiency. The power conversion unit includes the thermal receiver and the engine/generator. It absorbs the concentrated solar energy beams, converts them to heat, and transfers them to the engine/generator. A thermal receiver can be a bank of tubes with a cooling fluid—usually hydrogen or helium—typically the heat-transfer medium and an engine's working fluid. Alternate thermal receivers are heat pipes, where the boiling and condensing of an intermediate fluid transfers the heat to the engine. The most common type of heat engine used today in dish/engine systems is the Stirling engine (conceived in 1816 and mentioned earlier). This system uses the fluid heated by the receiver to move pistons and create mechanical power. Mechanical power runs a generator or alternator to produce electricity (USDOE, 1998).

Disk/engine systems use dual-axis collectors to track the sun. As mentioned, the ideal concentrator shape is parabolic, created by a single reflective surface, multiple reflectors, or facets. Many options exist for receiver and engine type, including the Stirling cycle, microturbine, and concentrated photovoltaic modules.

Because of the high concentration ratios achievable with parabolic dishes and the small size of the receiver, solar dishes are efficient (as high as 30% efficient) at collecting solar energy at very high temperatures (900–2,700°F; 500–1,500°C).

USDOE (1998) points out that solar dish/engine systems have environmental, operational, and potential economic advantages over more conventional power generation options because they

- produce zero emissions when operating on solar energy
- operate more quietly than diesel or gasoline engines
- are easier to operate and maintain than conventional engines
- start up and shut down automatically
- operate for long periods with minimal maintenance

Because of their size and durability, solar dish/engine systems are well-suited and often used for non-traditional power generation. Individual units range in size from 10 to 25 kW. They can operate independently of power grids in remote, sunny locations, pumping water and providing power to people living in isolated places.

As mentioned, dish/engine systems produce high temperatures, which in turn can be used to produce steam that can be used either for electricity production or select high-temperature industrial processes (Chiras, 2002). Dish/engine systems can also be linked to provide utility-scale power to a transmission grid. Such systems could be located near consumers, substantially reducing the need for building or upgrading transmission capacity. Mainly because of their high efficiency, the cost of these systems is expected to be lower than that of other solar systems for these applications (USDOE, 1998).

The power tower or the central receiver system consists of a large field of flat, sun-tracking mirrors known as heliostats (silver-laminated acrylic membranes) focused on concentrating sunlight onto a receiver on the top of a tower. A heat-transfer fluid heated in the receiver generates steam, which, in turn, is used in a conventional turbine generator to produce electricity. Some power towers use water/steam as the heat-transfer fluid. Other advanced designs are experimenting with molten (liquefied) nitrate salt because of its superior heat transfer and energy storage (heat retention) capabilities, allowing continued electricity production for several consecutive cloudy days (NREL, 2010). The energy storage capability, or thermal storage, allows the system to continue dispatching electricity during cloudy weather or at night. Therefore, a power tower system comprises five main components: heliostats, receivers, heat transport and exchange, thermal storage, and controls (see Figure 12.3). Individual commercial plants can be sized to produce up to 200 MW of electricity.

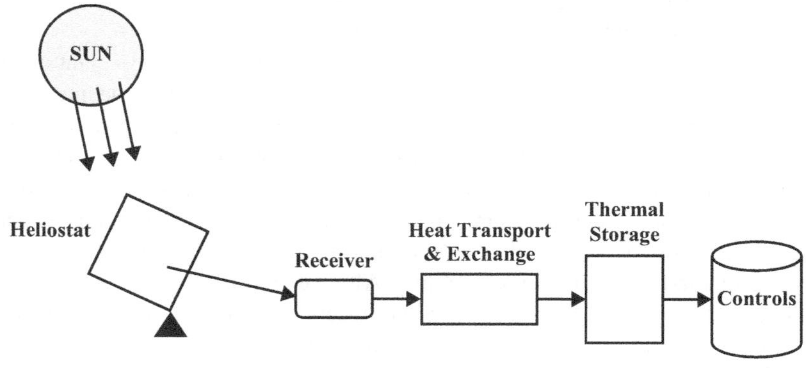

FIGURE 12.3 Solar power tower main components.

DID YOU KNOW?

Smaller CSP systems can be located directly where the power is needed. For example, a single dish/engine system can produce 3–25 kW of power and is well-suited for distributed applications. Larger, utility-scale CSP applications provide hundreds of megawatts of electricity for the power grid. Both linear concentrator and power tower systems can be easily integrated with thermal storage, helping to generate electricity during cloudy periods or at night. Alternatively, these systems can be combined with natural gas, and the resulting hybrid power plants can provide high-value, dispatchable power throughout the day.

DID YOU KNOW?

Concentrated solar power plants use water not only for lens cleaning but also for steam cycle production. Water consumption is an issue because they need water and they are most effective in locations where the sun is most intense, which in turn often corresponds to places like the Mohave Desert where there is little water (USDOE, 2008).

Thermal storage (TES) has become a critical aspect of any concentrating solar power (CSP) system deployed today (EERE, 2008c). One challenge facing the widespread use of solar energy is the reduced or curtailed energy production when the sun sets or is blocked by clouds. TES provides a workable solution to this challenge. In a CSP system, as mentioned, the sun's rays are reflected onto a receiver, creating heat that is then used to generate electricity. The thermal energy can be stored later if the receiver contains oil or molten salt as the heat-transfer medium. This allows CSP systems to be a cost-competitive option for providing clean, renewable energy. Presently, steam-based receivers cannot store thermal energy for later use.

An essential criterion in selecting a storage medium, such as the oil or molten salt referred to above, is the *substance's specific heat*. Specific heat (c) measures the heat energy required to increase a unit quantity of a substance by a unit of temperature. For example, at a temperature of 15°C, the heat energy needed to raise water's temperature by 1K (equal to 1°C) is 4,186 J/kg. Different materials absorb different amounts of heat in undergoing the same temperature increase. The relationship between a temperature change ΔT and the amount of heat Q added or subtracted is given by

$$Q = mc\Delta T \tag{12.1}$$

where c is the specific heat and m is the mass of the substance. Table 12.1 lists specific heats for different materials per pound and cubic foot. Another important parameter related to specific heat and TES materials is *heat capacity*, which is defined as the ratio of the heat energy absorbed by a substance to the substance's increase in temperature and is given by

$$\text{Heat Capacity} = \text{Specific Heat} \times \text{Density} \tag{12.2}$$

Another medium for thermal storage is *phase-change materials*, which are classified as latent heat storage units. Specifically, a phase material is a substance with a high heat of fusion (i.e., the amount of thermal energy that must be absorbed or evolved for 1 mol of a substance to change states from a solid to a liquid or vice versa without a temperature) which, melting and solidifying at a certain temperature, is capable of storing and releasing large amounts of energy. Heat is absorbed or

TABLE 12.1
Thermal Energy Storage Materials

Specific Heat		Density		Heat Capacity	
Material	**Btu/lb/°F**	**(lb/ft³)**	**kg/m³**	**(Btu/ft³–°F)**	**(kJ/m³–°C)**
Water	1.00	62	1,000	62	4,186
Iron	0.12	490	7,860	59	3,521
Copper	0.09	555	8,920	50	3,420
Aluminum	0.22	170	2,700	37	2,430
Concrete	0.23	140	2,250	32	2,160
Stone	0.21	170	2,700	36	2,270
White pine	0.67	27	435	18	1,220
Sand	0.19	95	1,530	18	1,540
Air	0.24	0.075	1.29	0.02	1.3

released when the material changes from solid to liquid. One common group of substances used as phase-change materials for solar applications (and elsewhere) are eutectic salts; salts are combined with water, such as sodium sulfate decahydrate, also known as Glauber's salt. This salt melts at 91°F with the addition of 108 Btu per pound. Conversely, when the temperature drops below 91°F, 108 Btu per pound of heat energy is released as the salt solidifies (Hinrichs and Kleinback, 2006).

Several TES technologies have been tested and implemented since 1985. These include the two-tank direct, two-tank indirect, and single-tank thermocline systems.

Solar thermal energy in this system is stored in the same fluid used to collect it. The fluid is stored in two tanks—one at high and the other at low temperatures. Fluid from the low-temperature tank flows through the solar collector or receiver, where solar energy heats it to a high temperature, and it then flows back to the high-temperature tank for storage. Fluid from the high-temperature tank flows through a heat exchanger, generating steam for electricity production. The fluid exits the heat exchanger at a low temperature and returns to the low-temperature tank. Two-tank direct storage was used in early parabolic trough power plants and at California's Solar Two power tower. The trough plants used mineral oil as the heat transfer and storage fluid; Solar Two used molten salt.

The two-tank indirect system functions similarly to the two-tank direct system, except that different fluids are used for heat transfer and storage. This system is used in plants where the heat-transfer fluid is too expensive or not suited for use as the storage fluid. The storage fluid from the low-temperature tank flows through an extra heat exchanger, which is heated by the high-temperature heat-transfer fluid. The high-temperature storage fluid then flows back to the high-temperature storage tank. The fluid exits this heat exchanger at a low temperature and returns to the solar collector or receiver, where it is heated back to the high temperature. Storage fluid from the high-temperature tank generates steam, like the two-tank direct system. The indirect system requires an extra heat exchanger, which adds cost. This system will be used in many of the parabolic power plants in Spain and has also been proposed for several U.S. parabolic plants. The plants will use organic oil as the heat-transfer fluid and motor salt as the storage fluid.

The single-tank thermocline system stores thermal energy in a single tank in a solid medium, most commonly silica sand. At any time during operation, a portion of the medium is at a high temperature, and a portion is at a lower temperature. A temperature gradient or thermocline separates the hot and cold temperature regions (see Figure 12.4). High-temperature heat-transfer fluid flows into the top of the thermocline and exits the bottom at a low temperature. This process moves the thermocline downward and adds thermal energy to the system for the stage. Reversing the flow pushes the thermocline upward and removes thermal energy from the system to generate steam and

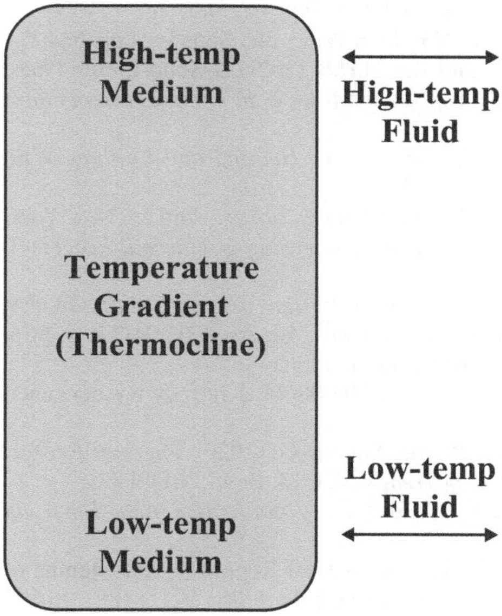

FIGURE 12.4 Single-tank thermocline thermal energy storage system.

electricity. Buoyancy effects create thermal stratification of the fluid within the tank, which helps to stabilize and maintain the thermocline.

Note: Using a solid storage medium and only needing one tank reduces the cost of the single-tank thermocline system relative to the two-tank systems.

++

12.5 THE BOTTOM LINE

TES refers to the energy stored in a material as a heat source or a cold sink and reserved for use at a different time. TES provides high energy density, relatively low-cost, indefinite storage time, no risk of freezing, efficient energy management, flexibility in charging and discharging, and high heat capacity. On the other hand, TES depends on the storage media type used, can lose energy during storage, has some high costs, and has some integrity and optimization challenges.

NOTES

1. Adapted from F.R. Spellman (2018). *The Science of Environmental Pollution*, 3rd ed. Boca Raton, FL: CRC Press.
2. Most of the information in this section is from DOE's (2001) *Concentrating Solar Power: Energy from Mirrors*. Washington, DC: U.S. Department of Energy; F.R. Spellman's (2021) *The Science of Renewable Energy*, 2nd ed. Boca Raton, FL: CRC Press.

REFERENCES

Alva, G., Lin, Y., and Fang, G. (2018). *An Overview of Thermal Energy Storage Systems*. Vol. 144, Amsterdam: Elsevier Ltd, pp. 321–378.

Alva, G., Liu, L., Huang, X., and Fang, G. (2017). *Thermal Energy Storage Materials and Systems for Solar Energy Association*. Vol. 68, Amsterdam: Elsevier Ltd, pp. 693–706.

Bauer, T., Odenthal, C., and Bonk, A. (2021). *Molten Salt Storage for Power Generation, in Chemie-Ingenieur-Technik*. Vol. 93, New York, NY: John Wiley and Sons Inc., pp. 534–546.

Bruno, T., Belusko, M., Liu, M., and Tay, N.H.S. (2021). 9-Solid-Liquid Phase Change Materials for Thermal Energy Storage. In Cabeza, L.F. (ed.). *Advances in Thermal Energy Storage System*, 2nd ed. Cambridge: Woodhead Publishing.

Chiras, D.D. (2002). *The Solar House: Passive Heating and Cooling*. White River Junction, VT: Chelsea Green Publishing Company, p. 33.

Dincer, I., and Rosen, M. (2011). *Thermal Energy Storage*, 2nd ed. New York, NY: Wiley.

Duffie, J. and Beckman, W. (1991). *Solar Engineering of Thermal Processes*. New York, NY: John Wiley & Sons, Inc.

EERE (2008a). *Solar Energy Technologies Program*. Accessed 02/04/24 @ www.eere.energy.gov.

EERE (2008b). *Linear Concentrator Systems*. Accessed 02.04/24 @ http://www.EERE.energy.gov/solar/printable_versions/linear_concentrators.html.

EERE (2008c). *Thermal Storage*. Accessed 02/08/24 @ http://www.eere.energy.gov/solar/printable_versions/thermal_storage.html.

Enescu, D., Chicco, G., Porumb, R., and Santan, G. (2020). *Thermal Energy Storage for Grid Applications: Current Status and Emerging Trends*. Vol. 13, Basel: MDPI AG.

Hinrichs, P. and Kleinback, M.H. (2006). *Energy and Its Use in the Environment*. Boston, MA: Brooks/Cole Publishing.

Ho, C.K. and Gerstle, W. (2021). Terrestrial Heat Repository for Months of Storage (THERMS). A Novel Radial Thermocline System. *Energy Sustainability*. Vol. 84881, Rockville, MD: American Society of Mechanical Engineers, p. V0006A003.

Kocak, B., Fernandez, A.I., and Paksoy, H. (2020). *Review on Sensible Thermal Energy Storage for Industrial Solar Applications and Sustainability Aspects*. Vol. 209, Amsterdam: Elsevier Ltd..

Ma, Z., Davenport, P., and Zhang, R. (2020). *Design Analysis of a Particle-Based Thermal Energy Storage System for Concentrating Solar Power or Grid Energy Stop Article-Based*. Vol. 29, Amsterdam: Elsevier Ltd.

Ma, Z., Wang, X., Davenport, P., Gifford, J., and Martinek, J. (2021). *Economic Analysis of a Novel Thermal Energy Storage System Using Solid Particles for Grid Electricity Storage*: Ed. Pre-print.

NREL (2009). *Concentrating Solar Power*. Accessed 06/13/24 @ http://www.nrel.gov/learning/re_csp.html.

NREL (2010). *CSP systems*. Accessed 02/07/24 @ http://www.nrel.gov/concentratingsolar.html.

Patel, M. (1999). *Thermal Energy and Solar Power Systems*. Boca Raton, FL: CRC Press.

Perez-Blanco, H. (2009). *The Dynamics of Energy: Supply, Conversion, and Utilization*. Boca Raton, FL: CRC Press.

Reilly, H.E. and Kolb, G.J. (2001). An Evaluation of Molten-Salt Power Towers, Including Results of the Solar Two Project.

Stull, R.B. (2000). *Meteorology for Scientists and Engineers*, 2nd ed. Belmont, CA: Brooks/Cole.

Trevisan, S., Wang, W., Guedez, R., and Laumert, B. (2022). *Experimental Evaluation of an Innovative Radial Flow High Temperature Packed Bed Thermal Energy Storage*. Vol. 311, Amsterdam: Elsevier Ltd.

USDOE (1998). *Solar Dish/Engine Systems*. Washington, DC: US Department of Energy.

USDOE (2008). *Solar Energy Technologies Program*. Accessed 06/13/24 @ http://www1.eere.Energy.gov/solar/printable_versions/about.html.

USDOE (2023). *Technology Strategy Assessment: Thermal Energy Storage Technology*. Washington, DC: United States Department of Energy.

Viswanathan, V., Mongrid, K., Franks, R., Li, X., Sprenkle, V., and Baxter, R. (2022). *2022 Grid Energy Storage Technology Cost and Performance Assessment*. Washington, DC: United States Department of Energy.

13 Supercapacitors

Note: Chapter 3 briefly discussed the basics of capacitance and capacitors and their circuit function in electricity and electronics. This chapter is concerned with the role of supercapacitors in storing energy.

13.1 CAPACITORS, SUPERCAPACITORS, ULTRACAPACITORS (USDOE, 2023)

Electrochemical capacitors, commercially referred to as supercapacitors or ultracapacitors or grouped under the general term "capacitors," are a family of energy storage devices with exceptionally high specific power compared with other electrochemical storage devices, such as lead-acid batteries, lithium-ion batteries, sodium-sulfur batteries, flow batteries, and Zn-air batteries. Supercapacitors do not require a solid dielectric layer between the two electrodes. Instead, they store energy by accumulating electric charge on porous electrodes filled with an electrolyte solution and separated by an insulating porous membrane. They can be charged and discharged quickly, offering an excellent cycle life, long operation life, and operation over a broad temperature range. All this is a plus. However, the major drawbacks of supercapacitors are low energy density and a high self-discharge rate. Battery University Group (2025) points out, for example, that a supercapacitor passively discharges from 100 to 50% in a month compared with only 5% for a lithium-ion battery. Note that supercapacitors' high capital cost and low energy density make the unit cost of stored energy (kWh) more expensive than alternatives such as batteries. Their attributes make them attractive when frequent small charges/discharges are required (e.g., ensuring power quality or providing frequency regulation).

13.1.1 Capacitor Groups

Electrical capacitors used for commercial applications are divided into two major groups—fixed and variable—and named according to their dielectric. The most common are air, mica, paper, ceramic, and electrolytic capacitors. Table 13.1 compares these types of electrical capacitors.

The fixed capacitor has a set capacitance value determined by its construction. The construction of the variable capacitors allows a range of capacitances. Within this range, the desired value of capacitance is obtained by some mechanical means, such as by turning a shaft (as in turning a radio tuner knob, for example) or adjusting a screw to adjust the distance between the plates.

TABLE 13.1
Comparison of Capacitor Types

Dielectric	Construction	Capacitance Range
Air	Meshed plates	10–400 pF
Mica	Stacked plates	10–5,000 pF
Paper	Rolled foil	0.001–1μF
Ceramic	Tubular	0.5–1,600 pF
	Disk	0.002–0.1 μF
Electrolytic	Aluminum	5–1,000 μF
	Tantalum	0.01–300 μF

DOI: 10.1201/9781003637851-15

The electrolytic capacitor consists of two metal plates separated by an electrolyte. The electrolyte, either paste or liquid, is in contact with the negative terminal, forming the negative electrode. The dielectric is a thin film of oxide deposited on the positive electrode, an aluminum sheet. Electrolytic capacitors are polarity sensitive (i.e., they must be connected in a circuit according to their polarity markings) and are used where a large amount of capacitance is required.

NREL (2009) points out that, like batteries, ultracapacitors, also known as supercapacitors, pseudocapacitors (PCs), electric double-layer capacitors, or electrochemical double-layer capacitors (EDLCs), are energy storage devices. They use electrolytes and configure various-sized cells into modules to meet various applications' power, energy, and voltage requirements. However, storage devices differ from batteries because ultracapacitors store energy electrostatically, whereas batteries store energy chemically. Ultracapacitors provide quick bursts of energy and improve capacitance by about two to three orders of magnitude (compared to an average capacitor) with a lower working voltage. Moreover, they can withstand hundreds of thousands of charge/discharge cycles without degrading. As an alternative energy source, ultracapacitors have proven themselves as reliable energy storage components used to power various electronic and portable devices such as AM/FM radios, flashlights, cell phones, and emergency kits. As ultracapacitor technology matures, it is being developed to function as batteries; for example, the vehicle industry is deploying ultracapacitors to replace chemical batteries. In the automotive industry, they have been integrated into electric vehicle (EV) and hybrid-electric vehicles (HEV) to help alleviate stress and extend the life of the batteries.

An ultracapacitor polarizes an electrolytic solution to store energy electrostatically. Though it is an electrochemical device, no chemical reactions participate in its energy storage mechanism. This highly reversible mechanism allows the ultracapacitor to be charged and discharged hundreds of thousands of times.

An ultracapacitor can be viewed as two nonreactive porous plates, or collectors, suspended within an electrolyte, with a voltage potential applied across the collectors. In an individual ultracapacitor cell, the applied potential on the positive electrode attracts the negative ions in the electrolyte, while the potential on the negative electrode attracts the positive ions. A dielectric separator between the two electrodes prevents the charge from moving between them.

Once the ultracapacitor is charged and energy is stored, a load can use this energy. The amount of energy stored is vast compared to a standard capacitor because of the enormous surface area created by the porous carbon electrodes and the slight separation (10 angstroms) made by the dielectric separation. However, it stores a much smaller amount of energy than a battery. Because the charge and discharge rates are determined solely by its physical properties, the ultracapacitor can release energy much faster (with more power) than a battery that relies on slow chemical reactions.

13.2 TYPES OF SUPERCAPACITORS

Based on the charge-storing mechanism, supercapacitors are divided into EDLCs, PCs, and hybrid electrochemical capacitors (HECs).

13.2.1 Electrochemical Double-Layer Capacitors (EDLCs)

EDLCs, called an electrical double layer (EDL), appear on an object's surface when exposed to fluid. They use two similar electrodes based on active carbon (or a derivative) to form two capacitors with the electrolyte. Charges are distributed on the surfaces by physical processes involving no chemical reactions. They exhibit a fast response to charge/discharge, good cycling stability, and a higher self-discharge rate. By changing the electrolyte, the performance characteristics of an EDLC can be adjusted for different applications.

13.2.2 Pseudocapacitors (PCs)

PCs use transition metal oxide-based electrodes to form highly reversible redox (faradaic) reactions, and energy is stored by transferring charge between the electrode and the electrolyte. These faradaic processes allow PCs to achieve greater capacitances and specific energy densities than EDLCs, involving only the electric double-layer effect at the electrode/electrolyte interface and not reactions at the electrodes.

13.2.3 Hybrid Electrochemical Capacitors (HECs)

HECs, called *asymmetrical capacitors*, use different anode and cathode (electrode) materials. The two electrodes have different electric potentials. The asymmetry implements faradaic and non-faradaic processes to store charge, leading to energy and power densities greater than EDLCs. These capacitors accomplish this without sacrificing the cycling stability and affordability that have limited the success of PCs (USDOE, 2023).

13.3 CURRENT USES

Practical uses of supercapacitors include the following:

- *Transportation*: Supercapacitors are used in automobiles, trucks, buses, trains, elevators, and cranes for regenerative braking, short-term energy storage, or burst-mode power delivery.
- *Medical*: Supercapacitors are used in medical implants (e.g., pacemakers), devices such as defibrillators, patient monitoring equipment, and other assorted equipment.
- *Automotive*: Supercapacitors are used in EVs for energy recovery during braking.
- *Critical infrastructure*: Supercapacitors provide ride-through power in critical infrastructure, typically backed up by large generators, which often take more than 15 seconds to start up.
- *Traction*: They are used in vehicles during acceleration, cruising, and deceleration phases.
- *Manufacturing and industrial*: Supercapacitors are often used in variable-frequency drives that operate critical manufacturing with constant voltage. They are also used in industrial applications that require quick peaking power, such as forklifts and seaport cranes.
- *Electronic and low-power applications*: They are used in solar power, battery backup, flashlights, and other applications.
- *Microgrids*: Microgrids and off-grid remote facilities can use battery energy storage to provide and absorb inrush currents during equipment startup (the current drawn during startup of some motors is six to ten times higher than running current) and during line faults. This reduces the discharge rate and extends battery life by maintaining ideal operating temperatures.
- *The Internet of Things (IoT)*: It is a network of physical objects embedded with sensors, software, and other technologies that connect and exchange data with other devices and systems via the Internet. Supercapacitors are often used in devices such as smart door cameras, security cameras, and portable point-of-sale devices to reduce battery cycling and extend the life of such devices, reducing maintenance.
- *Bulk power systems*: They are used in flexible AC transmission systems and high-voltage direct current transmission to alter the impedance of the line, regulating power factor and transmission capabilities by injecting or absorbing reaction power.
- *Renewable energy*: Supercapacitors are used in renewable systems integration to improve the power quality of fluctuating renewable generation (Bensmaine et al., 2015; Elmorshedy et al., 2021).

13.4 THE BOTTOM LINE

Supercapacitors are electrochemical devices with remarkably high specific power compared with other electronic storage devices. They feature unlimited cycle life, fast charge, and discharge rates, and no full charge circuit required. However, they have low energy density, high cost, and a high self-discharge rate, making them less attractive for long-duration energy storage.

REFERENCES

Battery University Group (2025). *BU-209: How Does a Supercapacitor Work?* Richmond, BC: Battery University Group.

Bensmaine, F., Bachelier, O., Slim Triani, G., and Mouni, C. (2015). LMI Approach of State-Feedback Controller Design for a STATCOM-Supercapacitors Energy Storage System Associated With Wind Generation. *Energy Conversion and Management*. vol. 96, pp. 453–472.

Elmorshedy, M.F., Elkadeem, K.M.K.M.R., Taha, I.B., and Mazzeo, D. (2021). Original Design and Energy Management of an Isolated Fully Renewable Energy System Integrating Batteries and Supercapacitors. *Energy Conversion and Management*. vol. 242.

NREL (2009). Concentrating Solar Power. Accessed 06/13/24 @ http://www.nrel.gov/learning/re_csp.html.

USDOE (2023). *Technology Strategy Assessment: Supercapacitors Energy Storage Technology*. Washington, DC: United States Department of Energy.

14 Hydrogen Storage/Fuel Cells

14.1 A PLAIN AND SIMPLE PROPOSAL

In casual conversations with energy experts, faculty, students, and a few associates, when energy cells (aka fuel cells) for vehicles are part of the conversation, most of them state that the day will come when we finally end the internal combustion engine's (ICE's) hundred-year reign as the dominant power source for personal transportation. They say it will be a winning situation. All consumers will get an efficient power source, communities will get zero emissions, and automakers will get another significant business opportunity—a growth opportunity. Energy/fuel cells are not just fodder for casual conversation, science fiction, or pie-in-the-sky technology. They are used in some applications today, and many wonder why they are not more commonly and universally used.

Fuel cells are like batteries in that they are composed of positive and negative electrodes separated by an electrolyte or membrane. Both are highly efficient. However, fuel cells do not need to be recharged like batteries, so they may run for extended periods if fuel and oxygen are provided.

When I lectured on energy and energy storage in my college-level environmental engineering classes, I focused on hydrogen and hydrogen storage. I presented the basics of hydrogen and then asked the students to tell me what they knew about hydrogen use and storage.

After listening to several responses (all excellent, by the way), I stated that I had a modest proposal about hydrogen fuel cells and others. I said that when the day arrives, we must rely on electric vehicles (EVs) for transportation: buses, trains, airplanes, and spaceships—those technologies have not yet been invented, produced, refined, and tested.

I also point out that, in my opinion, electricity is the future. When the well dries up for hydrocarbons, we can still produce electricity by solar, nuclear, wind, and running water. Whether running water is a natural run-of-the-river pool via dams to make and store hydroelectric energy via conventional run-of-the-river dams or pumped hydro energy storage is not the question; instead, it is the answer.

When hydrocarbons are no longer used or available, we must shift to electricity produced by other means and technologies. Hydrogen fuel cells may be the answer to many of our energy needs.

Usually, during my presentations, I pause momentarily and look for raised hands. There are always raised hands because my upper-class and grad students have exceptional IQs and have no fear of asking questions.

Asking questions—is this not how we all learn? That is if we are provided with the correct answers.

The students want a power-pack system that will get them from point A to B to C to wherever and maybe back to point A again. They want to accomplish this without a range of anxiety, high cost, or having to pray for an accessible and working charging station. We have found that even if remote charging stations are positioned along the highways, this does not mean they are operable. Many of the few recharge stations, even in city locations, are cannibalized for their contents, mainly copper. Copper demand is expanding because of the expanding data centers, computers, iPhones, and other devices that rely on copper in operating these devices.

So, my standard personal answer and the suggested proposal are to make vehicles so that their colossal battery packs, systems, stacks, or whatever configuration is too large and not easily changed when needed can be quickly replaced by a fuel cell.

Okay, what we are talking about here is a unique fuel cell. We need a fuel cell to produce electricity to power a traction motor in our vehicles and other transportation systems. This futuristic fuel cell will be a 4-ft-long and 4-inch-diameter cylinder designed to fit into the vehicle's slide-in/

DOI: 10.1201/9781003637851-16

slide-out factory-installed slot. Each car will have two fuel cell slots, one on each rear side. This is instead of the below carriage extensive battery package that many EVs have. The currently used huge under-carriage battery packages are not easily changed when necessary. However, the cylinder-shaped 4-ft length and 4-inch-in-diameter fuel cells can be changed when required and can be changed virtually by anyone, anytime, and anywhere.

Change-out stations (aka modified gas stations) located throughout the United States or in any country where cylindrical fuel cells are used are needed for cylindrical fuel cell change. These fuel cells can be hydrogen or some soon-to-be or later-developed cell type. These futuristic fuel cells range 800+ miles per cell, depending on how the vehicles are driven. High-speed driving, carrying, or towing heavy loads can lower the range to around 600 miles per cell. Because there are two fuel cells per vehicle, the range is doubled.

In a gasoline-driven vehicle, when the gas supply is low and runs out, the driver can carry a 5-gallon can of spare gas in the trunk of the vehicle. Or the driver can carry an empty 5-gallon can, and when and if he or she runs out of gas, they can hike or get a hitched ride to the nearest gas station, purchase gas, return to the vehicle, and fuel it enough to get a full tank at the same gas station.

The problem with today's EVs is that when the batteries are discharged, you can't remove them and replace them (my students call this a real "bummer"); replacement is complex and must be done at the dealership or repair shop. The 5-gallon can of spare gas is worthless in this situation. One of my students and his fellow students in the class stated that running out of battery charge is not a problem if you also carry a portable gasoline battery charger in your vehicle for emergencies. Since EVs are meant to reduce pollution, using a portable gas-driven generator in the car is nonsensical.

So, my modest proposal is that easily changed-out hydrogen energy/fuel cells are the future and should be discussed. Also, drivers can carry a spare fully charged fuel cell in the trunk instead of a useless 5-gallon container. However, both sides—for and against this technology—should be considered, and in this book, they are.

The bottom line: Instead of installing new fuel cell stations throughout the country, we should convert gas stations into change-out stations.

14.2 ENERGY/FUEL CELLS: A REALISTIC VIEW

14.2.1 Hydrogen[1]

Hydrogen, the chemical symbol H (see Figure 14.1), is the simplest element on Earth and the most abundant element in the universe. It contains only one electron and one proton. Hydrogen is a diatomic molecule—each molecule has two atoms of hydrogen (which is why pure hydrogen is commonly expressed as H_2). Although abundant on Earth as an element, hydrogen combines readily with other components and is almost always found as part of another substance, such as water, hydrocarbons, or alcohols. Hydrogen is also found in biomass, which includes all plants and animals. The following provides quick facts about hydrogen:

- Hydrogen is present in water, sugar, methane, and other common substances.
- Hydrogen is an energy carrier, not an energy source. Hydrogen can store and deliver usable energy, but it doesn't typically exist by itself; it must be produced from compounds containing it.

1
Hydrogen
1.008

FIGURE 14.1 Periodic table symbol for hydrogen.

- Hydrogen has very high energy by weight and is three times more energy-dense than gasoline.
- Hydrogen can be produced using diverse domestic resources, including nuclear, natural gas, coal, biomass, and other renewables, including solar, wind, hydroelectric, or geothermal energy. This diversity of domestic energy sources makes hydrogen a promising energy carrier essential to our nation's energy security. It is expected and desirable for hydrogen to be produced using a variety of resources and process technologies (or pathways).
- Department of Energy (DOE) focuses on hydrogen production technologies that result in near-zero net greenhouse gas emissions and use renewable energy sources, nuclear energy, and coal (when combined with carbon sequestration). Energy efficiency is also essential to ensure sufficient clean energy for our overall energy needs.
- Hydrogen can be produced via various process technologies, including thermal (natural gas reforming, renewable liquid and bio-oil processing, and biomass and coal gasification), electrolytic (water splitting using a variety of energy resources), and photolytic (splitting water using sunlight via biological and electrochemical materials).
- Hydrogen can be produced in extensive, central facilities (50–300 miles from the point of use), smaller semi-central facilities (located within 25–100 miles of use), and distributed facilities (near or at the point of use). Learn more about distributed versus centralized production.
- For hydrogen to be successful in the marketplace, it must be cost-competitive with the available alternatives. This competitive requirement in the light-duty vehicle transportation market means hydrogen must be available untaxed at $2 and $3/gge (gasoline gallon equivalent). This price would result in hydrogen fuel cell vehicles (FCVs) having the exact cost to the consumer on a cost-per-mile driven basis as a comparable conventional ICE or hybrid car.
- DOE is engaged in researching and developing various hydrogen production technologies. Some are further along in development than others—some can be cost-competitive for the transition period (beginning in 2015), and others are considered long-term technologies (cost-competitive after 2030).
- Hydrogen can be used to make fertilizer and steel and as a fuel for transportation.
- Hydrogen can be produced from multiple abundant fuel sources in the United States.
- Hydrogen can be used to store energy and make electricity, with only water as a byproduct.

Infrastructure is required to move hydrogen from where it's produced to the dispenser at a refueling station or a stationary power site. Infrastructure includes pipelines, trucks, railcars, ships, barges that deliver fuel, and the facilities and equipment needed to load and unload them.

Delivery technology for hydrogen infrastructure is currently available commercially, and several U.S. companies deliver bulk hydrogen today. Some of the infrastructure is already in place because hydrogen has long been used in industrial applications. Still, supporting widespread consumer use of hydrogen as an energy carrier is insufficient. Because hydrogen has a relatively low volumetric energy density, its transportation, storage, and final delivery to the point of use comprise a significant cost and result in some energy inefficiencies associated with using it as an energy carrier.

DID YOU KNOW?

The hydrogen fuel cell paradox: energy must first be used to produce hydrogen only to gain less energy back when hydrogen is used to generate electricity. Electrolyzers are used to make hydrogen with electricity and water in and hydrogen and oxygen out making hydrogen using electricity.

Options and trade-offs for hydrogen delivery from central, semi-central, and distributed production facilities to the point of use are complex. The choice of a hydrogen production strategy greatly affects the cost and method of delivery.

For example, more extensive, centralized facilities can produce hydrogen at relatively low costs due to economies of scale. Still, the delivery costs for centrally produced hydrogen are higher than those for semi-central or distributed production options (because the point of use is farther away). In comparison, distributed production facilities have relatively low delivery costs. Still, the hydrogen production costs are likely higher for lower volume production, which means higher equipment costs per unit of hydrogen.

Key challenges to hydrogen delivery include reducing delivery costs, increasing energy efficiency, maintaining hydrogen purity, and minimizing hydrogen leakage. Further research is needed to analyze the trade-offs between the hydrogen production and delivery options taken together as a system. Building a national hydrogen delivery infrastructure is a big challenge. It will take time to develop and will likely include combinations of various technologies. Delivery infrastructure needs and resources will vary by region and type of market (e.g., urban, interstate, or rural). Infrastructure options will evolve as hydrogen demand grows and delivery technologies develop and improve.

14.2.2 Hydrogen Storage

Storing enough hydrogen on board a vehicle to achieve a driving range of greater than 300 miles is a significant challenge. On a weight basis, hydrogen has nearly three times the energy content of gasoline (120 MJ/kg for hydrogen versus 44 MJ/kg for gas). However, on a volume basis, the situation is reversed (8 MJ/l for liquid hydrogen versus 32 MJ/l for gasoline). On-board hydrogen storage in the 5–13 kg H_2 range is required to encompass the entire platform of light-duty vehicles.

Hydrogen can be stored in various ways, but for it to be a competitive fuel for vehicles, the hydrogen vehicle must be able to travel a comparable distance to conventional hydrocarbon-fueled vehicles, without range anxiety. Hydrogen can be physically stored as either a gas or a liquid. Storage as a gas typically requires high-pressure tanks (5000–10,000 psi tank pressure). Storage as a liquid requires cryogenic temperatures because the boiling point of hydrogen at one-atmosphere pressure is 252.8°C. Hydrogen can also be stored on the surfaces of solids (by adsorption) or within solids (by absorption). In adsorption, hydrogen is attached to the material's surface as hydrogen molecules or hydrogen atoms. In absorption, hydrogen is dissociated into H-atoms, and then the hydrogen atoms are incorporated into the solid lattice framework.

Hydrogen storage in solids may make it possible to store large quantities of hydrogen in smaller volumes at low pressures and temperatures close to room temperature. It is also possible to achieve volumetric storage densities greater than liquid hydrogen because the hydrogen molecule is dissociated into atomic hydrogen within the metal hydride lattice structure.

Finally, hydrogen can be stored by reacting hydrogen-containing materials with water (or other compounds such as alcohol). In this case, hydrogen is effectively stored in both the material and the water. The term "chemical hydrogen storage" or chemical hydrides describes this form of hydrogen storage. It is also possible to store hydrogen in the chemical structures of liquids and solids.

14.2.3 Hydrogen Fuel Cell

The fuel cell uses hydrogen's chemical energy to cleanly and efficiently produce electricity, with water and heat as byproducts. Fuel cells are unique in terms of the variety of their potential applications; they can provide energy for systems as large as a utility power station and as small as a laptop computer.

Fuel cells have several benefits over conventional combustion-based technologies currently used in many power plants and passenger vehicles. They produce much smaller quantities of greenhouse gases and none of the air pollutants that create smog and cause health problems. If pure hydrogen is used as a fuel, fuel cells emit only heat and water as byproducts.

DID YOU KNOW?

Hydrogen FCVs emit approximately the same amount of water per mile as vehicles using gasoline-powered ICEs.

A *fuel cell* is a device that uses hydrogen (or hydrogen-rich fuel) and oxygen to create electricity by an electrochemical process. A single fuel cell comprises an electrolyte and two catalyst-coated electrodes (a porous anode and cathode). While there are different fuel cell types, all fuel cells work similarly:

- Hydrogen, or a hydrogen-rich fuel, is fed to the anode, where a catalyst separates hydrogen's negatively charged electrons from positively charged ions (protons).
- At the cathode, oxygen combines with electrons and, in some cases, with species such as protons or water, resulting in water or hydroxide ions.
- For polymer electrolyte membrane and phosphoric acid fuel cells, protons move through the electrolyte to the cathode to combine oxygen and electrons, producing water and heat.
- For alkaline, molten carbonate, and solid oxide fuel cells, negative ions travel through the electrolyte to the anode, combining hydrogen to generate water and electrons.
- The electrons from the anode cannot pass through the electrolyte to the positively charged cathode; they must travel around it via an electrical circuit to reach the other side of the cell. This movement of electrons is an electrical current.

14.2.4 The Color of Hydrogen

If we were to lay out each hydrogen color spectrum, we would form quite a rainbow. This spectrum of colors represents hydrogen's various production methods. These colors, ranging from green to pink, serve as a classification system within the energy industry to differentiate between the types of hydrogen. However, remember that not every country has adopted the color code as used in the United States and some other countries.

Is the color-coding system for hydrogen just a fancy naming convention? No, it is not. It provides insights into the production method, the resources consumed, and the emissions generated during the process. In its natural state, hydrogen is an invisible, colorless gas. The point is that the colors don't represent any visible difference but rather the production and environmental implications—how it is produced and what, if there are any, pollutants it produces. Some think the hydrogen industry should move away from the color designation and adopt policies focusing on the carbon intensity score, which they feel will lead us along the way. Hydrogen production is used for industrial processes rather than energy storage, and the carbon intensity score measures how clean our electricity is. It refers to how many grams of carbon dioxide are released to produce a kilowatt-hour (kWh) of electricity.

So, what are the colors of hydrogen? What fuel source is used to produce different hydrogen colors? What process is used to create the colors of hydrogen? What products are made using hydrogen color processes?

Okay, all these questions are good ones, and the best way to answer them is to present the colors, the fuel used to produce them, the process involved, and the end products produced in the following list:

Green Hydrogen: It is fueled by renewable energy by electrolysis with end product $H_2 + O_2$.
Yellow Hydrogen: It is fueled by solar power via electrolysis with end product $H_2 + O_2$.
Purple/Pink Hydrogen: Purple hydrogen is produced by nuclear power and pink by solar energy $H_2 + O_2$.
Red Hydrogen: It is fueled by atomic power/catalytic splitting from biomass, ending with $H_2 + O_2$.
Turquoise Hydrogen: It is fueled by natural gas via methane pyrolysis, ending $H_2 + C_{(solid)}$.
Blue Hydrogen: It is fueled by natural gas via steam reforming, ending $H_2 + CO_{2(\%\ captured\ and\ stored)}$.
Gray Hydrogen: It is fueled by natural gas via steam reforming, ending $H_2 + CO_{2(released)}$.
Brown/Black Hydrogen: It is produced from coal gasification via steam reforming ending $H_2 + CO_{2(released)}$.
White Hydrogen: It is produced from water, naturally occurring ending H_2.

Again, it is essential to note that the colors don't represent any visible difference but rather the production and environmental implications.

14.3 AI APPLICATIONS FOR FUEL CELLS

Promising materials must be found to manufacture a capable and efficient fuel cell. Artificial intelligence (AI) incorporation in developing and producing these materials is ongoing. Machine learning (ML) can enable performance predictions, service-life predictions, and fault diagnosis in fuel cells. AI can be used to increase hydrogen production, storage, and transport using algorithms and models. Moreover, AI can simulate, control, and optimize fuel cell operation.

DID YOU KNOW?

More than 95% of hydrogen production is used for industrial processes rather than energy storage (Bhandari et al., 2014).

14.4 THE BOTTOM LINE

The hydrogen fuel system must consist of an electrical hydrogen production device, a hydrogen storage unit, a device to generate electrical energy from the stored hydrogen, and power conversion and control equipment.

The discussion about fuel cells is often expected to be chained to powering EVs. Fuel cells can be used in various applications, providing power across multiple sectors, including transportation, industrial/commercial/residential buildings, and long-term energy storage for the grid in reversible systems. Another truth is that fuel cells have several benefits over conventional combustion-based technologies currently used in many power plants and vehicles. For example, fuel cells can operate at higher efficiencies than combustion engines and convert the chemical energy in the fuel directly to electrical energy with efficiencies that exceed 60%. Fuel cells have lower to zero emissions than combustion engines; regarding global climate change, hydrogen fuel cells emit only water, no carbon dioxide emissions, and no smog or health problems at the point of operation. The bottom-line truth about fuel cells is that they have very few moving parts, making them quiet while operating (USDOE, 2024).

The real bottom line is that hydrogen fuel cells will massively emerge as an alternative energy option (and could become the conventional option) in transportation and domestic use over the next several years. Looking into the future of progress in energy storage, hydrogen will be used as a fuel, produced from renewable or nuclear energy, offering an environmentally acceptable and efficient source of power and energy.

Some envision a hydrogen economy in which hydrogen is a primary energy source, decreasing carbon emissions, and helping to achieve global decarbonization goals.

The key challenges involved with hydrogen fuel cells include their high cost, the need for energy or fuel to produce them, their difficulty in storage and transport, and the limited infrastructure needed to move and use them.

NOTE

1 Information in this section from USDOE 2008, Hydrogen, Fuel Cells & Infrastructure Technologies Program. Accessed 12/12/23 @ http://www1.eere.energy.gov/hydrogenandfuelcells/production/basics.html.

REFERENCES

Bhandari, R., Trudewind, C.A., and App, P. (2014). Life Cycle Assessment of Hydrogen Production via Electrolysis—A Review. *J Clean Prod.* vol. 85, pp. 151–163.

USDOE (2024). *Fuel Cells*. Washington, DC: U.S. Department of Energy.

15 The Water Battery

When we speak of water, and its many manifestations, we speak of that endless quintessential cycle that predates all other cycles. Water is our most precious natural resource; we can't survive without it. There is no more water today than yesterday; that is, no more this calendar year than 100 million years ago. The water present today is the same water used by all the animals that ever lived, by cave dwellers, Caesar, Cleopatra, Christ, da Vinci, John Snow, Teddy Roosevelt, and the rest of us—again there is not one drop more or one drop less of water than there has always been. Although this life-giving cycle is a unique blend of thermal and mechanical aspects, it depends on solar energy and gravity for its existence. Nothing on earth is truly infinite in supply, but the energy available from water sources, in practical terms, comes closest to that ideal.

F.R. Spellman (2008)

Given enough time, there is nothing, absolutely nothing, in this world that is safe from water.

Frank R. Spellman (2014)

15.1 THE HOURGLASS EFFECT

Water batteries, also known as pumped storage hydropower (PSH), are made of two big pools of water, one high above the other, that act like an hourglass to provide power. Note that we are talking about some of the biggest batteries on Earth. This is why USDOE (2021) in Washington, DC, the Office of Energy Efficiency and Renewable Energy has listed ten reasons to love water batteries.

In addition to the ten reasons to love water batteries, USDOE (2021) points out that we use PSH daily to power our lives. The paradox is as essential as PSH is to our daily lives; it remains strange, mysterious, hidden, and unknown to most of those who depend on it. This is rather amazing, considering that PSH water batteries are some of the biggest batteries on Earth.

I agree with the USDOE and its ten reasons to love water batteries, which, in my view, are important now and will be the key energy producers of the future. The ten reasons are presented below.

1. Because of water batteries, grid overload is rare.
2. Water batteries fill energy gaps on still and cloudy days.
3. PSH provides 93% of U.S. energy storage.
4. In the United States, 21 states use water batteries to store energy.
5. In the United States, PSH can store up to 553 GWh of energy.
6. Pumped storage is the most efficient extensive energy storage system available at 70–80%.
7. Pumped storage facilities provide 40% of the power needed to jumpstart a grid blackout.
8. Hydropower and pumped storage provide flood control and irrigation benefits, helping absorb extra water during natural disasters.
9. PSH has grown almost as much as all other types of energy storage combined.
10. Water batteries are almost a century old.

15.2 TYPES OF HYDROPOWER FACILITIES

When and if the subject of hydropower ever comes up in people's conversations, those involved might immediately associate a dam or dams with hydropower. The truth is that there are more than 90,000 dams in the United States, and less than 3% of them produce power.

DOI: 10.1201/9781003637851-17

Okay, what are the other 97% of the dams used for?

Most dams are used for recreation, stock/farm ponds, water supply, irrigation, and flood control.

Hydropower systems range from small systems suitable for a single home or village to large projects producing electricity for utilities.

Let's learn more about the sizes of hydropower plants.

15.2.1 Impoundment

The most common type of hydroelectric power plant is an impoundment facility. An impoundment stores river water in a reservoir using a dam, typical of large hydropower systems. Water released from the reservoir flows through a turbine, spinning it, which activates a generator to produce electricity. The water may be released to meet changing electricity needs or other needs, such as flood control, recreation, fish passage (via a fish ladder), and other environmental and water quality needs.

15.2.2 Diversion

A diversion, also known as a "run-of-river" facility, channels a portion of a river through a canal and/or a penstock to utilize the natural decline of the riverbed elevation to produce energy. Penstock is a sluice, gate, or closed intake structure usually made of steel that channels water flow to turbines, with water flow regulated by the gates, valves, and turbines. Note that a diversion may not require the use of a dam.

15.2.3 Pumped Storage

PSH works like a giant battery. A PSH is like a seesaw, up and down, with water pumped up and stored at a higher level than below. When the grid needs more electricity, the water flows downward and turns into a turbine, producing electricity. When electricity demand is low, PSH pumps water up for storage when needed. Later in this chapter, PSH is discussed in greater detail. For now, discussing the fundamentals of hydropower in general is essential.

15.3 SIZES OF HYDROELECTRIC POWER PLANTS

Hydropower plants range in size from larger plants, which supply many consumers with electricity, to small and even "micro" plants, which are operated by individuals for their own energy needs and sell power to utilities.

15.3.1 Large Hydropower

Large hydropower plants have a capacity of more than 30 MW. Examples include the Grand Coulee Dam in Washington State, which uses 33 hydroelectric generators to produce 6,809 MW; Hoover Dam, which produces only 2,080 MW; and China's Three Gorges Dam, which produces 22,500 MW.

15.3.2 Small Hydropower

Small hydropower plants generate between 100 kW and 10 MW.

15.3.3 Micro Hydropower

A micro hydropower plant has a capacity of up to 100 kW. A small or micro-hydroelectric power system can produce enough electricity for a single home, farm, ranch, or village.

15.4 FUNDAMENTALS OF CONVENTIONAL HYDROELECTRIC DAMS

Okay, let's talk about the fundamentals of dams. First, dams are facilities that include some of all of the following components to perform their intended purposes safely: conventional dam or retaining structure (the structural sections that hold back the water), reservoir (the body of water impounded by the dam), spillways (the structure that passes normal and/or flood flows in a manner that protects the structural integrity of the dam), outlet works (the combination of structures and equipment required for safe operation and control of water released from a reservoir), powerhouse (the structure that houses turbines, generators, and associated control equipment for the production of hydroelectricity), and canal/aqueduct (a constructed channel, usually open, that conveys water by gravity or is used for navigation).

Dams are facilities that include some or all of the following components to perform their intended purposes: a conventional dam or impounding structure (structural sections that hold back the water), a reservoir (a body of water impounded by the dam), spillways (structures that facilitate the discharge of normal and/or flood flows in a manner that protects the structural integrity of the project), outlet works (a combination of structures and equipment required for safe operation and control of water released from a reservoir), a powerhouse (a structure that houses turbines, generators, and associated control equipment for the production of hydroelectricity), penstocks (pipelines or conduits used to convey water under pressure to turbines or pump units), and canals/aqueducts (constructed channels, usually open, that convey water by gravity and also may be used for navigational purposes).

In general, canals convey water with an open channel from a source to a destination and serve various functions. Some canals, constructed as part of water supply or transportation projects, have an embankment section forming one of the sides of the water conveyance structure and acting as a water retention structure. Water can be conveyed for various uses (e.g., domestic, municipal, industrial, or agricultural water supply) or as a means of transportation for boats, barges, or other watercraft. In some cases, dams also include navigation locks.

As pointed out above, one of the significant components of a hydroelectric dam is the area behind the dam, its reservoir. The water temporarily stored there is called gravitational potential energy. The water is stored above the rest of the dam facility to allow gravity to carry the water down to the turbines. Because this higher altitude is different than where the water would naturally be, the water is at an altered equilibrium. The result is stored energy of position: gravitational potential energy. The water has the potential to do work because of its position. As shown in Figure 15.1, gravity will force the water to fall to a lower position through the intake and the control gate. Installed within the dam, the water from the reservoir goes through the intake when the control gate is opened. It becomes translational kinetic energy as it falls through the following central part of the system, the penstock. Translational kinetic energy is the energy generated due to motion from one location to another. The water is moving (falling) from the reservoir towards the turbines through the penstock;

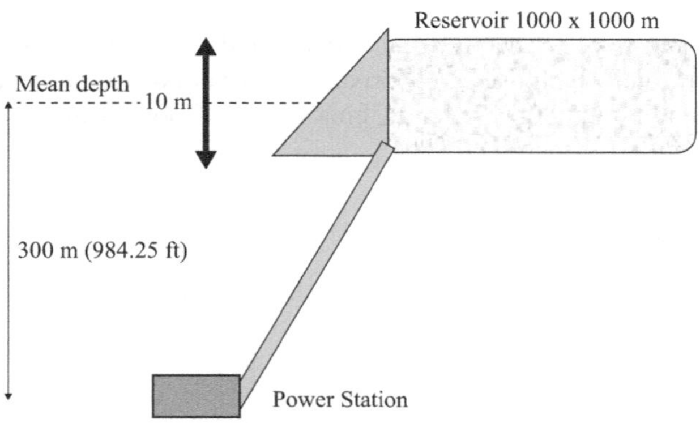

FIGURE 15.1 Schematic representation of a hydro scheme. (Adapted from N.K. Tovey 2005. *Energy Resources*. Norwich, Norfolk, UK: University of Anglia, pp. 109–110.)

the water is carried through the penstock's long shaft towards the turbines, where the kinetic energy becomes mechanical energy. The force of the water is used to turn the turbines that turn the generator shaft. The generators convert the energy of water into electricity, and then step-up transformers increase the voltage produced to higher voltage levels.

Dams may be classified according to the type of construction materials used, the construction methods, the slope or cross-section, how the structure resists the forces of the water pressure behind it, the means used for controlling the seepage, the purpose, and the hazard potential. Materials used for construction include earth, rock, tailings from mining or milling, concrete, masonry, steel, timber, miscellaneous materials such as plastic or rubber, or any combination of these materials. Today's two most common types of dams are embankment and concrete dams (Association of State Dam Safety Officials (ASDSO), 2015).

Embankment dams are the most common type of dam in use today. Depending on whether it is composed of compacted earth or mostly compacted or dumped rock, an embankment dam is termed an "earth fill" or "rockfill" dam. An embankment dam's ability to resist reservoir water pressure is primarily a result of its mass weight and the type and strength of the materials from which it is made.

Concrete dams may be categorized as gravity or arch dams. In concrete gravity dams, the most common form of concrete dam, the width of the concrete, and the friction force along the foundation are used to resist the reservoir water pressure. Gravity dams are constructed of vertical concrete blocks with flexible seals in the joints between the blocks. A buttress dam is a specific type of gravity dam in which the large mass of concrete is reduced, and the forces are diverted to the dam foundation through vertical or sloping buttresses. Concrete arch dams are characterized by thinner cross-sections in which the reservoir water pressure is carried laterally into the abutments. The shape of the arch may resemble a segment of a circle or an ellipse, and the arch may be curved in the vertical plane as well. Such dams are usually constructed of thin vertical blocks that are keyed together; barriers to stop water from flowing are provided between the blocks. Variations or arch dams include multi-arch dams in which more than one curved section is used and arch-gravity dams that combine some features of the two types of dams.

The National Inventory of Dams (NID) lists 82,642 dams in the United States; more than 70,000 are state-regulated, Federal agencies regulate 4,639, and any government agency does not restrict the remainder. The ownership of dams in the United States makes them a unique part of the national infrastructure. Although most infrastructure facilities, such as roads, bridges, and sewer (wastewater) systems, are owned by public entities, most dams (almost 65%) in the United States are owned by private entities. Table 15.1 illustrates the ownership structure, and Table 15.2 shows the primary purpose of dams in the United States.

Although Table 15.2 shows that only about 2% of the dams within the United States are used for hydroelectric production purposes, it is essential to point out that, regarding critical infrastructure applications, the production of electrical power and hydropower (i.e., the harnessing of water to perform work) in general is just as important as flood control and water supply. The power of falling water has been used in industry for thousands of years, as listed in Table 15.3.

TABLE 15.1
Ownership of U.S. Dams

Owner	Percentage of Owners
Private	65
Local government	20
State government	5
Unknown	4
Federal government	4
Public utility	2

TABLE 15.2
Dams by Primary Purpose

Primary Purpose	Percentage of Purpose
Recreation	34
Flood control and storm water management	17
Fire protection, stock, or small farm pond	15
Irrigation	10
Water supply	7
Other	7
Unknown	4
Hydroelectric	~2
Fish and wildlife pond	>2
Tailings	>1.5
Debris control	>1.0
Navigation	<0.5

TABLE 15.3
History of Hydropower

Date	Hydropower Event
B.C.E.	Hydropower was used by the Greeks to turn water wheels for grinding wheat into flour, more than 2,000 years ago.
Mid-1770s	French hydraulic and military engineer Bernard Forest de Belidor wrote a four-volume work describing vertical- and horizontal-axis machines.
1775	U.S. Army Corps of Engineers founded, with establishment of Chief Engineer for the Continental Army.
1880	Michigan's Grand Rapids Electric Light and Power Company, generating electricity by a dynamo belted to a water turbine at the Wolverine Chair Factory, lit up 16 brush-arc lamps.
1881	Niagara Falls city street amps powered by hydropower.
1882	World's first hydroelectric power plant began operation on the Fox River in Appleton, Wisconsin.
1886	About 45 water-powered electric plants in the United States and Canada.
1887	San Bernardino, CA. opens first hydroelectric plant in the west.
1889	Two hundred electric plants in the United States use water power for some or all generation.
1901	First Federal Water Power Act.
1902	Bureau of Reclamation established.
1907	Hydropower provided 15% of U.S. electrical generation.
1920	Hydropower provided 25% of U.S. electrical generation. Federal Power Act establishes Federal Power Commission authority to issue hydro-development licenses on public lands.
1933	Tennessee Valley Authority established.
1935	Federal Power Commission authority extended to all hydroelectric projects built by utilities engaged in interstate commerce.
1937	Bonneville Dam, the first Federal dam, begins operation on the Columbia River, and the Bonneville Power Administration is established.
1940	Hydropower provided 40% of electrical generation. Conventional capacity tripled in the United States since 1920.
Since 1980,	Conventional capacity has nearly tripled in the United States 1900.
2003	About 10% of U.S. electricity comes from hydropower. Today, there is about 80,000 MW of conventional capacity and 18,000 MW of pumped storage.

Source: EERE (2008).

DID YOU KNOW?

The potential energy in a specific slug or parcel of water is expressed in newton-meters (N · m). The newton is the standard unit of force, equivalent to a meter per second squared (1 m/s^2). This is the product of the mass of the slug in kilograms, the acceleration of gravity in meters per second squared (about 9.8 m/s^2), and the elevation of the parcel in meters (the vertical distance it falls as its energy is harnessed). The equivalent kinetic-energy unit is the joule (J), which is in effect equal to a watt-second (W · s).

Table 15.4 highlights solar energy's quadrillion Btu ranking in current renewable energy source use. As pointed out in the table, the energy consumption by energy source computations is from 2007. Thus, the 0.080 solar/PV quadrillion Btu figure is expected to increase steadily to a higher level, and this should be reflected in the 2008–2015 figures when they are released. Table 15.5 lists the largest U.S. hydroelectric power stations.

TABLE 15.4
U.S. Energy Consumption by Energy Source, 2007 (Quadrillion Btu)

Energy Source	2007
Total	**101.605**
Renewable	6.830
Biomass (biofuels, waste, wood, and wood-derived)	3.615
Biofuels	1.018
Waste	0.431
Wood-derived Fuels	2.165
Geothermal	0.353
Hydroelectric Conventional	**2.463**
Solar/PV	0.080
Wind	0.319

Source: EIA (2007). *U.S. Energy Consumption by Energy Source.* Accessed 06/12/25 @ http://www.eia.doe.gov/enaf/alternate/page/renew_energy_conump/table1.html.

TABLE 15.5
Largest U.S. Hydroelectric Power Stations

Name	Year of Completion	Total Capacity (MW)
Grand Coulee	1942/1980	6,809
Bath County PSP	1945	3,003
Chief Joseph Dam	1958/73/79	2,620
Robert Moses Niagara PP	1961	2,515
John Day Dam	1949	2,160
Hoover Dam	1936/1961	2,080
The Dalles Dam	1981	2,038

Source: Bureau of Reclamation. (http://www.usbr.gov/pn/news/02new/gcrecrd.html) Retrieved 5/5/25.

DID YOU KNOW?

One of the problems encountered in a hydraulic system is storing the liquid. Unlike air, which is readily compressible and is capable of being stored in large quantities in relatively small containers, a liquid such as water cannot be compressed. Therefore, it is not possible to store a large amount of water in a small tank—62.4 pounds of water occupies a volume of 1 cubic foot, regardless of the pressure applied to it.

$$1 \text{ foot}^3 \ H_2O = 62.4 \text{ pounds}$$

Both cubic feet and pounds are used to describe the volume of water. There is a defined relationship between these two methods of measurement. The specific weight of water is defined relative to a cubic foot. One cubic foot of water weighs 62.4 pounds. This relationship is valid only at a temperature of 4°C and at a pressure of one atmosphere [known as standard temperature and pressure (STP)—14.7 pounds per square inch at sea level containing 7.48 gallons]. The weight varies so little that, for practical purposes, this weight is used from temperature 0–100°C. One cubic inch of water weighs 0.0362 pounds. Water 1 foot deep will exert a pressure of 0.43 pounds per square inch on the bottom area (12 inches × 0.0362 pound/inch^3). A column of water 2 feet high exerts 0.86 psi, one 10 feet high exerts 4.3 psi, and one 55 feet high exerts 55 feet × 0.43 psi/feet = 23.65 psi.

A column of water 2.31 feet high will exert 1.0 psi. To produce a pressure of 50 psi requires a water column

$$50 \text{ psi} \times 2.31 \text{ feet / psi} = 115.5 \text{ feet}$$

Remember: The important points being made here are

1. 1 foot^3 H_2O = 62.4 pounds.
2. A column of water 2.31 feet high will exert 1.0 psi
   Another relationship is also important:
   1 gallon H_2O = 8.34 pounds

15.4.1 Flow/Discharge Rate: Water in Motion

The study of fluid flow is much more complicated than that of fluids at rest. Still, understanding these principles is essential because the water used in hydropower applications is nearly always in motion (e.g., the water used to propel turbine blades).

Discharge (or flow) is the quantity of water passing a given point in a pipe or channel during a given period. Stated another way for open channels: the flow rate through an open channel is directly related to the liquid's velocity and the liquid's cross-sectional area in the channel.

$$Q = A \times V \tag{15.1}$$

where

Q = Flow—discharge in cubic feet per second (cfs)
A = Cross-sectional area of the pipe or channel (foot^2)
V = water velocity in feet per second (fps or feet/second)

EXAMPLE 15.1

Problem:

The channel is 6 feet wide, and the water depth is 3 feet. The speed in the channel measures 4 feet per second. What is the discharge or flow rate in cubic feet per second?

SOLUTION:

$$\text{Flow, cfs} = 6\,\text{feet} \times 3\,\text{feet} \times 4\,\text{feet / second} = 72\,\text{cfs}$$

Discharge or flow can be recorded as gallons/day (gpd), gallons/minute (gpm), or cubic feet (cfs). Flows treated by many hydropower systems are large and often called a million gallons per day (MGD). Using appropriate conversion factors, the discharge or flow rate can be converted from cfs to other units such as gallons per minute (gpm) or million gallons per day (MGD).

15.4.2 Reservoir Stored Energy

One of the significant components of a hydroelectric dam is the area behind the dam, its reservoir. The water temporarily stored there is called gravitational potential energy. The water is stored above the rest of the dam facility to allow gravity to carry the water down to the turbines. Because this higher altitude is different than where the water would naturally be, the water is at an altered equilibrium. The result is stored energy of position, that is, gravitational potential energy. The water has the potential to do work because of its position. Gravity will force the water to fall to a lower position through the intake and the control gate. Installed within the dam, the water from the reservoir goes through the intake when the control gate is opened. It becomes translational kinetic energy as it falls through the following central part of the system, the penstock. Translational kinetic energy is the energy due to motion from one location to another. The water moves (falling) from the reservoir towards the turbines through the penstock; the water is carried through the penstock's long shaft towards the turbines, where the kinetic energy becomes mechanical energy. The force of the water is used to turn the turbines that turn the generator shaft. The generators convert the energy of water into electricity, and then step-up transformers increase the voltage produced to higher voltage levels.

15.4.3 Hydro Turbines

According to EERE (2008), there are two main types of hydro turbines: impulse and reaction. The type of hydropower turbine selected for a project is based on the height of standing water, as mentioned, this is referred to as "head," and the flow, or volume of water, at the site. Other deciding factors are the cost, the efficiency desired, and how deep the turbine must be set.

Impulse Turbine: Impulse turbines use the water's velocity to move the runner and discharge to atmospheric pressure. The water stream hits each bucket on the runner. There is no suction on the turbine's downside, and water flows out of the bottom of the turbine housing after hitting the runner. An impulse turbine is generally suitable for high-head, low-flow applications.

Reaction Turbine: Reaction turbines develop power from pressure and moving water combined. The runner is placed directly in the water stream, flowing over the blades rather than striking each individually. Unlike impulse turbines, reaction turbines are typically used for locations with low head and high flow.

15.4.4 Navigation Locks

The term "navigation lock" is used to denote navigation projects that include the locks themselves (single or multiple chambers), as well as the associated dam (unless located in a canal), and several appurtenant structures such as approach walls, control rooms, and administration and maintenance buildings. The U.S. waterway system includes 236 lock chambers at 192 lock sites owned and/or operated by the U.S. Army Corps of Engineers (USACE). Many navigation projects may serve multiple purposes besides the primary navigation function. For example, 46 lock-associated dams currently produce hydropower.

These facilities make many inland waterways viable, year-round transportation corridors. Natural riverbeds are often long, uneven downhill slopes with shallow areas and deep pools. Water depths vary with seasonal rainfall and snow melting; dry periods can significantly affect navigation. Locks on navigable rivers enable commercial and recreational river traffic to move safely from one pool level to another. The lift of the locks—the difference between raising or lowering from one pool to the next—can range from just a few feet to 110 feet (e.g., John Day Lock and Dam on the Columbia River, Oregon, and Washington).

JOHN DAY DAM

The John Day Dam is a concrete gravity run-of-the-river dam spanning the Columbia River on the Washington/Oregon border (BPA, 2010). The dam features a navigation lock plus fish ladders on both sides and has the highest lift (110 feet) of any U.S. Lock (USGS, 2015). Lake Umatilla is the reservoir impounded by the dam, and it runs 76.4 miles (123 km) up the river channel to the foot of McNary Dam (USGS, 2015).

Multiple locks may be part of one navigable river system. The Tennessee River system is managed through dams owned by the United States and operated by Tennessee Valley Authority (TVA) and the USACE. The system's nine main and five auxiliary locks allow smooth movement on the river as it drops 513 feet in elevation from its beginning to the point at which it joins the Ohio River. The Mississippi River is made navigable through 30 main and four auxiliary locks.

A principal value of the inland and Intracoastal navigation systems is the ability to convey large volumes of bulk commodities moving long distances efficiently. Towboats push barges lashed together to form a "tow." A 15-barge tow is standard on larger rivers with locks, such as the Ohio, upper Mississippi, Illinois, and Tennessee rivers. Such tows are a highly efficient mode of transportation, moving about 22,500 tons of cargo as a single unit. A single 15-barge tow is the equivalent of about 225 railcars or 870 tractor-trailer trucks. If the cargo transported on the inland waterways each year had to be moved by another mode, it would take an additional 6.3 million railcars or 25.2 million trucks to carry the load. The ability to move more cargo per shipment makes barge transport both fuel-efficient and environmentally advantageous. On average, a gallon of fuel allows 1 ton of freight to be shipped 155 miles by truck, 413 by railway, and 576 miles by barge. As a result, cargo transported by barge results in an average transportation savings of about $11 per ton over the cost of shipping by alternative modes. Environmental advantages can also be measured by comparing the greenhouse gas (GHG) emissions produced during fuel consumption. On average, the GHG emissions produced per million ton-miles are 71.61 by truck, 26.88 by railway, and 19.27 by barge.

Coal is the largest commodity by volume moving on the inland waterways. The Nation's electric industry depends on inland waterways for more than 20% of the coal that it consumes to produce electricity. Petroleum is the next largest group, including crude oil, gasoline, diesel fuel, heavy fuel, and asphalt. Another large group includes grain and other farm products, most of which move by waterway to ports on the Lower Mississippi or Columbia Rivers for export overseas. Sixty percent of the country's farm exports travel through inland waterways. Other major commodities include aggregates, such as stone, sand, and gravel used in construction; chemicals, including fertilizers; metal ores, minerals, and products, such as steel; and many other manufactured products.

There is more to a dam than retaining water to eventually flow through a dam's infrastructure to store and produce electricity from a run-of-the-river. The U.S. Department of Homeland Security (2010) points out that dam infrastructure is critical to the economy of the 21st century. The sector is a vital and beneficial part of the Nation's infrastructure. It continuously provides a wide range of economic, environmental, and social benefits, including hydroelectric power, river navigation, water supply, wildlife habitat, waste management, flood control, and recreation, to name a few. Some examples of the benefits derived from the dam sector assets are discussed below:

- *Water Storage and Irrigation*: Dams create reservoirs throughout the United States that supply water for many industrial, municipal, agricultural, and recreational uses. Ten percent of American cropland is irrigated by water stored behind dams, and thousands of jobs are tied to producing crops grown with irrigation water.
- *Electricity Generation*: The United States is one of the largest producers of hydropower in the world, second only to Canada. Dams in the United States produce more than 270,000 GWh, contributing 7% of the Nation's electricity and representing 70% of the nation's renewable energy generation.
- *"Black Start" Capabilities*: 4,316 MW of "incremental" hydropower are available at sites with existing hydroelectric facilities. Incremental hydropower is defined as existing hydro projects' capacity additions or improved efficiency. During the August 2003 blackout in the Northeast, hydropower projects in New York and several other states were able to quickly start generating electricity, leading the way to restoring power to millions of Americans.
- *Recreation*: Dams and other sector assets provide prime recreational facilities throughout the United States. In 2002, 105.7 million recreation users were supplied with days and nights at hydropower projects licensed by the Federal Energy Regulatory Commission (FERC). In addition, about 400 million people visit a project of the USACE annually, and about 90 million visit the Bureau of Reclamation project in a year.
- *Navigation*: Navigational projects constitute an essential component of the U.S. water system, including 236 lock chambers at 192 lock sites owned and/or operated by USACE (1993). A principal value of the inland and Intracoastal navigation system is the ability to convey large volumes of bulk commodities moving long distances efficiently. If the cargo transported on the inland waterways each year had to be moved by another mode, it would take an additional 6.3 million railcars or 25.2 million trucks to carry the load. The ability to move more cargo per shipment makes barge transport both fuel-efficient and environmentally advantageous.
- *Flood Risk Reduction*: Many dams and levees function as flood control projects, reducing the potential human health and economic impacts of flooding. Reservoirs and levees built by USACE reportedly prevented more than $19 billion in possible damage during the 1993 Midwest Flood. USACE lever systems currently provide a 6:1 return on flood damage prevented compared to initial costs, and robust levee systems provide a 24:1 return on investments. Levees and hurricane barriers mitigate flood damage in both rural and metropolitan areas.
- *Sediment Control*: Some dams enhance environmental protection by controlling detrimental sedimentation.
- *Impoundment of Mine Tailings and Industrial Waste Materials*: More than 1,500 mine tailings and industrial waste impoundments controlled by dams in the Nation facilitate mining and processing of coal and other vital minerals and manufacturing while protecting the environment.

Regardless of causation, asset failures within the dam sector pose considerable potential risks and could result in significant destruction, including loss of life, massive property damage, and severe long-term consequences. Health and welfare are threatened without a stable dam infrastructure sector, and the U.S. economy cannot function effectively.

As mentioned, the dam infrastructure sector includes subsections or elements such as dam projects, hydropower plants, navigation locks, levees, dikes, hurricane barriers, mine tailing and other industrial waste impoundments, and similar water retention and water control facilities. It is important to note that these subsections or elements of the dam sector encompass physical, human, cyber, and security elements.

15.4.5 Dams Sector Elements

The Dams Sector comprises physical, human, cyber, and security elements. The multiplicity and complexity of sector asset functions dictate the importance of these elements, as described below.

15.4.5.1 Physical Elements

Dams in the United States range in height from a few feet to more than 700 feet and in length from a few tens of feet to more than 1 mile. Normal reservoir storage (i.e., the volume of water stored in a reservoir for a majority of the time) behind U.S. dams can range from a few acre-feet to 30 million acre-feet. Dams in the U.S. Territories range from 30 to 274 feet high, with reservoir sizes that range from 100 to 55,000 acre-feet. Dams are also located near or along the borders with Canada and Mexico.

In addition to the main impounding structure, one or several appurtenant structures may be associated with the multiple purposes of the facility. Examples of the types of structures that could be appurtenant to a dam include free overflow and gated spillway structures, outlet works, water supply features, penstocks, power generation features, navigation locks, embankments associated with the main structure (also known as saddle dams, dikes, and freeboard dams), and different types of control structures. Each dam type and each appurtenant structure has its purpose and varying degrees of vulnerability. In general, each functional component may be associated with different protective configurations.

Levees (and flood risk reduction systems in general) may feature multiple components, including embankment sections, as well as floodwall sections, pumps and pumping stations, interior drainage works, flood damage reduction channels, closure structures, penetrations (e.g., utilities and pipes), and transitions (i.e., sections where the materials constituting the levee system change from, for example, a conventional floodwall section to an earthen section).

15.4.5.2 Human Element

The safe operation and security of the nation's critical infrastructure depend heavily on human elements. Dams Sector facilities can be either staffed or unstaffed. Many smaller or remote assets have minimal requirements for operation and are operated without personnel onsite for lengthy periods of time (e.g., a week). Some facilities are operated remotely from central operations centers and may or may not require onsite personnel on a daily basis. Some are staffed on a "business day schedule" (i.e., 8- or 10-hour scheduled for five to seven days a week). Some larger or more complex facilities require staff to be present on a 24/7 basis.

DID YOU KNOW?

Because many of the Dams Sector operations require small workforces, owners and managers are usually well acquainted with their workers and schedules. As a consequence, the potential for criminal infiltration is very limited in the imposters or infiltrators or inside threats would be easier to recognize that is the case for other types of infrastructure. In addition, due to this smaller and more easily recognized workforce, the work environment is significantly easier to monitor and unusual activity, such as surveillance, terrorist attacks, and other criminal activities, is much easier to distinguish and respond to in the early stages of a potential situation. However, this may not be necessarily applicable to other types of facilities, such as levees, for example, which are spatially distributed and require easy and open access for maintenance purposes.

The workforce for sector assets such as hydropower facilities or navigation locks is relatively stable compared to other industries, and each facility has a relatively small workforce.

Concerning hydropower dams, the number of personnel required to operate and maintain the asset can vary significantly. The need for staff depends greatly on the complexity of the dam and the benefits derived from it. Larger hydropower dams may require adjustments in operations several times per hour to respond to generation needs and economics. They may require several operators for each shift: a crew of maintenance and monitoring personnel; specialized staff that respond to hourly, daily, or weekly adjustments to power demands; marketing specialists; transmission line crews; and security personnel and guards. Simple, less complex facilities may require only one person to operate the structure and perform monitoring and maintenance duties. For example, recreational dams and stock ponds have minimal staffing needs. Staffing at levees and flood control projects is somewhat similar to that at dams but is more focused on operations and maintenance activities and will vary depending on water levels and weather conditions.

Since many of these positions are highly technical and specialized, they require personnel with three or more years of apprentice work or advanced training. Operations at some dams may also require full-time or corporate sharing of management and IT personnel. Personnel succession plans and continuity-of-operations plans help to ensure ongoing operation in the event of work stoppages and extended periods of illness, such as during pandemics. In addition, the schedule for noncritical operations can be modified to respond to situations that may be understaffed (e.g., establishing a deferred maintenance schedule).

15.4.5.3 AI Element

The efficiency and safety of dam sector operations and maintenance are constantly improving as technology in industrial control systems advances. Control systems vary widely based on age and generation; thus, they differ in complexity and sophistication. Some systems are closed and use isolated networks and proprietary communication protocols, while others use open architecture and standard communication protocols and rely on the Internet. In addition, the control systems at dams, which may also be connected to the electric power grid. Most dam owners, if not all, use various computer security methods for mastering terminal units, data servers, and historians, such as authentication procedures, encryption, firewalls, virus software, and anti-spyware.

Note that many industrial control systems used today were designed for operability and reliability during an era when security received low priority, before 9/11 and the rise in worldwide terrorism. These systems operated in relatively isolated environments and typically relied on proprietary software, hardware, and/or communications technology. Infiltrating and compromising these systems often required specific knowledge of individual systems' architecture and physical access to system components.

In contrast, newer control systems are highly network-based and use every day and open standards for communication protocols, such as addressable Internet protocols. Many controllers are Internet protocol addressable. Asset owners and operators have gained immediate benefits by extending the connectivity of their systems and have increasingly adopted commercial off-the-shelf technologies that provide the higher levels of interoperability required in today's modern infrastructure. Standard operating systems such as Windows or UNIX are increasingly used in central supervisory stations, which are now typically connected to remote controllers via public-private networks provided by telecommunications companies. Common telecommunications technologies, such as the Internet, public-switched telephone networks, or cable or wireless networks, are often used.

DID YOU KNOW?

As a general proposition, levee operations are conducted by smaller entities and are considerably less dependent on information technology than dams. They rely on basic telecommunications and information infrastructures that can be supplemented with manual operations.

Again, it is essential to point out that the dam sector (including the different types discussed earlier) provides services that are essential to the U.S. economy and a major contributor to the so-called good life (standard of living) we presently enjoy. The dam sector knows its vulnerabilities and is leading a significant voluntary effort to increase its planning and preparedness. Cooperation through industry groups has resulted in substantial information sharing of practical and best practices across the sector. Many sector owners and operators have extensive infrastructure protection experience and have recently focused on cybersecurity. In addition to the economic consequences of a successful homegrown or foreign terrorist attack against dam sector facilities, there is also the potential of a threat to public health and safety and the environment.

Pumped hydro energy storage (PHES) is not dependent on renewable energy sources like solar, wind, and tide. So long as we can confine an upper storage reservoir of water that, via gravity, flows downhill to another storage basin, we can produce electricity. Whether we use renewable energy to pump water to the upper basin or other forms of energy is not the point. The point is that as long as we have some form of energy to pump, we can keep running water downhill to spin turbine generators to produce electricity.

15.5 DAMS INTERDEPENDENCIES

As mentioned, the dam sector comprises assets, including dam products, hydropower generation facilities, navigation locks, levees, dikes, hurricane barriers, mine tailings, other industries' waste impoundments, and similar water retention and water control facilities. The dam sector is a vital part of the nation's infrastructure. It provides various economic, environmental, and social benefits, including hydroelectric power, river navigation, water supply, wildlife habitat, waste management, flood control, and recreation.

The Dams Sector has interdependencies with a wide range of other factors, including

- *The Emergency Services Sector*: Emergency response sometimes relies on dam sector assets for firefighting water supply, emergency water supply, and waterway access during a significant disaster where access by land is impossible.
- *The Energy Sector*: Hydropower dams produce approximately 8–12% of the nation's power needs.
- *The Food and Agriculture Sector*: Dams sector assets are a water source for our nation's food and agriculture production.
- *The Transportation Systems Sector*: Dams and locks manage navigable waters throughout inland waterways for shipping and recreation.
- *The Water and Wastewater Systems Sector*: Water and wastewater management systems rely on dam sector assets to provide water to large, populated areas and commercial facilities.

15.6 DAMS SECTOR ASSETS

As previously described, the dam sector assets encompass a broad spectrum of facilities nationwide.

15.6.1 Levees

The term "levee" is used to denote human-made flood risk reduction systems (e.g., embankment sections, floodwalls, closure structures, pumps, interior drainage works, and flood damage reduction channels) whose primary purpose is to furnish flood protection from seasonal high water and which are therefore subject to water loading for periods of only a few days or weeks a year. Embankments subject to water loading for prolonged periods (longer than standard flood protection requirements) or water loaded permanently should be designed according to earth dam criteria rather than the levee criteria given in Table 15.6.

Even though levees are similar to small earth dams, they differ from earth dams in the following important respects: (1) a levee embankment may become saturated for only a short period beyond the limit of capillary saturation, (2) levee alignment is dictated primarily by flood protection requirements, which often results in construction on poor foundations, and (3) borrow is generally obtained from shallow pits or from channels excavated adjacent to the levee, which produce fill material that is often heterogeneous and far from ideal. Selection of the levee section is often based on the properties of the poorest material that must be used.

Numerous factors must be considered in levee design. These factors may vary from project to project, and no specific step-by-step procedure covering the details of a particular project can be

TABLE 15.6
Levee Major and Minimum Requirements

Step	Procedure
1.	Conduct a geological study based on a thorough review of available data, including analysis of aerial photographs. Initiate preliminary subsurface explorations.
2.	Analyze preliminary exploration data and, based on the analysis, establish preliminary soil profiles, borrow locations, and embankment sections.
3.	Initiate final exploration to provide a. Additional information on soil profiles. b. Undisturbed strengths of foundation materials. c. More detailed information on borrow areas and other required excavations.
4.	Using the information obtained in Step 3: a. Determine embankment and foundation soil parameters and refine preliminary sections where needed, noting all possible problem areas. b. Compute rough quantities of suitable material and refine the borrow area location.
5.	Divide the entire levee into reaches of similar foundation conditions, embankment height, assign a typical trial section to each reach and fill material.
6.	Analyze each trial section as needed for: a. Under seepage and through seepage. b. Slope stability. c. Settlement. d. Trafficability of the levee surface.
7.	Design special treatment to preclude any problems as determined from Step 6. Determine surfacing requirements for the levee based on its expected future use.
8.	Based on the results of Step 7, establish final sections for each reach.
9.	Compute final quantities needed, determine final borrow area locations.
10.	Design embankment slope protection.

established. However, it is possible to present general, logical steps based on successful past projects that can be followed in levee design and used as a base for developing more specific procedures for any particular project. Such a procedure is given in Table 15.6.

CAPILLARY ACTION

If we were to mention the term capillary action to the man or woman on the street, they might instantly nod their heads and respond that their bodies are of full of them—that capillaries are the tiny blood vessels that connect the smallest arteries and the smallest of the veins. This would be true, of course. But in the context of water science, capillary is something different than capillary action in the human body.

Even if you've never heard of capillary action, it is still important in your life. Capillary action is important for moving water (and all of the things that are dissolved in it) around. It is defined as the movement of water within the spaces of a porous material due to the forces of adhesion, cohesion, and surface tension.

Surface tension is a measure of the strength of the water's surface film. The attraction between the water molecules creates a strong film, which among other common liquids is only surpassed by that of mercury. This surface tension permits water to hold up substances heavier and denser than it. A steel needle carefully placed on the surface of that glass of water will float. Some aquatic insects such as the water strider rely on surface tension to walk on water.

Capillary action occurs because water is sticky; thanks to the forces of cohesion (water molecules like to stay closely together) and adhesion (water molecules are attracted and stick to other substances). So, water tends to stick together, as in a drop, and it sticks to glass, cloth organic tissues, and soil. Dip a paper towel into a glass of water and the water will "climb" onto the paper towel. In fact, it will keep going up the towel until the pull of gravity is too much for it to overcome.

DID YOU KNOW?

Levees, as opposed to dams, are geospatially extended systems that can run for multiple miles; however, the levee system may consist of several segments under different governance units. This creates opportunities for multiple failure modes and points throughout the system. Proper levee maintenance is often overlooked because it appears to be fairly basic—repair erosion damage, protect against unwanted vegetation growth or animal burrowing, and so forth—so it may be given low priority or ignored completely.

The method of construction must also be considered. In the past, levees were built using methods of compaction varying from none to carefully controlled compaction. The local economic situation also affects the selection of a levee section. Traditionally, levees have been built with relatively steep slopes using controlled compaction in areas with high property values, land use, and sound foundation conditions. In regions where property values are lower, foundation quality is poor, or construction occurs during periods of high rainfall, it is more common to find uncompacted or semi-compacted levees with flatter slopes. This is evident by comparing the steep slopes of levees along the industrialized Ohio River Valley with levees along the Lower Mississippi River, which have much broader sections with gentler slopes. Levees built with smaller sections and steeper slopes generally require more comprehensive investigation and analysis than levees with broad sections and flatter slopes, whose design is more empirical. Where conditions allow, levees are designed with steeper slopes. Levee maintenance is another factor that often influences the selection of a levee section.

15.6.1.1 Levee Types According to Location

Due to the varying needs of each, levees are categorized based on the areas they safeguard: urban levees and agricultural levees. As used in this book, urban and agricultural levees are defined as follows:

1. *Urban Levees*: These are designed to provide flood protection for communities, including industrial, commercial, and residential areas.
2. *Agricultural Levees*: These levees help to control water levels on agricultural lands.

15.6.1.2 Levee Types According to Use

Table 15.7 lists some of the more common terms used for levees serving a specific purpose in connection with their overall purpose of flood protection.

HEAD

Head is defined as the vertical distance water must be lifted from the supply tank to the discharge, or as the height a column of water would rise due to the pressure at its base. A perfect vacuum plus atmospheric pressure of 14.7 psi would lift the water 34 feet. If the top of the sealed tube is opened to the atmosphere and the reservoir is enclosed, the pressure in the reservoir is increased; the water will rise in the tube. Because atmospheric pressure is essentially universal, we usually ignore the first 14.7 psi of actual pressure measurements, and measure only the difference between the water pressure and the atmospheric pressure; we call this *gauge pressure*. For example, water in an open reservoir is subjected to the 14.7 psi of atmospheric pressure, but subtracting this 14.7 psi leaves a gauge pressure of 0 psi. This shows that the water would rise 0 feet above the reservoir surface. If the gauge pressure in a water main were 120 psi, the water would rise in a tube connected to the main:

$$120\text{ psi} \times 2.31\text{ feet / psi} = 277\text{ feet (rounded)}$$

The *total head* includes the vertical distance the liquid must be lifted (static head), the loss to friction (friction head), and the energy required to maintain the desired velocity (velocity head).

$$\text{Total Head} = \text{Static Head} + \text{Friction Head} + \text{Velocity Head} \tag{15.2}$$

TABLE 15.7
Classifications of Levees According to Use

Type	Definition
Mainline and tributary levees	Levees that lie along a mainstream and its tributaries, respectively.
Ring levees	Levees that completely encircle in "ring", an area subject to inundation from all directions.
Setback levees	Levees that are built landward of existing levees, usually because the existing levees have suffered distress or are in some way being endangered, as by river migration.
Sublevees	Levees built for underseepage control. Sublevees encircle areas behind the main levee which are subject, during high-water stages, to high uplift pressures and possibly the development of sand boils. They normally tie into the main levee, thus providing a basin that can be flooded during high-water stages, thereby counterbalancing excess head beneath the top stratum with the basin. Sublevees are rarely employed as the use relief wells or seepage berms make them unnecessary except in emergencies.
Spur levees	Levees that project from the main levee and serve to protect the main levee from the erosive action of stream currents. Spur levees are not true levees but training dikes.

15.6.1.3 Causes of Levee Failures

The principal causes of levee failure are

1. Overtopping
2. Surface erosion
3. Internal erosion (piping)
4. Slides within the levee embankment of the foundation soils.

DID YOU KNOW?

Levee ownership and governance patterns can sometimes be politically, socially, and economically more complex than those for dams. Levee activities may be cost-shared among agencies at several levels of government, which can contribute to delays in repairs, response, and mitigation actions. Actual ownership, especially for older levees, may require clear communication for levee safety and resilience.

15.7 HURRICANE BARRIERS

A *Hurricane, flood, surge*, or *storm surge barrier* is a specific type of floodgate, designed to restrict the inflow of storm surges and extratropical storms into bays, estuaries, rivers, and streams, and/or onto the adjacent lowlands. These barriers consist of gates closed during storms from a fully open position or a partially open position, depending on the degree of routine use they also serve.

Hurricane barriers in the Northeast that protect against flooding in large bays are normally left fully open and are closed only during a storm. Barriers in southern Florida serve a routine water control purpose and are completely closed during a hurricane event.

Hurricane barriers are typically large steel structures. Each movable gate spanning the Providence River in Providence, Rhode Island, is 40 feet wide and weighs 53 tons. Hurricane barriers usually comprise at least two gates—one on each channel bank—but more complex configurations may include multiple gates that can be independently raised or closed depending on the surge volume.

15.8 MINE TAILINGS AND INDUSTRIAL WASTE IMPOUNDMENTS

Mining, electric power generation, and manufacturing industries often use impoundment structures to store and dewater waste. In the coal industry, embankment structures used to dispose of coal waste are called slurry impoundments; in other extractive industries, the structures used to dispose acceptable mine waste are called tailings ponds. In other sectors, they may be called surface or industrial waste impoundments. The primary goal of these impoundments is to store tailings or mine waste, which are transported suspended in water and settle out in the impoundment.

15.8.1 Tailings Dams

Large quantities of rock are mined, crushed, pulverized, and processed to recover metal and other mineral values needed for industrial processes, fertilizers, homes, cars, and other consumer products. A fine grind is often necessary to release metals and minerals. Hence, the mining industry produces enormous quantities of fine rock particles, in sizes ranging from sand-sized down to as low as a few microns. These fine-grained wastes are known as "tailings."

Until recent decades, most mines were small underground operations with correspondingly modest requirements for tailings disposal. Since then, due to increasing demand, it has become economical to mine large lower-grade deposits by utilizing advances made by mining equipment manufacturers and developments in mining and mining technology. This has significantly increased the tailings and waste produced by mining projects and the industry (USEPA, 1994).

There are approximately 1,000 active metal mines in the United States (USGS, 2003). Many of these have at least one tailings impoundment, and often, several impoundments are grouped in cells. USEPA estimates that several thousand tailings impoundments may be associated with active non-coal mining, and tens of thousands of inactive or abandoned impoundments.

By far the more significant proportion of ore mined in most industry sectors ultimately becomes tailings that must be disposed of. In the gold industry, for example, only a few hundredths of an ounce of gold may be produced for every ton of dry tailings generated. Similarly, the copper industry and others typically mine relatively low-grade ores that contain less than a few percent of metal values; the residue becomes tailings. Thus, tailings disposal is a significant part of most hard rock mining projects' overall mining and milling operations. There are several methods used for tailings disposal. These include disposal of dry or thickened tailings in impoundments or free-standing piles, backfilling underground mine workings and open-pits, subaqueous disposal, and the most common method, the disposal of tailings slurry in impoundments. Modern tailings impoundments are engineered structures for permanently disposing of the fine-grained waste from mining and milling operations. At some projects, tailings embankments reach several hundred feet in height, and the impoundments cover several square miles.

Historically, tailings were disposed of where convenient and most cost-effective, often in flowing water or directly into drainages. As local concerns arose about sedimentation in downstream watercourses, water use, and other issues, mining operations began impounding tailings behind earthen dams, which were often constructed of tailings and other waste materials. The impoundments served the dual purpose of containing the tailings and, particularly in the arid west, allowing the reuse of scarce water.

Concerns about the stability and environmental performance of tailings dams and impoundments have been raised more recently. Stability concerns are raised partly by the use of tailings material in tailings dams/embankments; to mitigate these concerns, such embankments often rely on a certain amount of controlled seepage to enhance stability, affecting environmental performance. Ritcey (1989) has speculated that the need for sound impoundments in the uranium industry "probably" accounts for much of the attention paid to impoundment design in other facilities. Perhaps triggered by the initial attention to uranium impoundments, the increasing concern for environmental performance has led to better engineering design of tailings dams in other mining industry sectors for both stability and environmental performance. For instance, experience gained with leach pad liners is being transferred to linings for tailings ponds, and synthetic lining materials are growing (although use of liners is still far from the industry norm). In addition, using cyanide and other toxic reagents in mill processes has raised unique concerns for some tailings. It leads to increased treatment before disposal and increased attention to containment. Finally, continuing concerns over acid mine drainage result in a growing body of research and emerging long-term control or mitigation concepts.

Inactive tailings impoundments are also receiving more attention due to the long-term effects of windblown dispersal, groundwater contamination, and acid drainage. Often, remediation costs surpass those of the tailings impoundments' initial design and operation.

The design depends on the quantity and the individual characteristics of the tailings produced by the mining and milling operation, as well as the climatic, topographic, geologic, hydrogeologic, and geotechnical characteristics of the disposal site and on regulatory requirements related to dam safety and environmental performance. What may work for one type of tailing or the same tailings at different sites may not work for another. Hence, each situation requires its own design process. The estimated quantity of tailings to be disposed of is particularly important given the evolving nature of most mining projects. Estimates of tailings quantities are based on continuously changing reserve estimates as mine development progresses. Accordingly, the final size and design of tailings impoundments can differ substantially from initial projections. This presents significant challenges to federal land managers and state permit writers, who are faced with reviewing and overseeing tailings impoundment planning, design, and performance, and to the general public, who may ultimately pay for miscalculations resulting in environmental damage.

15.9 PUMPED HYDRO ENERGY STORAGE (PHES)

Depending on whom you are talking to, one might say it is a "gravity battery," and others might call it a "water battery," but it is commonly known as pumped energy storage hydropower. These two monikers are correct when used to describe what this author calls nicknames for mechanical PHES. And the nicknames are applicable because PHES acts similarly to a giant battery. After all, it can store energy and then release it when needed. Between 1960 and 1980, pumped storage plants for hydroelectric power in the United States were built. It is interesting to note that PHESs are the largest source of electricity storage technology in the United States, both in capacity and number of plants. All remaining commercial-sized electrical storage uses thermal energy or batteries.

Simply put, PHES is made of two large pools (aka reservoirs; see Figure 15.2) of water, one higher than the other, acting like an hourglass to provide energy; the system stores energy in the form of gravitational potential energy of water. The process involves pumping water into an elevated pool or storage reservoir during low electricity demand and low electricity prices, such as at night. When electricity demand is high, water flows downhill (gravity flow = water or gravity battery stored energy) from the reservoir through piping (penstock) to hydroelectric generators at a dam. Compared to conventional hydro plants, PHESs are more expensive in construction costs; moreover, operating costs are high because the water must be pumped uphill.

There are two types of PSH systems. An *open-loop system* has a continuous source of downstream water pumped uphill to an upper storage reservoir. More often or not, an open-loop plant pumps water that has already passed through the dam up to the storage reservoir above the dam. By contrast, *closed-loop systems* pump water from a lower storage reservoir, which is not continuously filled with water and is not connected to a flowing source.

How important and influential are these water batteries or gravity batteries? First, because of these units, grid overload is rare. When other energy sources like wind and solar power make more electricity than nearby industries and homes, the extra energy or power pushes water into the top pool of water, where it sits, waiting and charging the water battery. Then, during a grid-overloading heat wave, the water is released from the top pool and flows down, acting as a prime mover (like a motor or diesel engine, etc.), spinning a turbine that creates electricity to energize lights, refrigerators, and other essentials.

We commonly think or feel that renewable energy is crucial for a clean energy future, but our needs go beyond environmental needs; the fact is, we need electrical power. This is not complex or mysterious; it is crucial for maintaining a good life. Water batteries can fill energy gaps when the weather limits the production of wind and solar energy.

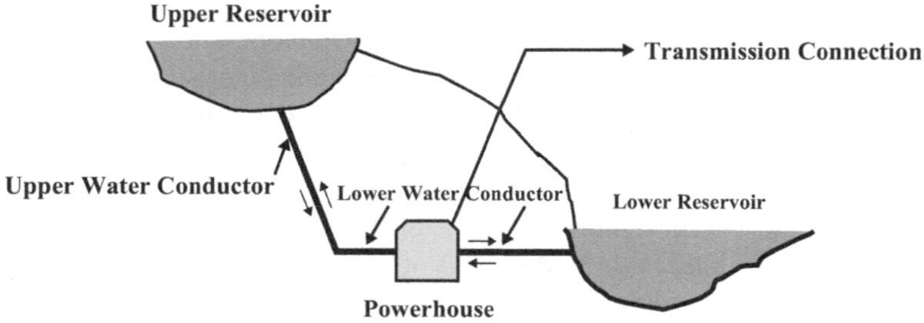

FIGURE 15.2 Typical configuration of PSH plant.

PSH provides more than 90% of U.S. energy storage. These projects are some of the most significant long-term energy storage systems today. These projects are placed in remote locations, and people living nearby generally know little about their importance until the lights go out.

In 2021, 18 states and all major regions of the country will use PSH to store energy. Currently, South Carolina, Virginia, and California use most of these clean energy pools, and other states have projects to help grow the presence and usage of these water batteries.

In the United States, PSH can store more than 550 GWh of energy. To give an example of the power generated, consider that American video gamers use about as much energy as 85 million refrigerators or 5 million cars yearly. The 550 GWh of PSH in the United States provides this energy.

Presently, pumped hydro storage is the most efficient and extensive energy storage system currently available, with a capacity of up to 80% and the ability to be turned on quickly. However, remember that energy storage, even the typical lead-acid storage battery, is not 100% efficient. Pumping water into a water battery's top reservoir requires a surge of energy. Another factor affecting efficiency in pumped hydro storage is evaporation.

Pumped hydropower storage sites provide more benefits to their locations than just storing water energy as a prime mover of electricity turbines. For instance, during a driving rainstorm that typically causes severe flooding in certain areas, the water pools accept excessive rainwater, which absorbs excess water and helps keep flooding to a minimum. This excess rainwater also adds to the pools' total water content, which reduces the need to use energy and pumps to pump up water levels in the pools. The pooled water can also provide water to fight wildfires and protect the area from extensive damage. During severe drought conditions, the pools can provide clean water for irrigating crops and a backup supply of safe drinking water.

In the United States, retrofitting existing PSH allows growth and increased energy output capacity without extensive new construction. This means that, within the capacity of U.S. pumped storage, without any new construction, it can grow pumped storage as much as all other types of energy storage combined.

15.9.1 Reversible Pumps/Turbines

Many existing PSH plants worldwide use reversible pumps/turbines, which the TVA first applied in 1956 at the Hiwassee PSH plant (Koritarov et al., 2022). Typically, Francis-type turbines are designed for both generating and pumping. Most PSH plants use motor generators that can operate as motors and generators. Before reversible pumps and turbines were invented, PSH plants employed a pump and motor on one shaft and a turbine and generator on another. Separate pumps and turbines are still used for some PSH configurations, such as ternary, quaternary, and pump-back PSH plants with separate pumping stations. A pump-back PSH plant can utilize natural inflows to the upper reservoir to produce electricity as a conventional hydropower plant. Still, it can also pump the water back to the upper reservoir for additional storage as a PSH plant.

The most used types of PSH technology are fixed-speed, adjustable, and ternary technologies. Practically all of the PSH plants in the United States and the majority of the plants around the world use fixed-speed technology, which allows the PSH unit to vary its power output in the generating mode; however, the unit continuously pumps with full power (i.e., it cannot vary the pumping power). First developed in Japan in the early 1990s, adjustable-speed units use doubly fed induction or converter-fed synchronous machines, which can vary the pumping power and have a wider operating range in the generating mode (Koritarov et al., 2014). The motor generators are adjustable speed types where the rotational speed is adjusted using either partial- or full-size frequency converters. A three-component configuration is used in ternary PSH technology, whereby a motor or generator is coupled with a separate pump and a turbine.

A quaternary PSH technology is being proposed, which uses separate pumps, turbines, motors, and generators. Both ternary and quaternary technologies can operate when the pump and the turbine can operate simultaneously; this operation is known as a hydraulic short circuit. It's all about flexibility, which allows for excellent operation flexibility and a complete operating range—from −100 to 100% of PSH capacity. Regarding the typical round-trip efficiency of new PSH plants, they currently come in at approximately 80%. Current PSH plants provide a large amount of dispatchable capacity (plant sizes are typically several hundred megawatts) and energy storage. This is positive because it helps balance grid operations and store surplus generation from variable renewables like solar and wind. PSH also provides numerous grid services, such as inertial response, frequency regulation, operating reserves, voltage support, and black start (USDOE, 2023).

15.10 CHALLENGES TO DEPLOYMENT

PSH deployment is not without its challenges. Some of the main challenges facing PSH developers include the following:

- *Licensing and Permitting Process*: We are talking about several years before the permitting and licensing of new PSH projects can be approved. In a move to reduce the duration of this process, the FERC established an accelerated three-year procedure for licensing closed-loop PSH projects.
- *Risk*: PSH projects are very capital-intensive investments, requiring significant money for project development and long payback periods. This is the case even though larger projects benefit from economies of scale; the investment risk is also higher because the developers must commit large amounts of capital for project development.
- *Financing*: For new PSH projects, there is a need to consider many uncertainties that may affect the project's financial viability. Rules, prices in the long term, the electricity market structure, including the potential impacts of long-term prices of natural gas and other fuels, the penetration of variable renewable energy resources and their generation and price impacts, government incentives, and regulatory policy are some of the market uncertainties that must be considered.
- *Market Uncertainties*: In a word, competition is or can be an additional challenge or risk. This is because many proposed PSH projects will operate in competitive wholesale electricity markets, which experience rapid changes in generation plant mixes. These changes result in uncertainties in the long-term market prices for capacity, energy, and grid services, and they estimate the potential costs and revenue of PSH plants in the long term.
- *Public Acceptance*: Getting the public to buy in and accept large-scale energy projects is no slam-dunk. The public also may assume that the development of a new PSH project will involve the construction of a large dam on a river and will disturb fish and other habitats. Most proposed new PSH projects in the United States, however, are closed-loop type, typically using two manufactured reservoirs that are not connected to any natural bodies of water and are devoid of fish and other aquatic life.
- *Long Construction Period*: The building of PSH projects takes time; this is not an overnight project and takes about five years to complete. These projects are designed to benefit from economies of scale and provide large energy storage capacities. Several construction techniques are being tested to accelerate PSH projects. In light of the need for more rapid construction, small modular PSH designs that could be constructed in less time are being developed.
- *Regulations and the Environment*: Several issues may affect new PSH projects, including environmental concerns, water rights, land acquisition, state and local energy policies, and site-specific regulations (e.g., related to the use of brownfield sites and abandoned mines).
- *Ignorance, General Lack of Understanding*: The prolonged absence of comprehension of energy storage (LDES) values and the services it can provide to the grid during extreme weather events and other disturbances, and concerns regarding the rules about domestic production will be defined.
- *Return on Investment*: These are the issues concerning the long payback period for return on investment.

15.11 THE BOTTOM LINE

The bottom line of pumped hydro storage systems, unlike variable renewable sources like wind, tides, and solar, is that they can respond continuously to fluctuations in demand and sudden surges and provide energy at a low cost. However, the effects of downstream development, limited maintenance funding, aging infrastructure, previous design practices, and exposure to extreme events from natural hazards or deliberate actions create a complex risk scenario that may impact millions of people in the event of a dam or levee failure. Artificial intelligence is used in hydroelectric power by activating functions, data gathering and analysis, accelerating clean energy deployment, measuring energy demand, and optimizing power production and maintenance scheduling.

REFERENCES

Association of State Dam Safety Officials (ASDSO) (2015). *Introduction to Dams*. Accessed 05/10/15 @ www.damsafety.org.

BPA (2010). *The Columbia River System Inside Story*. Accessed 5/11/15 http://www.bpa.gov/power/pg/columbia_river_inside_sotry.pdf.

EERE (2008). *History of Hydropower*. Accessed 06/15/09 @ www.eere.energy.gov/windandhydro/printable_versions/hydro_hisotry.html.

Koritarov et al. (2014). *Modeling and Analysis of Advanced Pumped Storage Hydropower in the United States*. Washington, DC: U.S. Department of Energy.

Koritarov, V., Ploussard, Q., Kwon, J., and Balducci, P. (2022). *A Review of Technology Innovations for Pumped Storage Hydropower*. Washington, DC: U.S. Department of Energy.

Ritcey, R.C. (1989). *Design and Evaluation of Tailings Dams*. Washington, DC: US Environmental Protection Agency.

Spellman, F.R. (2008). *The Science of Water*, 2nd ed. Boca Raton, FL: CRC Press.

Spellman, F.R. (2014). *The Science of Water*, 3rd ed. Boca Raton, FL: CRC Press.

U.S. Department of Homeland Security (2010). *Dam Sector-Specific Plan: An Annex to the National Infrastructure Protection Plan*. Washington, DC: USDHS.

USACE (1993). *Impoundments*. Washington, DC: US Army Corps of Engineers.

USDOE (2021). *10 Reasons to Love Water Batteries*. Washington, DC: U.S. Department of Energy.

USDOE (2023). *Technology Strategy Assessment: Supercapacitors Technology*. Washington, DC: United States Department of Energy. Supercapacitors Technology.

USEPA (1994). *Design and Evaluation of Tailings Dams*. Washington, DC: U.S. Environmental Protection Agency.

USGS (2003). *Active Mines and Mineral Processing Plants in the United States*. Washington, DC. United States Geological Survey.

USGS (2015). *Lake Umatilla*. Accessed 5/11/25 http://geonames.usgs.gov/apex/f?p=gnispq:3:::NO::{#_P3_FID:1513298.

16 Geo-Energy
Geothermal Energy = Geo (Earth) + Thermal (Heat)

Note: This chapter discusses not only the vast natural geothermal energy storage but also basic Earth morphology, geothermal history, and the double benefits of geothermal energy storage and mineral extraction from the brine. While teaching various environmental-based college-level courses at Old Dominion University from 1988 to 2009, I taught courses that discussed geothermal energy storage. Then, following my lecture on geothermal energy storage, my next lecture was about thermal energy storage. Before presenting the two separate lectures, I learned I needed to define the difference between the two subjects.

Why?

It is not difficult to mix the two topics, as they are both based on heat energy and storage. However, there is a very distinct difference between them. First, geothermal energy is energy derived from the heat stored in the Earth. Second, the difference is stark because thermal energy storage (TSE) manifests itself from a temperature change, basically a phase or transition change.

16.1 INTRODUCTION[1]

It is commonly called "geo-energy," "Earth Heat energy," or "geothermal," and these terms are a perfect fit for this discussion about energy storage. This is the case because there is nothing new about Earth-generated heat energy. Energy from geothermal resources has benefited humankind from its earliest origins. Prehistoric civilizations used hot springs and fumaroles (steam discharges) for cooking, heating, and therapeutic bathing. Today, we classify this usage as geothermal *direct-use applications*. Note that geothermal energy has provided affordable, reliable, and renewable energy in the United States since the 1890s. Mink (2017) points out that Boise, Idaho, began directly using geothermal energy to heat commercial and residential buildings. Subsequently, the use of geothermal energy in the United States has expanded to include utility-scale electricity production, distributed heating and cooling applications, and the augmentation of various industrial processes.

The U.S. Geological Survey has calculated that the heat energy in the upper 10 km of the Earth's crust in the United States equals over 600,000 times the country's annual non-transportation energy consumption. Probably no more than a tiny fraction of this energy could ever be extracted economically. However, just one-hundredth of the total is equal to half the country's current non-transportation energy needs for over a century, with only a fraction of the pollution from fossil-fueled energy sources (McLarty et al., 2000). Geothermal heat is the only renewable energy source created naturally by the Earth. Suppose we utilize waste biomass, solar (passive and thermal), wind (onshore), photovoltaic, geothermal, and other renewable resources available to us in the United States. In that case, we will exceed the demand (what we need) by at least five times as much energy as we need, all from clean, renewable sources.

Approximately 4,000 miles below the Earth's surface is the core, where temperatures can reach 9000°F. This heat, geothermal energy (geo, meaning Earth, and thermos, meaning heat), flows outward from the core, heating the surrounding area. This can form underground reservoirs of hot water and steam. These reservoirs can be tapped for various uses, such as generating electricity or heating buildings.

DOI: 10.1201/9781003637851-18

Note: Renewable energy consumption in the United States was approximately 9%. In 2023, geochemical energy accounted for 1% of energy used in the United States and about 0.4% of total electricity produced.

The geothermal energy potential in the uppermost 6 miles of the Earth's crust amounts to 50,000 times the energy of all oil and gas resources globally. In the United States, most geothermal reservoirs are in the Western states, Alaska, and Hawaii. However, geothermal heat pumps (GHPs), which uses the shallow ground's stable temperature to heat and cool buildings, can be used almost anywhere.

Again, it is essential to note that there is nothing new about renewable energy. Humans have used renewable resources since time immemorial, from solar power to burning biomass (wood) in caves and elsewhere. For example, hot springs have been used for bathing since Paleolithic times or earlier (DOE, 2009). The early Romans used hot springs to feed public baths and underfloor heating. France's world's oldest geothermal district heating system has operated since the 14th century (Lund, 2007). The history of geothermal energy use in the United States is lengthy. In the following, a brief chronology of major geothermal events in the United States is provided (DOE, 2006).

GEOTHERMAL TIMELINE

8,000 B.C. (and earlier)
Paleo-Indians used hot springs for cooking, refuge, and respite. Hot springs were neutral zones where members of warring nations bathed together in peace. Native Americans have a history with every major hot spring in the United States.

1807
As European settlers moved westward across the continent, they gravitated toward these springs of warmth and vitality. In 1807, the first European to visit the Yellowstone area, John Colter (c. 1774–c. 1813), widely considered to be the first mountain man, probably encountered hot springs, leading to the designation "Colter's Hell." Also, in 1897, settlers founded the city of Hot Springs, Arkansas, where, in 1830, Asa Thompson charged 1 dollar each for three spring-fed baths in a wooden tub, and the first known commercial use of geothermal energy occurred.

1847
William Bell Elliot, a member of John C. Fremont's survey party, stumbles upon a steaming valley just north of what is now San Francisco, California. Elliot calls the area The Geysers—a misnomer—and thinks he has found the gates of Hell.

1852
The Geysers is developed into a spa called The Geysers Resort Hotel. Guests include J. Pierpont Morgan, Ulysses S. Grant, Theodore Roosevelt, and Mark Twain.

1862
At springs located southeast of The Geysers, businessman Sam Brannan pours an estimated half million dollars into an extravagant development dubbed "Calistoga," replete with a hotel, bathhouse, skating pavilion, and racetrack. Brannan's was one of many spas reminiscent of those of Europe.

1864
Homes and dwellings have been built near springs through the millennia to take advantage of the natural heat of these geothermal springs, but the construction of the Hot Lake Hotel near La Grande, Oregon, marks the first time that the energy from hot springs is used on a large scale.

1892
Boise, Idaho, provided the world's first district heating system, piping water from hot springs to town buildings. Within a few years, the system served 200 homes and 40 downtown businesses. Today, four district heating systems in Boise provide heat to over 5 million square feet of residential, business, and governmental space. There are now 17 district heating systems in the United States and dozens more around the world.

1900
Hot springs water is piped to homes in Klamath Falls, Oregon.

1921
John D. Grant drills a well at The Geysers to generate electricity. This effort was unsuccessful, but Grant met with success across the valley at another site one year later, and the United States' first geothermal power plant went into operation. Grant uses steam from the first well to build a second one, and several wells later, the operation produces 250 kW, enough electricity to light the buildings and streets at the resort. The plant, however, is not competitive with other power sources, and it soon falls into disuse.

1927
Pioneer Development Company drills the first exploration at Imperial Valley, California.

1930
The first commercial greenhouse use of geothermal energy is undertaken in Boise, Idaho. The operation uses a 1000-foot well drilled in 1926. In Klamath Falls, Charlie Lieb develops the first downhole heat exchanger (DHE) to heat his house. Today, more than 500 DHEs are in use around the country.

1940
Nevada's first residential space heating began in the Moan area in Reno.

1948
Geothermal technology moved east when Carl Nielsen developed the first ground-source heat pump for his residence. J.D. Krocker, an engineer in Portland, Oregon, pioneered the first commercial building use of a groundwater heat pump.

1960
The country's first large-scale geothermal electricity-generating plant begins operation. Pacific Gas and Electric operates the plant located at The Geysers. The first turbine produces 11 MW of net power and has operated successfully for over 30 years. Today, 69 generating facilities operate at 18 resource sites around the country.

1978
Geothermal Food Processors, Inc. opened the first geothermal food-processing (crop-drying) plant in Brady Hot Springs, Nevada. The Load Guaranty Program provided $3.5 million for the facility.

1979
The first electrical development of a water-dominated geothermal resource occurs at the East Mesa field in the Imperial Valley in California. The plant is named for B.C. McCabe, the geothermal pioneer who, with his Magma Power Company, did field development work at several sites, including The Geysers.

1980
TAD's Enterprises of Nevada pioneered using geothermal energy for the cooking, distilling, and drying processes associated with alcohol fuel production. UNOCAL built the country's first flash plant, generating 10 MW at Brawley, California.

1982
Economic electrical generation begins at California's Salton Sea geothermal field using crystallizer-clarifier technology. The technology resulted from a government/industry effort to manage the high-salinity brines at the site.

1984
A 20-MW plant begins generating power at Utah's Roosevelt Hot Springs. Nevada's first geothermal electricity is generated with a 1.3-MW binary power plant that begins operation.

1987
Geothermal fluids were used in the first geothermal-enhanced heap leaching project for gold recovery near Round Mountain, Nevada.

1989
The world's first hybrid (organic Rankine/gas engine) pressure-geothermal power plant begins operation at Pleasant Bayou, Texas. It uses both the heat and the methane of a pressured resource.

1992
Electrical generation begins at the 25-MW geothermal plant in the Puna field of Hawaii.

1993
A 23-MW binary power plant is completed at Steamboat Springs, Nevada.

1995
Integrated Ingredients dedicates an annual food-dehydration facility at Empire, Nevada, that processes 15 million pounds of dried onions and garlic. A DOE low-temperature resource assessment of 10 Western states identifies nearly 9000 thermal wells and springs and 271 communities collocated with a geothermal resource greater than 50.

2002
GeoPowering the West organized geothermal development working groups in five states: Nevada, Idaho, New Mexico, Oregon, and Washington. Group members represent all stakeholder organizations. The working groups identify barriers to geothermal development in their state and bring together all interested parties to arrive at mutually beneficial solutions.

2003
The Utah Geothermal Working Group is formed.

16.2 GEOTHERMAL ENERGY: THE BASICS

As mentioned, geothermal energy processes evolve around the Earth's natural heat, which can be used for beneficial purposes when the heat is collected and transported to the surface. To properly understand geothermal energy as it is currently used, it is essential to define enthalpy, the heat content of a substance per unit mass. The term "enthalpy" is used because temperature alone is not sufficient to specify the applicable energy content of a steam/water mixture. A mass of steam at a given temperature and pressure can provide much more energy than the same mass of water under the same conditions. Enthalpy is a function of pressure, volume, and temperature:

Pressure + Volume + Temperature = Enthalpy

Geothermal practitioners usually classify geothermal resources as "high enthalpy" (water and steam at temperatures above 180–200°C), "medium enthalpy" (about 100–180°C), and "low enthalpy" (<100°C). For this text, it is sufficient to think of temperature and enthalpy going hand in hand.

DID YOU KNOW?

The temperature at the center of the Earth, about 6,500 m (4,000 miles) deep, is about the same as the sun's surface (nearly 6,000°C, about 10,800°F).

16.2.1 Earth's Layers

Earth comprises three main compositional layers: crust, mantle, and core (shown horizontally in Figure 16.1). The crust has variable thickness and composition: the continental crust is 10–50 km thick, while the oceanic crust is 8–10 km thick. Silicon, oxygen, aluminum, and iron make up the Earth's crust. Like the shell of an egg, the Earth's crust is brittle and can break. Earth's continental crust is 35 km thick. The oceanic crust is 7 km thick.

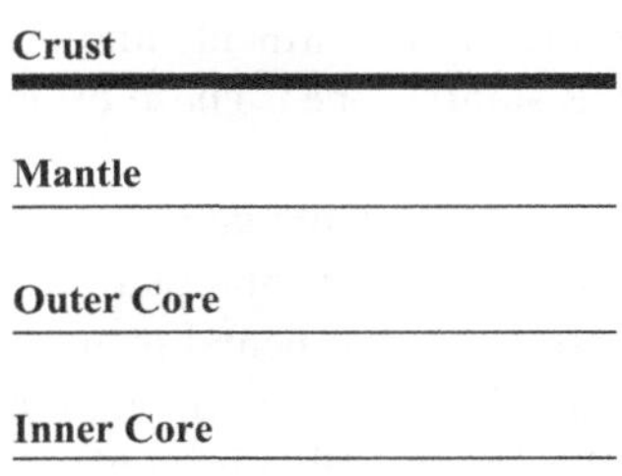

FIGURE 16.1 Earth's horizontal layers.

Based on seismic (earthquake waves) that pass through the Earth, we know that below the crust is the mantle, a dense, hot layer of semi-solid (plastic-like liquid) rock approximately 2,900 km thick. The mantle, which contains silicon, oxygen, aluminum, iron, magnesium, and calcium in more significant amounts than the crust, is hotter and denser because temperature and pressure inside the Earth increase with depth; as a comparison, the mantle might be thought of as the white of a boiled egg. The Moho layer is the 30-km-thick transitional layer between the mantle and crust. The temperature at the top of the mantle is 870°C. The temperature at the bottom of the mantle is 2,200°C.

DID YOU KNOW?

The Energy Independence and Security Act of 2007 (Pub. L. No 110–140) defines geothermal energy as renewable. Although distinct from solar and wind, which tap an instantly renewable energy source, geothermal is a renewable resource with life cycles and timescales more like those of sustainable forests.

At the center of the Earth lies the core, which is nearly twice as dense as the mantle because its composition is metallic [iron (Fe)-nickel (Ni) alloy] rather than stony. Unlike the yolk of an egg, however, the Earth's core comprises two distinct parts: a 2,200-km-thick liquid outer core and a 1,250-km-thick solid inner core. The liquid outer core spins as the planet rotates, creating the Earth's magnetic field. Several essential points related to Earth's structure and geothermal properties include the following:

- Heat flows outward from the center because of radioactive decay.
- The crust (about 30–60 km thick) insulates us from the interior heat.
- A solid inner core followed by a liquid outer core, with the semi-molten mantle.
- Temperature at the base of the crust is about 1000°C, increasing slowly into the core.
- Hot spots are located 2–3 km from the surface.

16.2.2 Crustal Plates

Within the past 45- or 50 years, geologists have developed the theory of plate tectonics (tectonics: Greek, "builder"). The theory of plate tectonics deals with the formation, destruction, and large-scale motions of significant segments of Earth's surface (crust), called *plates*. This theory relies heavily on the older concepts of continental drift (developed during the first half of the 20th century) and seafloor spreading (understood during the 1960s), which help to explain the cause of earthquakes and volcanic eruptions and the origin of fold mountain systems.

Regarding crustal plate activity and geothermal energy, and as mentioned, large quantities of heat (that are economically extractable) tend to be concentrated in places where hot or even molten magma exists at relatively shallow depths in the Earth's outermost layer (the crust). Such "hot" zones generally are near the boundaries of the dozen or so slabs of rigid rock (or plates) that form the Earth's lithosphere.

These crustal plates are composed of great slabs of rock (lithosphere), about 100 km thick, and cover many thousands of square miles (they are thin compared to their length and width); they float on the ductile asthenosphere, carrying both continents and oceans. Many geologists recognize at least eight main plates and numerous smaller ones. These **main** plates include

- African plate covering Africa—continental plate
- Antarctic plate covering Australia—continental plate
- Australian plate covering Australia—continental plate
- Eurasian plate covering Asia and Europe—continental plate
- The Indian plate covering the Indian subcontinent and a part of the Indian Ocean—continental plate
- Pacific plate covering the Pacific Ocean—oceanic plate

- North American plate covering North America and northeast Siberia—continental plate
- South American plate covering South America—continental plate

The **minor** plates include

- Arabian plate
- Caribbean plate
- Juan de Fuca plate
- Cocos plate
- Nazca plate
- Philippine plate
- Scotia plate

The plates ride in the asthenosphere, the upper mantle's ductile, soft, plastic-like zone. Crustal plates move about one another at one of three types of plate boundaries: convergent (collision boundaries), divergent (spreading boundaries), or transform boundaries. These boundaries between plates are typically associated with deep-sea trenches, significant faults, fold mountain ranges, and mid-oceanic ridges.

16.2.3 Convergent Boundaries

Convergent boundaries (or active margins) develop where two plates slide toward each other, commonly forming either a subduction zone (if one plate subducts or moves underneath the other) or a continental collision (if the two plates contain continental crust). One plate is deformed and slips below the other to relieve the stress created by the colliding plates.

16.2.4 Divergent Boundaries

Divergent boundaries occur when two plates slide apart. Oceanic ridges are examples of these divergent boundaries. They are where new oceanic, melted lithosphere materials well up, resulting in basaltic magmas that intrude and erupt at the oceanic ridge, creating new oceanic lithosphere and crust (new ocean floor). Along with volcanic activity, the mid-oceanic ridges are also areas of seismic activity.

16.2.5 Transform Plate Boundaries

Transform or shear/constructive boundaries do not separate or collide; instead, they slide past each other horizontally with a shearing motion. Most transform boundaries occur where oceanic ridges are offset on the sea floor. The San Andreas Fault in California is an example of a transform fault.

DID YOU KNOW?

- Plates are in constant motion (several centimeters/year).
- When collision or grinding occurs, it can create mountains, volcanoes, geysers, and earthquakes.
- Near the junction of these plates is where heat travels rapidly from the interior.

16.3 ENERGY CONVERSION

To generate power from geothermal systems, three elements are needed:

- *Heat:* Plentiful heat is found in rocks deep underground, varying by depth, geology, and geographic location.
- *Fluid:* Enough fluid to carry heat from the rocks to the Earth's surface.
- *Permeability:* Small pathways that facilitate fluid movement through the hot rocks.

The conversion of heat to electricity is standard for most power plants. This is the case even though the energy source may be coal, gas, nuclear, wind, solar, water, or geothermal. Powering the non-transportation section of our economy is essential, of course, so using any fuel source to convert to electrical power for industrial use is one of our premium objectives in our constant and insatiable appetite for energy. Because of our increasing need for more electricity generation, one might surmise that we should focus solely on electricity production to power our economy, industrial complex, and everything else—one genie in one bottle to accomplish everything. Liquid fuels are the fuels of choice now because they are accessible, available, and relatively inexpensive. Thus, though our ongoing research for other energy sources persists, we still do not have that absolute pressing need to develop a liquid fuel replacement—not yet, we don't.

When our energy-needs-focus shifts due to necessity, absolute or otherwise, geothermal is available, and we need to continue our research in this critical area. Moreover, energy conversion occurs at present with all forms of energy in use, and with geothermal conversion, it is nothing new; only the procedure and methodology differ. In geothermal energy, "conversion" refers to the power plant technology that converts the hot geothermal fluids into electric power.

Even though geothermal power plants have much in common with traditional power-generating stations, turbines, generators, heat exchanges, and other standard power-generating equipment, there are essential differences between geothermal and other power-generating technologies. For example, each geothermal site has its own unique set of characteristics and operating conditions. For instance, geothermal sites are essentially site-specific; these conditions must be considered. That is, the fluid produced from a geothermal well can be steam, brine, or a mixture of the two, and the temperature and pressure of the resource can vary substantially from site to site. The chemical composition of the resource can contain dissolved minerals, gases, and other hard-to-manage substances. Because these site-specific conditions can profoundly affect efficiency, productivity, and economic viability, engineers strive to fine-tune geothermal conversion technology, precisely matching plant design to the site-specific conditions.

16.4 GEOTHERMAL POWER PLANT TECHNOLOGIES

Geothermal power plants fall into one of three conversion categories: direct steam, flash, and binary plants. The type of conversion used depends on the state of the fluid (whether steam or water) and its temperature. Dry steam power plants were the first type of geothermal power generation plants built. They use the steam from the geothermal reservoir from wells and route it directly through turbine/generator units to produce electricity. Flash steam plants are the most common type of geothermal power generation plants in operation today. They use water temperatures greater than 360°F (182°C) pumped under high pressure to the generation equipment at the surface. Binary-cycle geothermal power generation plants differ from dry steam and flash steam systems in that the water or steam from the geothermal reservoir never meets the turbine/generator units.

DID YOU KNOW?

The Earth's temperature increases, on average, by about 28°C for every kilometer, or 80°F for every mile, of depth below the surface for the first several kilometers down.

16.4.1 Dry Steam Power Plants

Location-specific, dry-steam power plants take advantage of subterranean rocks that are so hot that the water from a geothermal power plant vaporizes on its way up through the production

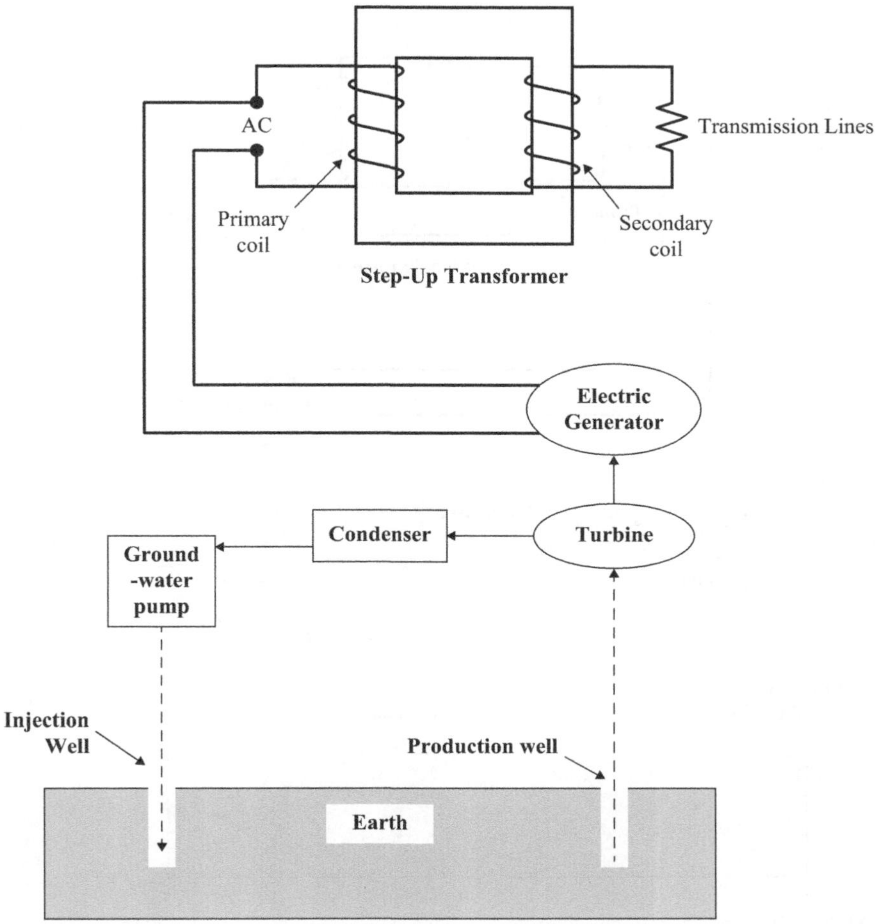

FIGURE 16.2 Simplified functional diagram of a dry steam geothermal power plant.

well (see Figure 16.2). This geothermal power plant can use steam directly from a geothermal reservoir to drive the turbine. This is the oldest type of geothermal power plant. It was first used at Larderello in Italy in 1904 and is still very effective. These geothermal plants emit only excess steam and a few gases (USDOE, 2008). Steam technology is still relevant today and is currently in use in Northern California at The Geysers, the world's largest single source of geothermal power.

16.4.2 Flash Steam Power Plants

Flash steam plants are a common type of geothermal power plant in operation today. Figure 16.3 is a functional diagram of a *flash steam geothermal power plant.* Fluids at temperatures greater than 182°C/360°F, pumped from deep underground, travel under high pressure to a low-pressure tank at the Earth's surface. Fluid is sprayed into a tank held at a much lower pressure than the fluid, causing some fluid to vaporize or "flash rapidly." The vapor then drives a turbine, which drives a generator, which, in turn, produces electricity. If any liquid remains in the tank, it is

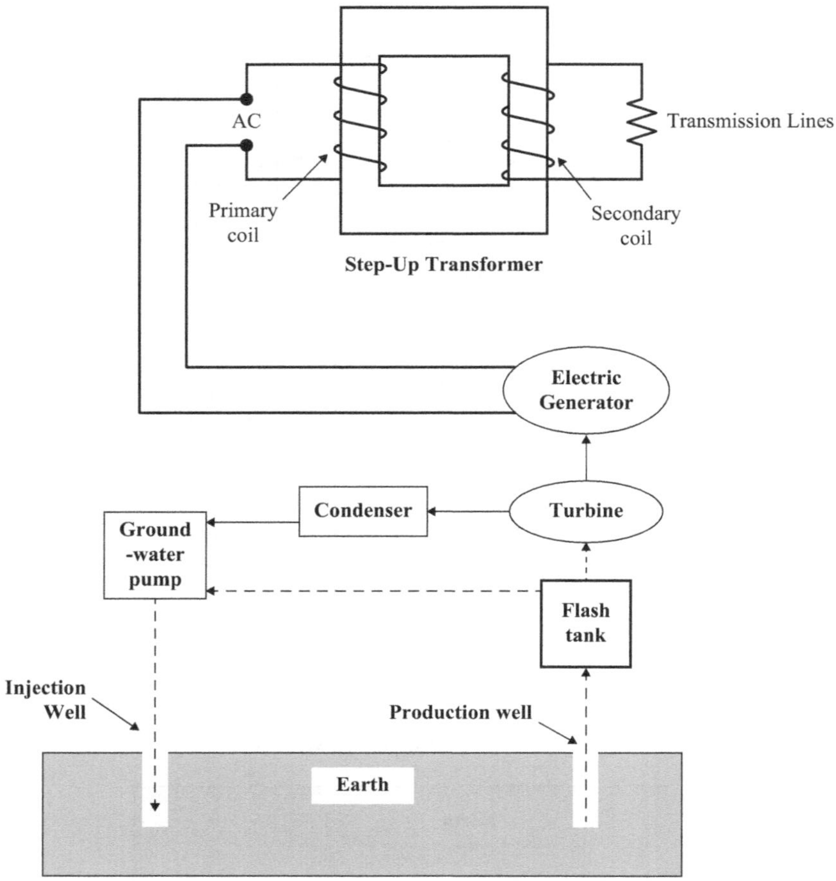

FIGURE 16.3 Simplified functional diagram of a flash steam geothermal power plant.

returned to the groundwater pump to be forced down into the Earth again and can be flashed again to extract more energy (USDOE, 2008).

DID YOU KNOW?

Most geothermal power plants are flash steam plants.

16.4.3 Binary-Cycle Power Plants

Binary-cycle geothermal power plants can use lower-temperature geothermal resources, making them a vital technology for deploying geothermal electricity production in more locations. In a binary-cycle geothermal power plant (see Figure 16.4), water is pumped into the earth and back up hot, just as in the flash steam system. Instead of going into a flash tank, the hot water enters a *heat exchanger* (see Figure 16.4), where most of its energy is transferred to another fluid called a *binary liquid*. This fluid can be water, but more often, it is a volatile liquid resembling a refrigerant that boils quickly into vapor at a lower temperature than water. The liquid-to-vapor conversion occurs in a particular low-temperature boiler. Then the vapor leaves the turbine, is cooled back into liquid

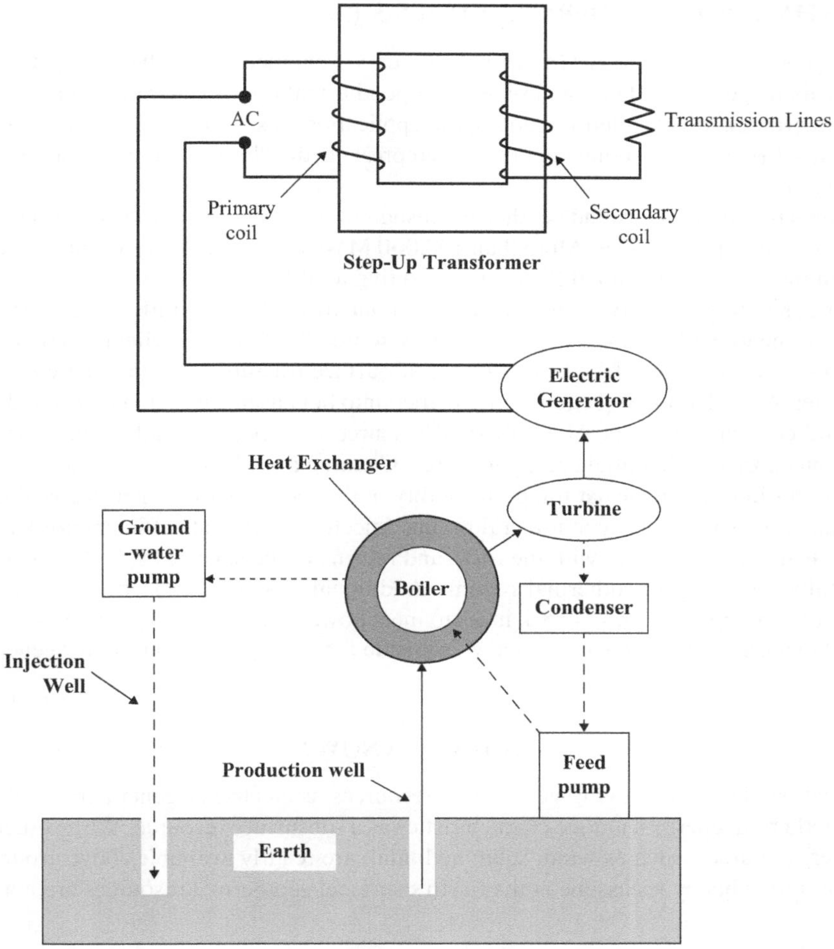

FIGURE 16.4 Simplified functional diagram of a binary-cycle geothermal power plant.

by a condenser, and is recirculated to the boiler. Because this is a closed-loop system, virtually nothing is emitted to the atmosphere. Moderate-temperature water is by far the most common geothermal resource, and most geothermal power plants in the future will be binary-cycle plants (USDOE, 2008).

DID YOU KNOW?

Geothermal sources should have the highest temperatures (150°C or greater) to produce electricity. However, GHPs can provide efficient residential and commercial heat and cooling when located in a shallow-Earth environment where ground temperatures are relatively constant. Considering this, it is essential to note that geothermal power production can occur at resource temperatures below 150°C (~300°F) but such projects tend to be the exception and require a combination of technical and access factors, and not to forget economics, that enable development.

16.5 ENHANCED GEOTHERMAL SYSTEMS (EGS)

Enhanced geothermal systems (EGS), sometimes called engineered geothermal systems, can dramatically expand geothermal energy use. Present geothermal power generation comes from hydrothermal reservoirs and is limited in geographic application to specific ideal places in the Western United States. EERE (2006) points out that this represents the "lower-hanging fruit" of geothermal energy potential.

EGS offers the chance to extend geothermal resources to large areas of the Western United States and into new geographic regions. More than 100,000 MWe of economically viable capacity may be available in the continental United States, representing a 40-fold increase over the present geothermal power generating capacity. This potential is about 10% of the overall U.S. electrical capacity today and, as mentioned, means a domestic energy source that is clean, reliable, and proven.

The EGS concept extracts heat by creating a subsurface fracture reservoir where injection wells can add water. A production-injection well is drilled into hot basement rock with limited permeability and fluid content. This type of geothermal resource is sometimes called "hot, dry rock" and represents an enormous potential energy resource. Creating an enhanced or engineered geothermal system requires improving the natural permeability of the rock. Rocks are permeable due to minute fractures and pore spaces between mineral grains. Injected water is at sufficient pressure to ensure fracturing, is heated by contact with the rock, and returns to the surface through production wells, as in naturally occurring hydrothermal systems. Additional production-injection wells are drilled to extract heat from large volumes of rock mass to meet power generation requirements. A previously unused but extensive energy resource is now available for clean geothermal power generation.

DID YOU KNOW?

Technical barriers to deploying geothermal resources as electricity generators result mainly from geothermal energy's unique characteristics as a subsurface resource. While other renewable energy sources, such as wind, solar, and tidal, are readily available above ground, making them immediately accessible at the Earth's surface, geothermal resources are not.

16.6 GEOTHERMAL HEAT PUMPS

GHPs (sometimes referred to as GeoExchange, earth-coupled, ground-source, or water-source heat pumps) have been in use since the late 1940s. GHPs use the constant temperature of the Earth as the exchange medium instead of the outside air temperature. This allows the system to reach high efficiencies (300%–600%) on the coldest winter nights, compared to 175%–250% for air-source heat pumps on cool days. GHPs are used for space heating, cooling, and water heating. Its great advantage is that it concentrates naturally existing heat rather than producing heat through fossil fuel combustion. The system includes three principal components (see Figure 16.5):

- *Geothermal Earth connection subsystems:* Using the Earth as a heat source/sink, a series of pipes, commonly called a "loop," is buried near the building to be conditioned. The loop can be buried either vertically or horizontally. It circulates a fluid (water or a mixture of water and anti-freeze) that absorbs heat from or relinquishes heat to the surrounding soil, depending on whether the ambient air is colder or warmer than the soil.
- *Geothermal heat pump subsystem:* For heating, a GHP removes heat from the fluid in the Earth connection, concentrates it, and then transfers it to the building. For cooling, the process is reversed.
- *Geothermal heat distribution subsystem:* Conventional ductwork distributes heated or cooled air from the GHP throughout the building.

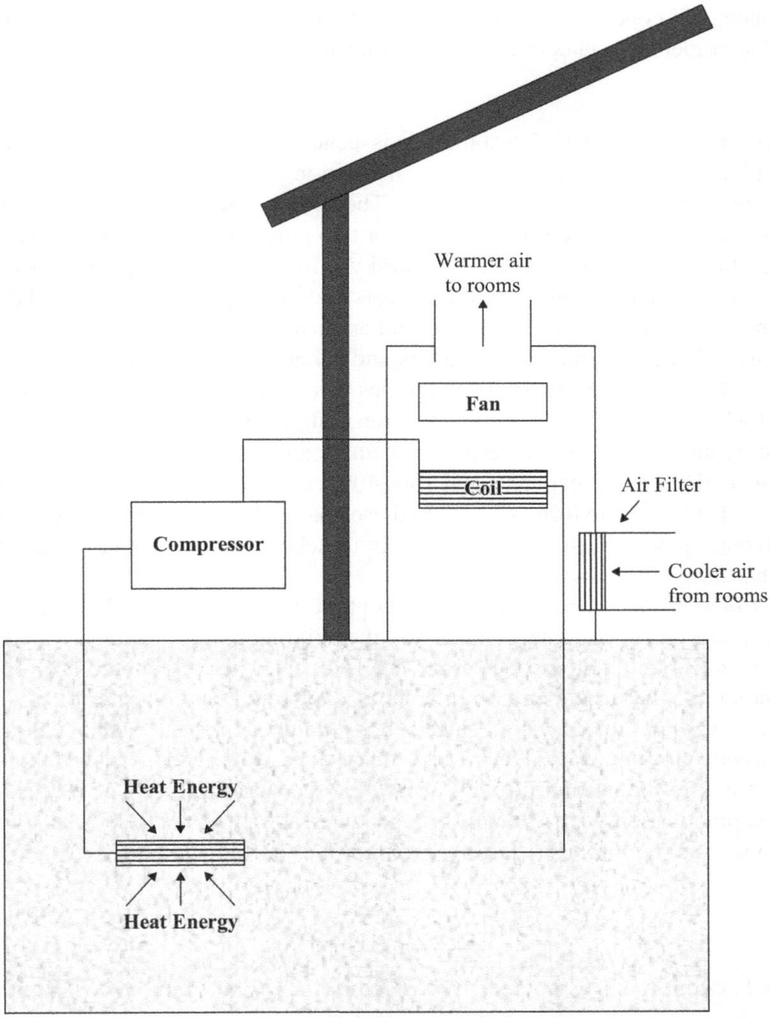

FIGURE 16.5 Geothermal heat pump system.(Adapted from Gibilisco (2007), *Alternative Energy Demystified*, pg. 54. New York: McGraw-Hill.)

In addition to space conditioning, GHPs can provide domestic hot water when the system is operating. Many residential systems now have desuperheaters that transfer excess heat from the GHP's compressor to the house's hot water tank. A desuperheater provides no hot water during the Spring and Fall when the GHP system is not operating; however, because the GHP is so much more efficient than other means of water heating, manufacturers are beginning to offer "full demand" systems that use a separate heat exchanger to meet all a household's hot water needs. These units cost-effectively provide hot water as quickly as any competing system (EERE, 2009).

16.6.1 Types of Geothermal Heat Pumps

Energy Savers (2009) lists four basic ground-loop GHP system types. Three closed-loop systems are horizontal, vertical, and pond/lake. The fourth type is the open-loop option. The best depends on the site's climate, soil conditions, available land, and local installation costs. All these approaches can be used for residential and commercial building applications.

- *Closed-Loop Systems*
  - *Horizontal:* This type of installation is generally most cost-effective for residential installations, particularly for new construction where sufficient land is available. It requires trenches at least 4 feet deep. The most common layouts use two pipes, one buried at 6 feet and the other at 4 feet, or two pipes placed side-by-side at 5 feet in the ground in a 2-foot-wide trench. The Slinky™ method of looping pipe allows more pipe in a short trench, cutting installation costs and making horizontal installation possible in areas without conventional horizontal applications.
  - *Vertical:* Large commercial buildings and schools often use vertical systems because the land area required for horizontal loops would be prohibitive. Vertical loops are also used where the soil is too shallow for trenching, and they minimize the disturbance to existing landscaping. For a vertical system, holes (approximately 4 inches in diameter) are drilled about 20 feet apart and 100–400 feet deep. Two pipes connected at the bottom with a U-bend to form a loop into these holes. The vertical loops are connected with horizontal pipe (i.e., manifold), placed in trenches, and connected to the heat pump in the building.
  - *Pond/lake:* This may be the lowest-cost option if the site has an adequate water body. A supply line pipe runs underground from the building to the water and coiled into circles at least 8 feet under the surface to prevent freezing. The coils should only be placed in a water source that meets minimum volume, depth, and quality criteria.
- *Open-loop system:* It uses well or surface body water as the heat exchange fluid that circulates directly through the GHP system. Once it has circulated through the system, the water returns to the ground through the well, a recharge well, or surface discharge. This option is practical only when there is an adequate supply of relatively clean water and all local codes and regulations regarding groundwater discharge are met.

DID YOU KNOW?

Geothermal system life is estimated at 25 years for the inside components and 50+ years for the ground loop. Approximately 50,000 GHPs are installed in the United States each year. Even with the large number of annual installations, it is essential to remember that heat pumps are relatively expensive. This is especially true of the deep-ground source type. A new system may take a long time to pay for itself.

16.7 GREATER THAN THE SUM OF PARTS

Geothermal energy can provide additional value beyond electric or non-electric applications; it can compensate for our mineral shortages. For example, converting geothermal energy into electricity creates byproducts that can provide additional economic value streams. The fluids, geothermal brines processed through a geothermal power plant, may contain minerals whose extraction and refinement could, under appropriate market conditions, add revenue beyond the sale of electricity.

Geothermal brine is a potentially significant source of valuable minerals and metals. It is water heated by natural heat flow from the depths of the Earth. The hotter brine generates electricity, typically with temperatures greater than 120°C. Lower-temperature brine is used to supply thermal energy to space heating, aquaculture, and agriculture applications. Note that the geothermal waters had intimate and lengthy contact with the layers of the Earth's crust that they flow through, resulting in the dissolution of minerals and metals from the rocks and solution into the hot water. These brine solutions can be processed to recover minerals and metals. Potential products include lithium, silica, zinc, and other materials. Solution mining, or in-situ leaching or recovery, involves extracting minerals and metals from geothermal brine. This procedure is followed by applying established or new hydrometallurgical techniques for isolation and purification.

Note that although brine was often considered waste, a nuisance more than an asset, geothermal brine contains significant concentrations of potentially valuable mineral resources. Even though it is often considered just a waste, more interest and focus are on improving the economics of geothermal energy by producing and marketing some of the dissolved constituents. Note that the costs associated with resource extraction from geothermal fluids are relatively low for several reasons (Bourcier et al., 2005):

- There are no costs associated with mining and physical processing of the ore, and no harmful environmental impacts.
- Plant costs are split between power and mineral production. The brine is already pumped and processed by the geothermal power plants.
- Because the minerals are already in solution, there are no costs associated with the dissolution of ore materials.
- Even though the mineral resources are relatively low, the mass of water and brine is large.

Geothermal brine is water that percolates through and is heated by hot rocks. The water source may be meteoritic, filling the pores of stones (connate), or a mixture. In some cases, a magmatic component is present from the devolatilization of hot magma.

The chemical components of geothermal brine are determined by their source (e.g., meteoric, seawater, magmatic), the rock types with which they have reacted along their flow path, the temperature of that interaction, and the chemistry of the brine. Mixing and boiling within the reservoir also impact chemistry. The chemistry of brine sampled at the surface reflects its chemical and physical history. Some of the elements may be especially indicative of their source. Examples are lithium, cesium, and rubidium, which are often enriched in fluids hosted by silica-rich volcanic rocks.

The key to mining geothermal brine for essential elements is separation and purification. Critical components are retrieved from the brine and solid materials, such as sludge or scale, that precipitate. Two other keys to mining the brine are economics, for-profit, and guarding against detrimental environmental impact.

Extraction of resources from geothermal brine has focused on a wide variety of materials, including platinum, zinc, lead, copper, manganese, rare earth, silica, lithium, arsenic, antimony, boron, bromine, iodine, strontium and barium sulfate, precipitated calcium carbonate, calcium chloride, potash, table salt, sodium sulfate, and others such as boric acid, sulfur, and ammonium salts (Gallup, 1998; Harper et al., 1992). Recently, many processes have been reported to recover other valuable materials from geothermal brine (Duyvesteyn, 1992; Lin et al., 2000, 2001).

16.7.1 Recovery from Sludges and Scales

Various processes are used to recover from solid geothermal sludge and scale. These recovery methods are discussed below (Bourcier et al., 2005):

- *Acid leaching:* In the Salton Sea area in the Imperial Valley of Southern California, the geothermal power plants produced solid waste separated as filter cake from the clarifier containing a mixture of iron-bearing silica, salts, and heavy metals. Hydrochloric acid has been employed to leach out the iron and other metals so that the remaining silica becomes pure as a pozzolanic additive for different types of cement.
- *Biochemical leaching:* It is a biochemical process that employs microorganisms to extract metal from their solid waste ashes. Biochemical leaching used for mining low-grade

copper, uranium, and gold ores (Ehrlich and Brierley, 1990) has been modified to treat the solid waste separated as filter cake from the clarifiers at Salton Sea plants. In the process, acid-loving acidophilic bacteria leach out most toxic heavy metals to make the treated sold safe for landfill disposal. In other cases, the byproduct is pure enough to be used as feedstock for further processing (Premuzic et al., 1995b).

16.7.2 Recovery of Metals and Salts from Geothermal Fluids

- *Sorption:* It is the process by which one substance becomes attached to another. As such, sorption is employed with synthetic ion-exchange resins and bacteria known to adsorb ions selectively from the solution. The selectivity and capacity of the adsorption are pH, temperature, and ionic strength dependent. Research has been reported, including laboratory studies on the use of the adsorptive property of cell wells to recover U, Co, Zn, Mn, and Li ions in the solution (Premuzic et al., 1995b). However, it is essential to note that no biological process has yet been commercialized. Commercial ion-exchange resins have been used to recover zinc from high-salinity brines.
- *Evaporation:* Solar evaporation ponds were used to recover salt minerals from saline brines, and this was ongoing throughout the early 1970s in Imperial Valley, CA. Note that evaporation is not without a cost; it is an energy-intensive process but can be employed under rare circumstances when energy costs and the need for water reinjection are of no concern. Additionally, reinjection of fluid after slat and potash removal could negatively impact the injection zone (Premuzic et al., 1995a).
- *Precipitation as sulfides:* Schultze and Bauer (1985) point out that hydrogen sulfide (H_2S) was added to geothermal fluids to precipitate most heavy metals as insoluble metal sulfides. The advantage of this treatment is its near quantitative efficiency. However, there is a potential problem because if the geothermal brine is rich in many metals, quantitative precipitation gives rise to a complex metal sulfide mixture that requires further purification (SRI International, 1980). Note that precipitation, such as hydroxides induced by raising the pH, gives rise to similar problems (Hazen Research Inc., 1978).

DID YOU KNOW?

A ubiquitous component of geothermal brine is silica. It must be removed or reduced in concentration to allow other components to be removed. Most hydrothermal systems equilibrate with quartz (SiO_2) causing the brine to have silica concentrations that reflect the temperature of the reservoir—the hotter the reservoir, the higher the silica concentration of the brine.

16.8 GEOTHERMAL ENERGY STORAGE AND AI

In Chapter 3, it was pointed out that machine learning (ML) uses various algorithms (clustering, neural networks, penalized regression, and Gaussian process regression) to compress the time needed to characterize better battery performance, lifetime, and safety. ML provides a gateway to monitoring and achieving efficient and quality operation. Generative artificial intelligence (AI) algorithms analyze data streams from geothermal wells to detect anomalies and predict failures. Remote sensing (RS) is also used to detect abnormalities.

ML has a significant ability to learn. The machine "learns" as the algorithm running the program is taught to forecast, categorize, and find key insights while data mining. ML uses advanced algorithms to identify patterns and make inferences from data. Geothermal investigation, identifying and predicting the locations of new geothermal resources, research, resource exploration, and power plant operations contain vast amounts of data. This is where AI and ML come into play. ML and AI provide impactful insights on large, complicated datasets, including those used to analyze geothermal energy production and storage.

16.9 THE BOTTOM LINE

The supply of geothermal energy is vast and can be considered renewable if each site is engineered correctly and operated to ensure that excessive water is not pumped into the earth in one location in too short a time. Geothermal energy can be accessed by drilling water or steam wells in a process like drilling for oil. Geothermal energy is an enormous, underused heat and power resource that is clean (emits little or no greenhouse gases), reliable (average system availability of 95%), and homegrown (making us less dependent on foreign oil).

Geothermal resources range from shallow ground to hot water and rock several miles below the Earth's surface and even farther down to the boiling molten rock called magma. Mile-or-more-deep wells can be drilled into underground reservoirs to tap steam and boiling water to be brought to the surface for various applications. In the United States, most geothermal reservoirs are in the Western states, Alaska, and Hawaii. However, before geothermal electricity can be considered a vital element of the U.S. energy infrastructure, it must become cost-competitive with traditional forms of energy.

AI has been leveraged for various energy purposes, including geothermal, and has played a transformative role in the geothermal industry. It can improve the efficiency of geothermal technologies through preventive maintenance, cost reduction, output resource allocation, energy storage optimization, advanced control systems, and grid integration and stability. AI is used in geothermal production, storage, and usage research to make them better performers.

Geothermal energy developments offer environmental and efficiency advantages relative to other energy sources.

NOTE

1 Information from DOE (2001). *Renewable Energy: An Overview.* Washington, DC: U.S. Department of Energy; F.R. Spellman's (2025). *The Science of Energy.* Boca Raton, FL: CRC Press.

REFERENCES

Bourcier, W.L., Lin, M., and Nix, G. (2005). *Recovery of Minerals and Metals from Geothermal Fluids.* Presented at SME Annual Meeting, Cincinnati, OH, United States—February 24, 2003 through February 26, 2003.

DOE (2006). *Geothermal Technologies Program: A History of Geothermal Energy in the United States.* Accessed 06/18/24 @ http://www1.eere.energy.gov/geothermal/printableversison/history.html.

DOE (2009). *Fossil Fuels.* U.S. Dept of Energy. Accessed 06/10/24 @ http://www.energy.gov/energysources/fossilfuels.htm.

Duyvesteyn, W.P.C. (1992). Recovery of Base Metals From Geothermal Brines. *Geothermics.* vol. 21, pp. 773–799.

EERE (2006). *How an Enhanced Geothermal System Works.* Accessed 03/29/24 @ http://www1.eere.energy.gov/gelotherm/printable_versions/egs_animation.html.

EERE (2009). *Geothermal Heat Pumps.* Accessed 03/27/24 @ http://www1.eere.energy.gov/geothermal/heatpumps.html?print.

Ehrlich, H.L. and Brierley, C.L., eds. (1990). *Microbial Mineral Recovery.* New York: McGraw-Hill Publishing Co., 434 p.

Energy Savers (2009). *Types of Geothermal Heat Pump Systems.* Washington, DC: US Department of Energy. Accessed 12/12/24 @ http://www.energysavers.gov/your_home/space_heating_cooling/index.

Gallup, D.L. (1998). Geochemistry of Geothermal Fluids and Healthy Scales, and Potential for Mineral Recovery. *Ore Geol Rev.* vol. 12, pp. 225–236.

Gibilisco, S. (2007). *Alternative Energy Demystified*. New York: McGraw-Hill.

Harper, R.T., Thain, I.A., and Johnson, J.H. (1992). Towards the Efficient Operation of Geothermal Resources. *Geothermics*. vol. 21, no. 5/6, pp. 641–651.

Hazen Research Inc. (1978). *Magmamax No. 1 Geothermal Brine Minerals Recovery and Bulk Precipitation Pilot Plants. Final Report on USBM Contract No. L3881964*. Golden, CO: Hazen Research Inc.

Lin, M.S., Bohenek, M., Premuzic, E.T., and Johnson, S. (2000). Silica Production from Low-Salinity Geothermal Brines. *Geotherm Res Counc Trans*. vol. 24, pp. 671–674. San Francisco, California. Sept. 24–27.

Lin, M.S., Premuzic, E.T., Wei, M. Z., and Johnson, S. (2001). Mineral Recovery: A Promising Geothermal Power Production Co-Product. *Geotherm Res Counc Trans*. vol. 25, pp. 497–500. San Diego, California, August 26–29.

Lund, J.W. (2007). Characteristics, Development, and Utilization of Geothermal Sources. Klamath Falls, Ore: Oregon Institute of Technology. *Geo-Heat Centre Quarterly Bulletin*. vol. 28, no. 2, pp. 1–9.

McLarty, L., Grabowski, P., Entingh, D., and Robertson-Tait, A. (2000). Enhanced Geothermal Systems R&D in the United States. *WGC 2000*. pp. 3793–3796.

Mink, L.L. (2017). The Nation's Oldest and Largest Geothermal District Heating System. *Geotherm Res Counc Trans*. vol. 41, pp. 205–212.

Premuzic, E.T., Lin, M.S., Jin, J.-Z., and Hamilton, K. (1995b). Geothermal Waste Treatment Biotechnology. *Proceedings of the World Geothermal Congress*. vol. 4, p. 2769–2772.

Premuzic, E.T., Lin, M.S., Lian, H., and Miltenberger, R.P. (1995a). Geothermal Brines and Sludges: A New Resource. *Geotherm Res Counc Trans*. vol. 19, p. 77–80.sp. Oct. 1995.

Schultze, L.E., and Bauer, D.J. (1985). *Recovering Lead-Zinc Sulfide from a Geothermal Brine*. USBM (US Bureau of Mines) Report of Investigation 8922. Washington, DC: U.S. Department of Interior.

SRI International (1980). *Recovery of Heavy Metals from High Salinity Geothermal Brines*. Final Report on USBM (U.S. Bureau of Mines) Contract J0188076. Washington, DC: U.S. Department of Interior.

USDOE (2008). *Hydrothermal Power Systems*. Accessed 03/26/24 @ www1.eere.energy.gov.geothermal/printable_versions/powerplants.html.

17 Gravity Batteries

17.1 INTRODUCTION

Earlier in Chapter 15, the terms "water battery" and "gravity battery" were mentioned regarding *pumped storage hydroelectricity* (PSH). It is true that PSH falls into either of these two titles or is titled "energy battery," which fits PSH as the most widely used and highest-capacity form of grid energy storage. Recall that in PSH, water is pumped from a lower reservoir to a higher reservoir, which can then be released through turbines to produce energy (electricity).

In addition to PSH, other types of gravity batteries are being built, planned, or already in operation. *Energy storage-by-rail* is a type of gravity battery in which excess renewable energy is used to run heavy train cars uphill during low energy **demand**. The potential energy is released later using regenerative braking (an energy recovery system) as they roll downhill, acting as a gravity battery (Massey, 2017). Hebrock (2020) points out that a utility-scale (50 MW) facility known as Advanced Rail Energy Storage (ARES) GravityLine (i.e., a fixed motor, chain-drive system that converts electricity into potential mechanical energy of a mass raised to higher elevations) began construction in 2020 by ARES, located in the Pahrump Valley, Nevada; this system is designed to deliver up to 15 minutes of service at full capacity (Hebrock, 2020).

Lift Renewable Energy uses another form of gravity battery. To store energy, buoyant gas containers are pulled down in water by a winch, lifting the water hundreds of meters. The cycle is then reversed, generating electricity as the gas containers rise. These systems can be sited near significant cities, requiring very little infrastructure. The round-trip efficiency exceeds 85% and can be constructed at a gigawatt-hour scale.

In Lifted Weight Storage (LWS) technology, a pulley system mechanically lifts solid weights vertically using surplus energy. When backup energy is needed, the mass is lowered and the pulley turns a generator (Gardiner, 2020).

An LWS system has been designed using a tower built from 32-ton concrete blocks stacked with 120-m cranes. One commercial unit is expected to store 20 MWh of energy, enough to power 2,000 Swiss homes daily (EnergyVault, 2020).

Another LWS system uses an underground shaft housing and an underground electrical winch to lift a 500- to 5,000-ton weight, which, when lowered, turns the winch motor into a generator. This system generates 10 MWh, enough to power more than 13,000 homes. The weight can be dropped quickly for a small burst of power (Gravitricity, 2020; Kropotin, 2023).

17.2 GRAVITY BATTERY OPERATION

The gravity battery is an innovative solution to the challenge of using gravity to store energy. In this fast-paced world, where technology plays a crucial role in our daily lives, and with the increasing use of artificial intelligence (AI), large data centers are required to power up technology and AI as needed, on demand. Gravity batteries are unique energy storage systems that have the potential to revolutionize the way we store and utilize renewable energy.

Again, gravity batteries operate on the principle of storing and releasing potential energy through the force of gravity. The gravity battery system consists of essential components, including a heavy mass, a lifting mechanism, and a generator to convert the falling mass into

DOI: 10.1201/9781003637851-19

electricity. In the process of storing and releasing energy in a gravity battery, this takes place in the following steps:

First: Energy Generation

Excess energy from renewable sources such as solar and/or wind power is generated. The excess energy is used to lift the heavy mass to a certain height, which is stored as potential energy.

Second: Energy Storage

After the heavy mass is raised to the desired height, it is locked in place using a hydraulic or mechanical system to prevent premature fall. To calculate the energy stored in the mass, use the formula $E = mgh$, where E is the potential energy, m is the object's mass, g is the acceleration due to gravity, and h is the object's height.

Third: Energy Release

17.2.1 Gravity Battery Benefits and Challenges

Gravity batteries offer several key advantages over traditional energy storage technologies, making them an attractive option for storing renewable energy. The benefits of gravity batteries include the following:

- High energy storage capacity
- Long lifespan
- Cost-effective
- Provide grid stability
- Environmentally friendly

Along with the benefits of gravity energy storage, challenges/disadvantages include the following:

- Decreased storage capacity
- Conversion efficiency problems
- High initial costs

17.3 THE BOTTOM LINE

Gravity batteries are designed to be connected with renewable energy sources like wind, solar, and tides. The limiting factor, at present, is cost.

REFERENCES

EnergyVault (2020). *EnergyVault: Enabling a Renewable World.* Accessed 4/20/25 @ https://energyvault.com.

Gardiner, L. (2020). *What Are Gravity Batteries?* Accessed 4/15/25 @ https://www.thegreenage.co.uk/what-are-gravity-batteries/.

Gravitricity (2020). *Fast, Long-Life Energy Storage.* Accessed 4/20/25 @ https://www.gravitricity.com.

Hebrock, R. (2020). Energy Storage Project Breaks Ground in Pahrump. *Pahrump Valley Times*, October 16, 2020.

Kropotin, P. (2023). *Gravity Energy Storage Systems with Weightlifting.* Accessed 4/20/25 @ https:www.theropedia.com/content/10359/.

Massey, N. (2017). *Energy Storage Hits the Rails Out West.* Accessed 4/18/25 @ https//www.scientificamerican.com//article/energy-storage-hits-the-rails-out-west.

Part III

Physical Security and Cybersecurity of Energy

18 Energy Storage Security

> We depend on critical infrastructure every day. Our ability to travel, to communicate with friends and family, to conduct business, to handle our finances, and even our ability to access clean, safe food and water are all reliant upon our Nation's critical infrastructure networks and systems. These essential services that underlie daily life in American society are increasingly being run on digital networks. People connect to the national grid daily without realizing it from smartphones, computers, and tablets. As a result, these critical systems are prime targets for cyberattacks from those seeking to cause our country harm.
>
> *DHS (2010)*

> In God we trust, all others we monitor.
>
> *Anonymous*

Note: In this book, we use the terms artificial intelligence (AI; aka "The Umbrella"), information technology (IT), cyberspace, digital network, and cybersecurity interchangeably because, although different, they are inherently and technologically closely related.

18.1 INTRODUCTION

Energy storage is a critical infrastructure. It provides vital services during natural disasters and contributes to maintaining a stable electrical grid. Dispersed energy storage can decrease downtime and diminish the impact of disruptions on residents and businesses. Energy storage systems are becoming an essential part of the power grid now and in the future. Our critical infrastructure energy storage systems are primary targets for physical and cyberattacks.

Physical security and cybersecurity technologies have become required to protect energy storage systems from adversarial actions that could, so to speak, put the lights out when the standard energy supply is intentionally disrupted and there are no backup energy storage systems. They have also been the victims of natural disasters or terrorist attacks. All elements within our energy supply systems must have high availability, confidentiality, and integrity to provide the systems with effective control and monitoring ability; thus, to ensure this, energy supply systems must coordinate with other grid operators, devices, or systems, which need reliable cybersecured communications.

The bottom line: Because the penetration of energy storage systems is increasing rapidly and nonstop, and grid operators rely more on energy storage system functionality, it is critical to protect those assets from terrorist physical or cyberattacks to maintain grid reliability and continuity of service.

18.2 A WEAPON AND A SHIELD[1]

Since 9/11, we have heard it said by many of our teachers, neighbors, security specialists, and students (and many others) that there is controversy about the definition of the politically charged word *terrorism*. Terrorism, like pollution, is a judgment call. For example, when defining pollution, if two neighbors live next door to an air-polluting facility, one neighbor with no personal connection with the polluting plant will likely label the plant's output as pollution.

DOI: 10.1201/9781003637851-21

The other neighbor, an employee of the plant, may see the plant's pollution as dollar bills—dollars that are his or her livelihood. Workers who dive into ponds full of raw sewage to find a leak in an effluent pipe say that the sewage in the pond and its associated odor is money in the bank … and long-term employment because toilets and their function are not likely to vanish anytime soon. Why a money-making enterprise? Because there are few people around who would join them in diving into trenches filled with raw sewage and doing their work. On the terrorism front, when someone deliberately spikes a tree to prevent loggers from cutting it down, the tree-spiker might feel he or she is a patriot, just a knight in shining green armor, and not a terrorist. On the other hand, the logger who has to take the tree down and puts his or her life and limbs at risk in taking down the spiked tree has little doubt in his or her mind on what to call the tree-spiker, and what the tree-spiker is called indeed has nothing to do with patriotism. Thus, what we are saying here in the example presented, pollution versus terrorism, is that along with attempting to define pollution, trying to define terrorism may be a judgment call, especially in the view of the terrorists. But there is never doubt about terrorism from those people who have experienced or witnessed it.

The best way to show how terrorism can and did affect an infrastructure is to provide a famous example of a terrorist action using cyber technology. Consider the following example as a real-time, real-life act of terrorism conducted on another critical infrastructure, wastewater treatment plant operations.

Barton Gellman (2002) points out that on April 23, 2000, police in Queensland, Australia, stopped a car on the road and found a stolen computer and radio inside. Using commercially available technology, a disgruntled former employee had turned his vehicle into a pirate command center of sewage treatment along Australia's Sunshine Coast. The former employee's arrest solved a mystery that had troubled the Maroochy Shire wastewater system for two months. Somehow, the system leaked hundreds of thousands of gallons of putrid sewage into parks, rivers, and the manicured grounds of a Hyatt Regency hotel—marine life died, the creek water turned black, and the stench was unbearable for residents. Until the former employee's capture—during his 46th successful intrusion—the utility's managers did not know why.

Specialists studied this cyber-terrorism case because, at the time, it was the only known one in which someone deliberately used a digital control system to cause harm. The former employee's intrusion shows how easy it is to break in and how unrestrained he was.

To sabotage the system, the former employee set the software on his laptop to identify itself as a pumping station and then suppressed all alarms. During his intrusions, the former employee was the "central control station" with unlimited command of 300 SCADA nodes governing sewage and drinking water.

The bottom line: As severe as the former employee's intrusions were, they paled compared to what he could have done to the freshwater system—he could have done anything he liked.

Hacking a digitally controlled system is common. Other reports of digital network exploits illustrate the debilitating effects of such attacks on the nation's security, economy, and public health and safety.

- In May 2015, media sources reported that data from 1.1 million health insurance customers in the Washington, DC. area were stolen in a cyberattack on a private insurance company. Attackers accessed a database containing customers' names, birth dates, e-mail addresses, and subscriber ID numbers.
- In December 2014, the Industrial Control Systems Cyber Emergency Response Team (ICS-CERT, which partners with law enforcement agencies to reduce risks within and across all critical infrastructure sectors) issued an updated alert on sophisticated malware compromising numerous industrial control system environments. Their analysis indicated that this campaign had been ongoing since at least 2011.

- In its monitor report, released from January 2014 to April 2014, ICS-CERT reported that a public utility had been compromised when a sophisticated threat actor gained unauthorized access to its control system network through a vulnerable remote access capability configured on the system. The incident highlighted the need to evaluate security controls employed at the perimeter and ensure that potential intrusion vectors are configured with appropriate security controls, monitoring, and detection capabilities.
- In December 2016, a Wisconsin couple was charged after allegedly defrauding Enterprise Credit Union in Brookfield out of more than $300,000. One of the defendants managed the bank's accounts and had her co-conspirator cash bank checks worth $980 several times each week beginning in May 2015. The charges allege that the couple used the money to buy drugs.

  Source: 2015. http://www.wauwatrosanow.com/story/news/cirme/2016/12/09/two-charged-allegedly-scamming-credit-unions-over-300k/95207718/.

In 2000, the FBI identified and listed threats to critical infrastructure. These threats are listed and described in Table 18.1. In 2015, the GAO described the sources of digital network-based threats. These threats are listed and described in detail in Table 18.2.

TABLE 18.1
Threats to Critical Infrastructure Observed by the FBI

Threat	Description
Criminal groups	Criminal groups that attack systems for monetary gain are increasingly using cyber intrusions.
Foreign intelligence services	Foreign intelligence services use cyber tools as part of their information gathering and espionage activities.
Hackers	Hackers sometimes crack into networks for the thrill of the challenge or bragging rights in the hacker community. While remote cracking once required a fair amount of skill or computer knowledge, hackers can now download attack scripts and protocols from the Internet and launch them against victim sites. Thus, while attack tools have become more sophisticated, they have also become easier to use.
Hacktivism refers to attacks on publicly accessible websites or e-mail servers that are motivated by political reasons. These groups and individuals overload e-mail servers and hack into websites to send a political message.	
Information warfare	Several nations are aggressively working to develop information warfare doctrine, programs, and capabilities. Such capabilities enable a single entity to have a significant and profound impact by disrupting the supply, communications, and economic infrastructures that support military power, impacts that, according to the Director of Central Intelligence, can affect the daily lives of Americans across the country.
Inside threat	The disgruntled organization insider is a principal source of computer crimes. Although insiders may not need much knowledge about computer intrusions, their knowledge of a victim system often allows them to gain unrestricted access to cause damage to the system or steal system data. The insider threat also includes outsourcing vendors.
Virus writers	Pose an increasingly serious threat. Several destructive computer viruses and "worms" have harmed files and hard drives, including the Melissa macro virus, the ExploreZip worm, the CIH (Chernobyl) virus, Nimda, and Code Red.

Source: FBI (2014).

TABLE 18.2
Common Digital Network Threat Sources

Source	Description
Non-adversarial-malicious	
Failure in information technology equipment	Failures in displays, sensors, controllers, and information technology hardware responsible for data storage, processing, and communications.
Failure in environmental controls	Failures in temperature/humidity controllers or power supplies.
Software coding errors	Failures in operating systems, networking, and general-purpose and mission-specific applications.
Natural or human-made disaster	Events beyond an entity's control such as fires, floods/tsunamis, tornadoes, hurricanes, and earthquakes.
Unusual or natural event	Natural events beyond the entity's control that are not considered to be disasters (e.g., sunspots).
Infrastructure failure or outage	Failure or outage of telecommunications or electrical power.
Unintentional user errors	Failures resulting from erroneous, accidental actions taken by individuals (both system users and administrators) in the course of executing their everyday responsibilities.
Adversarial	
Hackers or hacktivists	Hackers break networks for the challenge, revenge, stalking, or monetary gain, among other reasons. Hactivists are ideologically motivated actors who use cyber exploits to further political goals.
Malicious insiders	Insiders (e.g., disgruntled organization employees, including contractors) may not need a great deal of knowledge about computer intrusions because their position with the organization often allows them to gain unrestricted access and cause damage to the target system or to steal system data. These individuals engage in purely malicious activities and should not be confused with non-malicious insider accidents.
Nations	Nations, including nation-state, state-sponsored, and state-sanctioned programs, use cyber tools as part of their information-gathering and espionage activities. In addition, several nations are aggressively working to develop information warfare doctrine, programs, and capabilities.
Criminal groups and organize crime	Criminal groups seek to attack systems for monetary gain. Specifically, organized criminal groups use cyber exploits to commit identity theft, online fraud, and computer extortion.
Terrorist	Terrorists seek to destroy, incapacitate, or exploit critical infrastructures to threaten national security, cause mass casualties, weaken the economy, and damage public morale and confidence.
Unknown malicious outsiders	Unknown malicious outsiders are threat sources or agents that, due to a lack of information, agencies are unable to classify as being one of the five types of threat sources or agents listed above.

Source: GAO analysis of unclassified government and nongovernmental data. GAO 16–79.

DID YOU KNOW?

Presidential Policy Directive 21 defined "All hazards" as a threat to an incident natural or human-made that warrants action to protect life, property, the environment, and public health or safety and to minimize disruptions of government, social, or economic activities.

Threats to systems supporting critical infrastructure are evolving and growing. As shown in Table 18.2, cyber threats can be unintentional or intentional. Unintentional or non-adversarial threats include equipment failures, software coding errors, and the actions of poorly trained employees. They also include natural disasters and failures of critical infrastructure that the organization depends on but are outside of its control. Intentional threats include targeted and untargeted attacks from various sources, including criminal groups, hackers, disgruntled employees, foreign nations engaged in espionage and information warfare, and terrorists. These threat adversaries vary regarding the actors' capabilities, willingness to act, and motives, including seeking monetary gain or an economic, political, or military advantage (GAO, 2015).

All the threats listed in Tables 18.1 and 18.2 and threats against energy storage systems may manifest differently. Various attacks may involve the use of vehicles, handguns, rifles, pipe bombs, or other means. Cybersecurity attacks exploit vulnerabilities in communication or control systems to disrupt system operations or execute malicious actions. The attack surface has dramatically increased with the beginning of distributed energy resources (DERs), which include consumer-owned small energy storage systems frequently linked to public networks. AI is a double-edged sword in protecting against cyber threats, but on the other edge is a tool that cybercriminals leverage to create sophisticated attacks.

This means and points to the seismic shift in the cybersecurity landscape caused by the advent of AI.

Never underestimate the time, expense, blood, sweat, and effort a terrorist will expend to compromise the security of any energy-producing system and energy storage.

18.3 U.S. ENERGY INFRASTRUCTURE: KID IN CANDY STORE SYNDROME[2]

U.S. energy infrastructure is everywhere, in every direction, in almost endless combinations, interrelationships, and quantities. In making this simple observation, it matters little whether potential terrorists are of foreign or domestic origin. They only need to look around or surf the Internet to locate the components of the U.S. Energy Sector. In doing this, terrorists must be in utter fascination, with many good or tempting goodies (targets) around them.

Let's take a look at what anyone, including terrorists, can find out about our energy infrastructure.

- In the somewhat desolate Columbia-Snake River basin of Washington and Idaho states and surrounding the Hanford Site (aka Hanford Works, Hanford Project, Hanford Nuclear Reservation) with its nine decommissioned weapons production nuclear reactors and 53 million gallon (204,000 m^3) of high-level radioactive waste are several hydroelectric dams, strung like beads on the threads known as the Columbia and Snake Rivers. These dams include Grand Coulee, Chief Joseph, Wells, Rocky Reach, Rock Island, Wanapum, Priest Rapids, and Ice Harbor. Going westward along the Snake River are Lower Monumental, Little Goose, Lower Granite, and finally, to the south, the Columbia are McNary, John Day, The Dalles, and Bonneville dams. To the east, along the Snake River, are Lower Monumental, Little Goose, Lower Granite, and finally to the south, along the Snake River bordering Idaho-Oregon, Hells Canyon, Oxbow, and Brownlee dams.
- At the present time, 104 licensed nuclear power plants are located in many states in the United States. The power plants are listed by group depending on the region of the United States in which they are located. Region IV is the largest geographical region with the smallest number of power plants, 20. The region includes the states of North Dakota, South Dakota, Nebraska, Missouri, Arkansas, Louisiana, and all the states west of these. Region III comprises Minnesota, Wisconsin, Michigan, Iowa, Illinois, Indiana, and Ohio. This region includes 24 nuclear power stations. Region II consists of the southeastern states of West Virginia, Kentucky, Virginia, Tennessee, North Carolina, South Carolina, Georgia, Alabama, Mississippi, Florida, and the territories of Puerto Rico and the Virgin Islands. This region includes 34 nuclear power stations. Region I consists of the New England states of Maine, New Hampshire, Vermont,

New York, Massachusetts, Pennsylvania, Connecticut, New Jersey, Delaware, Maryland, and Washington, DC. This region includes 26 nuclear power plants.

- Hydroelectric dams and nuclear power plants produce electrical power, of course. However, remember that they are only the best-known (the crown jewels) and easily recognized electricity producers. In the United States, we also generate electrical power at lesser-known plants that use coal, petroleum liquids, petroleum coke, natural gas, and other gases.
- The production of electrical energy involves more than just its initial production. Along with generation and production, electrical power transmission and distribution are also part of the production process. Electricity infrastructure includes a North American power grid comprising long-distance transmission lines that move electricity from region to region and local distribution lines that carry electricity to homes and businesses.
- Petroleum products are pumped from the ground in domestic oil fields and conveyed through gathering lines to pipelines and refineries, transforming them into gasoline, diesel fuel, or heating oil. These finished products then travel through pipelines and tanker trucks to distribution outlets for consumers to purchase. Finished product pipeline systems involve complex infrastructure of their own, including compressor stations or plumbing stations and control systems that open and close valves and control product flow through pipes, often using computer technology.
- Facilities turn raw natural resources into useful energy products.
- Energy infrastructure includes rail networks, truck lines, and marine transportation.

Again, an individual terrorist or group of terrorists does not have to look far in the United States or on the Internet to find the location of energy infrastructure targets. Terrorists also have little trouble in recognizing that energy infrastructure is vulnerable to physical and cyber disruption that could threaten not only its integrity and safety but also the integrity and safety of large regions of the entire country. Moreover, one need not be a rocket scientist to determine that the U.S. power infrastructure is complex and interdependent. Consequently, any disruption of one element can have extensive consequences for others, if not all.

18.4 WE CALL IT 9/11; OTHERS CALL IT CANTOR FITZGERALD

Early in the morning of August 19, 2000, a Saturday, a 30-inch-diameter natural gas transmission pipeline operated by El Paso Natural Gas Company accidentally ruptured (due to severe internal pipe erosion) adjacent to the Pecos River near Carlsbad, New Mexico. The released gas ignited and burned for 55 minutes. Twelve people who were camping under a concrete-decked steel bridge that supported the pipeline across the river were killed, and their three vehicles were destroyed. Two nearby steel suspension bridges for gas pipelines across the river were extensively damaged. The primary safety issues identified in the post-investigation were the design and construction of the pipeline, the adequacy of El Paso Natural Gas Company's internal corrosion control program, the adequacy of federal safety regulations for natural gas pipelines, and the adequacy of federal oversight of the pipeline operation. Other examples of natural phenomena (events) include wildfires, flash floods, earthquakes, active volcanoes, droughts, and storms. These unpredictable natural events cannot yet be controlled or prevented.

Accidental incidents involving new types of emergencies began to surface in the 1940s. They were linked to the behavior of specific chemical and energy products (gas, oil, coal, nuclear, LPG, etc.), collectively called hazardous materials, whenever they were misused or involved in unintended mishaps and fires.

DID YOU KNOW?

Before the first World Trade Center bombing in 1993 and the 1995 Oklahoma City bombing, emergency incidents were primarily or generally thought to be caused by natural or accidental events.

Where there are people, there are energy sources/products—from the busiest metropolis to the most remote village. Energy is at the heart of our technology-based society in its myriad forms. Energy can make our lives longer, healthier, more productive, prosperous, and comfortable. Energy is essential to the U.S. economy and standard of living; its many forms allow us to live or experience the trappings of the "good life." The energy industry manufactures products that are fundamental elements of other economic sectors. For example, it produces electricity to light our homes and run our machines, energy supplied heat can warm our homes or fuel our industrial furnaces, and polymers that create plastics from petroleum for innumerable household and industrial products. The most common items are plastics, leather, paper, rubber, paints, textiles, pesticides, solvents, detergents, fuels, medicines, fertilizers, building materials, electronics, sporting equipment, and automobiles. In addition, it is essential to point out that an enormous amount of the energy industry's products go to industrial and domestic or household applications. To eliminate energy products from our society would not only be impractical, but it would also be undesirable. Instead, since it is impractical and undesirable to eliminate energy products from our culture, we must find ways to work and live safely with these products.

It was not that far in our distant past that we mainly dealt with the ramifications of natural disasters. Although never welcomed, these natural occurrences were expected, planned for, guarded against, and recovered from. Disasters were our primary concern. Of course, there have always been human-made or human-caused disasters that we have also had to contend with. The sinking of the *Titanic* prompted new safety regulations for passenger ships; a fire at New York City's *Triangle Shirtwaist Factory* in 1911 caused the deaths of 146 young women and was the catalyst for the development of workplace safety and fire regulations. However, the ultimate wake-up call for Americans came in the form of the events that we now summarize as "9/11" or, as some characterize, Cantor Fitzgerald 9/11 (because of the 658 employees of this business firm that were killed in the World Trade Center on 9/11/2001). Call it what you will; one fact remains, however, on September 11, 2001, terrorists struck at the heart of America—on American soil in an unforgettable way. Airplanes filled with people and fuel were turned into guided missiles of death and destruction.

What the 9/11 terrorists provided us with was a wake-up call. It was a wake-up call that alerted us to a new type of hateful venom, coldly delivered by groups or individuals to kill massive numbers of people, cause substantial property damage, and affect economic stability. Most importantly, for the public and emergency responders, 9/11 brought to the forefront knowledge and awareness that just about anything unthinkable is possible. *The bottom line*: Homeland security is vital to all of us.

Homeland Security is an ongoing problem that must be dealt with 24/7. Simply, there is no magic on-off switch that we can use to turn off the threat of terrorism in the United States or elsewhere. Consequently, there is no way we can turn off the security switch either. We must stay awake and be alert to the threat.

The threat to our security is not only ongoing but is also universal, including potential or actual threats from within—from our citizens (homegrown terrorism). Consider the American Timothy McVeigh, for example, who blew up the Government Building in Oklahoma City in 1995, killing almost 200 people, including several children. McVeigh, who bombed the building in revenge for the FBI's Waco, Texas raid, thought the army (he was a former decorated U.S. Army veteran) had implanted chip in his butt to track his movements, according to reports.

It is interesting to note that McVeigh, who was no doubt suffering from some severe disturbance, acted primarily alone. McVeigh is the exception that proves the rule—most terrorist acts on America are planned by a group beforehand.

Before 9/11, terrorism was not previously seen as a significant threat to the United States. However, the fact is that while the United States has not experienced continuous 9/11-type terror events in the past, the current climate suggests that the United States is at significant risk of further terror activity in the future. Forecasting the future is impossible, but we can look at the past and learn from it. Table 18.3 summarizes past acts of sabotage/terrorism carried out in the United States or against U.S. citizens overseas.

TABLE 18.3
Terror Attack Summaries (within United States or against Americans Abroad)[3]

Date/Location	Summary of Attacks
Sept. 16, 1920, New York City	Bomb exploded in New York City's Wall Street area, killing 40 and injuring hundreds. Perpetrators fled the country and were never apprehended.
1951–1956, New York City	Between 1951 and 1956, former Consolidated Edison employee set off series of bombs at New York City landmarks, including Grand Central Station and Radio City Music Hall. No deaths. Bomber was later caught and committed to a state mental institution.
Jan. 24, 1975, New York City	Bomb set off in historic Fraunces Tavern killed 4 and injured more than 50 people. Puerto Rican National Group (FALN) claimed responsibility, and police tied 13 other bombings to the group.
Nov. 4, 1979, Tehran, Iran	Iranian radical students seized the U.S. embassy, taking 66 hostages. Fourteen were released. The remaining 52 were freed after 444 days on the day of President Reagan's inauguration.
1982–1991, Lebanon	Thirty U.S. and other Western hostages kidnapped in Lebanon by Hezbollah. Some were killed, some died in captivity, and some were eventually released. Terry Anderson was held for 2,454 days.
April 18, 1983, Lebanon	U.S. embassy destroyed in suicide car-bomb attack; 63 dead, including 17 Americans. The Islamic Jihad claimed responsibility.
Oct. 23, 1983, Lebanon	Shiite suicide bombers exploded a truck near U.S. military barracks at Beirut airport, killing 241 marines. Minutes later, a second bomb killed 58 French paratroopers in their barracks in West Beirut.
Dec. 12, 1983, Kuwait City, Kuwait	Shiite truck bombers attacked the U.S. embassy and other targets, killing 5 and injuring 80.
Sept. 20, 1984, East Beirut, Lebanon	A truck bomb exploded outside the U.S. embassy annex, killing 14, including 2 U.S. military.
Dec. 3, 1984, Beirut, Lebanon	Kuwait Airways Flight 221, from Kuwait to Pakistan, hijacked and diverted to Tehran. Two Americans killed.
April 12, 1985, Madrid, Spain	Bombing at restaurant frequented by U.S. soldiers, killed 18 Spaniards and injured 82.
June 14, 1985, Beirut, Lebanon	TWA Flight 847 en route from Athens to Rome hijacked to Beirut by Hezbollah terrorists and held for 17 days. A U.S. Navy diver was executed.
Oct. 7, 1985, Mediterranean Sea	Gunmen attacked Italian cruise ship, *Achille Lauro*. One U.S. tourist was killed. Hijacking linked to Libya.
Dec. 18, 1985, Rome, Italy, and Vienna, Austria	Airports in Rome and Vienna were bombed, killing 20 people, 5 of whom were Americans bombing linked to Libya.
April 2, 1986, Athens, Greece	A bomb exploded aboard TWA flight 840 en route from Rome to Athens, killing four Americans and injuring nine.
April 5, West Berlin, Germany	Libyans bombed a disco frequented by U.S. servicemen, killing two and injuring hundreds.
Dec. 21, 1988, Lockerbie, Scotland	N.Y.-bound Pan-Am Boeing 747 exploded in flight from a terrorist bomb and crashed into Scottish village, killing all 259 aboard and 11 on ground.
Feb. 26, 1993, New York City	World Trade Center bombing in New York City. Six deaths and 1,000 injuries. Intended plan of total structural collapse did not occur. Several members of Middle Eastern extremist organizations convicted of roles in the bombing.
Dec. 7, 1993, New York City	Immigrant opened fire on commuters aboard Long Island Railroad commuter train, killing six and wounding 19. Gunman convicted of shootings dubbed "the Long Island Railroad Massacre."
April 19, 1995, Oklahoma City	Car bomb exploded outside federal office building, collapsing walls and floors. 168 people were killed, including 19 children and one person who died in rescue effort. Over 220 buildings sustained damage. Timothy McVeigh and Terry Nichols later convicted in the anti-government plot to avenge the Branch Davidian standoff in Waco, Tex., exactly two years earlier.

(*Continued*)

TABLE 18.3 (*Continued*)
Terror Attack Summaries (within United States or against Americans Abroad)[3]

Date/Location	Summary of Attacks
Nov. 13, 1995, Riyadh, Saudi Arabia	Car bomb exploded at U.S. military headquarters, killing five U.S. military servicemen.
Oct. 9, 1995, Hyder, Arizona	12-car Amtrak passenger train traveling from New Orleans to Los Angeles carrying 248 people derailed killing one and seriously injured 12. Investigations indicated that tracks had been deliberately tampered with. Incident classified as deliberate act of sabotage. Ongoing investigation is in progress.
June 25, 1996, Dhahran, Saudi Arabia	Truck bomb exploded outside Khobar Towers military complex, killing 19 American servicemen and injuring hundreds of others. 13 Saudis and a Lebanese, all alleged members of Islamic militant group Hezbollah, were indicted on charges relating to the attack in June 2001.
July 27, 1996, Atlanta, Georgia	Bomb exploded during concert at Summer Olympic killing one and injuring 118.
October 1997, San Francisco, CA	An outage at electrical substation in San Francisco affected 126,000 customers. Pacific Gas and Electric said at the time that someone might have deliberately manipulated equipment at the substation to break circuits.
Aug. 7, 1998, Nairobi, Kenya, and Dar es Salaam, Tanzania	Truck bombs exploded almost simultaneously near two U.S. embassies, killing 224 (213 in Kenya and 11 in Tanzania) and injuring about 4,500. Four men connected with al-Qaeda two of whom had received training at al-Qaeda camps inside Afghanistan were convicted of the killings in May 2001 and later sentenced to life in prison. A federal grand jury had indicted 22 men in connection with the attacks, including Saudi dissident Osama bin Laden, who remained at large.
May 2000, worldwide impact	Computer hacker in the Philippines unleashed "Love Bug" computer virus overloading corporate and government e-mail systems in many countries. Estimated $10–15 billion in damage. Charges against hacker eventually dropped in the Philippines because of inadequate laws against computer hacking.
July 25, 2000, Detroit, Michigan	Saboteur with knowledge of Detroit lighting system ripped wiring out of 100 street lights in downtown Detroit, leaving live wires exposed and forcing city officials to shut off power to more than 600 downtown street lights to ensure public safety. Damages estimated at $26,000. One person was arrested and charge.
September 11, 2001, New York City, Washington, DC, and Pennsylvania	Terrorist cells of Middle Eastern origination hijacked four commercial jet airlines. Two aircraft used against World Trade Center Towers in NYC, third against Pentagon. Fourth aircraft brought down by passenger intervention. Approximately 3,000 killed.
September 2001, The United States	The U.S. postal system used as delivery vehicle for Anthrax spores contained in letters. Numerous news organizations and government agencies were affected across the United States and in some overseas locations with strong U.S. affiliations. Several Killed.
On October 7, 2001, Alaska	Intoxicated man shot at the Alaska pipeline, which resulted in a 7,000-barrel crude oil spill, and the pipeline was shut for several days.

Case Study 18.1 When Whiskey Sours

My name is W.W. Williams III … call me Willy or Junior. I come from a long line of the same kind, patriots all, thank you very much! Usually, I would not be having this conversation with you … no, sir, I wouldn't. Like my father … you know, the one who blew up that sugar factory … I've never been about talking or jiving or other such dribble. No, sir! Talk is cheap and action is expensive … my kind of action anyway. Besides, you may have heard this introductory incantation or salutation (not sure which word to use here … in my genius, I do become confused; don't we all?) before … from my deceased father, the one I mentioned earlier; a true patriot, who unfortunately served a very short life sentence; he died a despicable death in a federal prison. So there went another good man, a true patriot, down the proverbial drain, so to speak.

Well, in my family, we keep it short, pointed, and irrefutable … so let's cut to the chase. Not that I have anything else more pressing to do. No, sir. It's just that I am not used to being out of my psychic ether mode … from which I perpetually surfed … until I got caught.

Now, at present, I am disgustingly sober and in total control of my wits and sanity. But I am not free. No, sir. I sit here on my cement bed in Supermax, Colorado … the ultimate prison for the world's ultimate prisoners. Yes, I am one of the worst … bad, harmful to the bone marrow, but, alas, sickeningly sober.

Regarding sobriety, I, like my family, am one of the Pied Piper's followers: Hunter Thompson's followers. Thompson described the good life … the physical Ethernet at its best! In the fog of soberness, I have memorized Thompson's famous description of his driving orgy through Las Vegas—my family's credo.

In the trunk, we "stow two bags of grass, seventy-five pellets of mescaline, five sheets of high-powered blotter acid, a saltshaker half full of cocaine and a whole galaxy of multicolored uppers, downers, screamers, laughers … A quart of tequila, a quart of rum, a case of Budweiser, a pint of raw ether and two dozen amyls."

As my daddy always said, doesn't it just torque your jaws that some of us know how to live … and the rest exist. Thompson is my hero.

Sigh! Let's get back to the topic and conversation at hand. I was given this chance to state my case … to explain why I put my gang together to attack the U.S. electrical grid and put it out of commission and kill a few of those bleeding-heart liberals (just call them bleeders) in the bargain. So, let's get to it. Let me tell you my story. I suppose it isn't pretty in some folks' way of thinking. But that … others' thinking has never bothered me too much …. No, sir. I do it … and did it my way. And I would be happy forever sitting on this cement bed speaking to you all if they would just let me have my ether. Oh well, what the hell … can't have it all … not anymore, anyway.

I joined the Greenway Coalition in 1970 during one of Ohio's first Earth Day celebrations. I was 11 years old at the time. I was involved in some glorious undertakings during my storied career with Greenway. For example, I was with the team that tree-spiked several old-growth trees on the Olympic Peninsula, Washington State. In my last incident there, I personally drove several metal rods into various tree trunks and saved the trees. Unfortunately, a couple of loggers died and others were injured when their saws struck the spikes and the saws kicked back and decapitated two loggers and seriously wounded the others … just collateral damage … necessary collateral damage, in my opinion.

As it turned out, spiking all those trees was great, but looking at all the blood and guts later was a bit overwhelming, so I shifted gears from tree spiking to monkeywrenching or ecotaging.

My monkeywrenching activities also dealt with saving old-growth forests. However, instead of spiking trees, I ecotaged logging equipment. This was an ideal practice because I could prevent tree destruction. After all, the logging equipment malfunctioned and, in some cases, massively destroyed itself when they tried to start it. The sabotaged equipment harmed only a few loggers, which I didn't mind. So, it was a win-win situation for a while, at least.

Over the years, Greenway Coalition grew in numbers and sophistication … and I grew along. In the late 1990s, we shifted gears from forestry to other areas of interest, including committing arson to preserve the pristine wilderness and energy storage facilities and systems. The Vail Mountain arson attack, for example, where we torched buildings that were going up … those buildings would have destroyed an outdoor paradise. Man, there are just some places on Earth where man should not intrude. Later, we returned to Oregon and torched the principal office of a logging company. In 1999, we even went to New Zealand to uproot fields of genetically engineered potatoes. The engineering of foodstuffs should be left to Mother Nature alone and not to humans. A few years later, we burned down a 200-unit condominium being built in southern California. We put out the word that "If you build it, we will trash it." Then again we shifted gears. We understand that one of the biggest mistakes mankind made was in developing the automobile. These horrendous machines are nothing but polluters of the environment, with low fuel efficiency and destroyers of things they happen to run into. So, we decided to attack several car dealerships, garages, and storehouses in Los Angeles. Most of the autos we attacked were those gas guzzling road hogging SUVs or Hummers.

Several years ago, we decided to redirect our focus once more. We also agreed that arson was a bit too damaging … we would take out the bad guys for sure, but in some cases, we burned down the surrounding forest or city blocks, etc. Collateral damage for sure, but we decided we had made our point in that area. We had put many loggers and car dealerships out of business and the spotted owl sleeps better at night in cleaner air because of our actions.

We also decided to remove those horrible solar and wind farms when we shifted focus. Those solar farms cook birds flying over and take up ground that desert turtles enjoy. Then we hit the wind farms. Those giant colossal structures kill birds, including bald eagles … not good for sure. Besides, I don't want another Quixote charging those windmill structures.

One of the things that surprised us, devastated us (if you really must know) was the distorted press and other news media venues out there in la-la land that had the nerve to address these heroic actions of ours as acts of vandalism and other forms of destruction. Some even dared to label our actions terrorism or the acts of terrorists. Can you believe that? One thing is sure: the new Homeland Security Secretary, Kristi Noem, would not be pleased with anyone, including me, being labeled a "terrorist."

So, here I vegetate in this cement block locked up like one of those so-called terrorists (no offense intended to the new Secretary of Homeland Security). One media outlet dared to call me crazy … a real nut case … a crank! Can you believe that man!

Well, speaking of crank, it (the crank) is partially to blame (I guess) for me being entombed in concrete. Why? Well, guess I ought to explain the rest of this fiasco so you will understand my plight … my situation, my travesty, so to speak.

It came to me one stormy night like lightning … illuminating the way for me and my followers. You see, because of my innate ability to plan, kill, and destroy, I became the leader of my so-called fringe group. We had no name or title … no, sir. We did not need any of that. Interference with and destruction of man's intrusion on nature was our goal and mantra.

Back to that stormy night.

I was tooting up some crack and crank, chased by several shots of my favorite straight whiskey … I always drink my whiskey straight … none of that mix to make it sour for me. We were camping in a tent in the Texas Panhandle one night. Now, when I say we, I mean me and a few of my closest associates … you know, the ones I could trust the most, women, of course … actually, they were the ones with the dope and whiskey. So, we sat inside the tent, combining a bit of crack and crank into smokes and chasing it all with America's finest whiskey when a tremendous storm came up and things got loud, windy, and wet.

After several toots and down the hatches, so to speak, I decided to step outside into the wind, downpour, and streaking lightning exploding like a giant kaleidoscope, reflecting dancing light in all directions … I shook my fist at the heavens. I demanded that someone up there bring this mess to a halt. Simply, enough was enough. And that is when it hit me!

No, not the lightning … though I could have used a battery recharge about then … no, sir, it was not the lightning that hit me. It was … it was … gee, how do I describe it? As I said, the wind was howling, the sky lit up with streaks of lightning, thunder pounding, and rain everywhere. But what got my attention … made me almost sober … was those long, ghastly, sharpened fingers slashing, whipping, lashing, whirling viciously though the air above me and off to the west as far as I could see in the intermittent light. What a scene. When is enough, enough, man? Of course, in time, I recognized those arms severing the air for what they were … human-made giant scissor-handed hulking grotesque monsters slashing and groping their way across the high plains … wind turbos from hell!

It might seem that my experience could be compared to tilting at windmills like Don Quixote, but this is inaccurate. While I acknowledge that I was somewhat intoxicated during the event, it does not reflect madness akin to Don Quixote. No. No way. Hell no! I can assure you that I do not have a crazy bone in my body. None!

For years after camping in one of America's largest wind farms, I have not been able to get those horrendous, grotesque mechanical images out of my mind. Alas, that particular spectral experience shaped the rest of my life and landed me here to vegetate in this cement garden.

After that night of stuporous manifestations from hell (there is no other way to describe it, thank you very much), I realized—my Eureka moment—that I had finally found my mission in life—to clear the land of wind turbines. I was so excited!

You might be wondering, what issues do wind turbines have? Aren't wind turbines ecologically more practical than fossil-fueled energy producers? Do wind turbines leave no carbon footprint compared to other fuel sources? Most importantly, is wind power renewable?

Sure. All that is true … but think about the land area those wind farms occupy. Think about the noise they produce. Consider the distorted impact these images have on typically scenic landscapes. Then there are those birds and bats whose turbine blades are hacked. Now, understand, I am not a fan of birds and bats … to me, they are nothing more than direct descendants of dinosaurs; they make good fossils, etc., and that is about all. Besides, all those avian would-be monsters do is eat and poop … mostly poop. Who needs all that? On the other hand, birds and bats have as much right to exist as we do … dinosaur-like or not, they are living things. Thus, they must be protected. Protecting them became my ultimate mission in life.

So, during the few intervening years before my ultimate coup de grace, I torched a few SUV dealerships here and there and planned my assault on those terrible wind turbine farms. However, being my genius, it was clear from the beginning of my planning stage that I could not attack thousands of wind turbines alone. No way, Jose!

I have consistently worked independently. Working alone, outside any command structure, works better for me. I am and always have been a part of the Greenway Coalition, of course. Still, as an appendage, I'm connected by deeds, actions, ideological and philosophical identification, but not by any direct group communication tie-in. I have found that my tactics and methods, conceived by me alone, are best directed and controlled by myself. I consider myself in a league far above my heroes, Timothy McVeigh, Theodore Kaczynski (the Unabomber), and Eric Robert Rudolph of 1996 Atlanta Olympics fame. These three all epitomize my lone wolf characteristics. The difference between us? They got caught and I never intended to follow suit … but you know what they say about the best laid plans of both … etc.

This time, I realized a committed team would destroy the wind turbine successfully. However, like I mentioned earlier, I like my whiskey straight and do not sour it by mixing; I do not like mixing with people … with a group. Too many nuts and bolts and, therefore, too many chances for something to go wrong … sour the operation. Each additional person added to the team is just another weak link, so to speak.

Then I had another revelation. Attacking individual wind turbines made little sense and was impractical. Attacking entire wind farms was also a problem because there was no way we could muster up enough explosives and correctly place them undetected to do the damage I wanted done. So, I thought through the matter and determined that the best bang for the buck could be obtained by attacking the electrical substations fed by electrical power from wind turbines. Such an attack by a group of activists had already occurred in 2005 in South Middleton, Pennsylvania. The activists shot out a transformer at the power substation, knocking out electrical power in three counties and affecting 20,000 customers.

I recruited a dedicated team of 40 men and women from all parts of the United States. All these folks were old hands at ecotage and all were members of Greenway Coalition … more than eager to participate in any destruction of U.S. infrastructure … my kind of people. I decided that we would meet in Sturgis, South Dakota … in August of that year, during the Sturgis Rally … you know the annual gathering of more than 400,000 bikers for a week of fun and disobedience on two wheels, usually resulting in around 450 individual arrests for one crime or another … don't you just love them, man!

Sturgis seemed the obvious choice. I told all my team members to wear blue jeans, leather jackets, unshaven, unkempt, and just plain ready to fit in with the biker crowd. And they did, and we did … we fit in … no one seemed to notice we even existed … which was what I wanted … flies on the wall preparing for the ultimate egg-laying attack.

My team met twice. We met in a picnic area outside Sturgis along I-90 … perfect location. The 40-member team was briefed on the plan. The plan? Well, it was relatively easy, definitely not complicated … I was working with a team of true patriots, but many were a bit short on brain power, a few were short a few microchips, and few had very little common sense … suitable killer types but not too bright, if you understand. Each team member was to work alone in a designated location where wind turbines fed electricity to electrical substations, which fed electricity to a town or city or a significant distribution network. These locations were in California, Arizona, New Mexico, Texas, and West Virginia. I planned to put the lights out in as many places as possible and silence those horrible air-cutting wind turbines. I knew the destruction would be temporary, but I also knew that we could and would strike again and again … each time making reconstruction just a bit too expensive in the long run. Thus, I would win!

Each team member was outfitted with a 30.06 rifle. We spent several hours in the Sturgis picnic area doing target practice. I outfitted each of the 40 shooters with 180-grain bullets. I knew we needed the high-speed, high-energy impact of those 180-grain bullets to do the damage required to put those substation transformers out of commission. Turned out that the shooters were pretty good at it … and it was funny, when you have 40 people shooting off high-power rifles, you would think someone would notice. People did notice … they just smiled and waved to us as they passed by. I guess in that part of the country all you need is a horse, a Harley, a meth lab or a gun to fit right in … fitting right in is what we are/were all about.

I planned the attack for Labor Day. I did this purposefully because it was a holiday … when those so-called working folks couldn't care less about anything. Also, symbolically, Labor Day was perfect … a sort of twisted protest on my part. Since I believe that the commie labor unions and their constant chant of "give me more, more, and more and we will give you less, less, and less in return" is what has destroyed this great country of ours, it torques my jaw. You think I am exaggerating? Look at the American auto manufacturing folks … how well are they doing? Why has almost all other manufacturing left this country and moved to commie-land? Why are so many red-blooded Americans unemployed? Unions … it is all about unions. They have destroyed the land I love. Well, now it is get-even time, my friends … the worm has turned.

Along with target practice with the rifles, I needed to brief each shooter on where to aim their rounds at an electrical substation for the most significant effect. Because my case is still undergoing another appeal in my cement heaven, I can keep certain legal documents in my cell. So, if you look at the photo of the substation, you will see the same image we used for training.

You might wonder why I freely admit all that happened, even showing you the substation training photo, especially when my case is under appeal. My appeal is based on police harassment, failure to be Mirandized appropriately, and also jailor brutality … you know how those brutal jailhouse pigs act, those same want-to-be pigs who can't cut the mustard as real cops, they are bullies … everyone. Unfortunately, this is my fourth appeal … so chances are slim I will win, anyway … though some of that ether right now would be nice, thank you very much!

Anyway, let's get back to the fiasco … and a fiasco it turned out to be … indeed. As I mentioned, I had set the date for the simultaneous attack on 40 different major city electrical substations for Labor Day. These attacks were to occur in different time zones, so I had each person set their watches to the same time and planned the attack at 10:00 pm. No problem there. The problem arose at a vast power substation outside Amarillo, Texas. Unfortunately, as it turned out the woman I assigned to this station, JoAnn Garner-Potts II, was the ultimate ding-a-ling-dong and meth head … an hour before the 10:00 pm attack time she was observed wobbling around with rifle in hand in front of the electrical substation, her target, scoping the place out. Well, needless to say, someone called the pigs and they showed up in force and threw her methed-up body to the ground, brutalized the hell out of her, etc., etc., etc. Within 5 minutes, JoAnn Garner Potts II told her captors everything

… meaning everything. Before any of the rest of us, 39 of us on station, could do our damage, we were all arrested, brutalized by the pigs, incarcerated by the bleeders that run the courts, and here I sit in this cement palace … surrounded by some of the worst killers known to modern man. What a waste! And that old axiom keeps ringing in my ears: All I want to know is where I am going to die so I will never go there—my problem? I am there—for life, and death.

So, now it is two years later and here I sit after one of the longest most ridiculous trials in criminal history … topped off by three losing appeals and, as mentioned, waiting on the outcome of another. Everyone was out to get me. I hadn't killed or injured anyone at the substations. I just don't get it!

Sitting here now and basking in the total disgrace of failure, only three things bother me: one, I got caught by totally incompetent, dysfunctional deadheads; two, during the trial and all the publicity that followed, they classified me as an American terrorist. "Terrorist?" No way. It is true I have killed … collateral damage … but the overall goal of saving the environment is all that mattered … the rest is poppycock … trivial to the bone. I am not a terrorist. I don't think the press, law, legal authorities, or the people know what a terrorist is. Maybe someone ought to write a book on the subject to inform everyone what a terrorist is. Indeed, no, sir, I am not a terrorist. Finally, the third item I am upset about is that I did not stick to my mantra. As I mentioned earlier, when you add mix to straight whiskey, you turn it sour. Likewise, you sour the operation when you trade lone wolf operations for mixed ops.

Before we move on with the rest of the text, it is essential to summarize a couple of key points from the preceding case study. First, the incident just related points to the purpose of this text; namely, to emphasize the importance of energy industry infrastructure and its protection and being ready for "any" contingency—"any" being the key word. Second, most of the news we listen to or read about at present dealing with terrorism or acts of terrorism is focused on incidents that take place in foreign countries and is perpetrated by foreigners. Yet, the sidebar above shows a different side of terrorism: domestic terrorism. As described in the case study, the destructive acts were committed by domestic terrorists. This is a touchy area of discussion in the United States. Americans do not want to visualize or even think about their fellow citizens being capable of destroying their own country and their people. However, we (the authors) feel that terrorist acts committed within the United States are most likely to be committed by homegrown individuals rather than by radical terrorists from other parts of the world. One only needs to study the history of homegrown terrorism within this country—the acts of the homegrown terrorists Timothy McVeigh, the Unabomber, and Eric Rudolph—to understand that danger to our country and its critical infrastructure may be more of an in-house problem, closer to home than a distant external problem. To retain our country as a free society, we must be on guard against all threats, no matter what their origin.

Keep in mind that many define "free society" in different ways. This is important to us in this text because, in our view, to be truly safe from the terrorist threat (if that is possible and we have our doubts), we must give up certain freedoms and accept closer scrutiny and vigilance regarding actions that we usually assume to be no one's business but our own. There are differing opinions on this topic, and we fully accept that. Again, we each define freedom in our own way, our right as free people.

++

Note: A Precise definition of key terms is required for us to understand the purview of this text clearly. Remember that the key terms and their definitions below may differ from those used in the U.S. Department of Homeland Security (DHS) National Infrastructure Protection Plan.

- *Vulnerability* is the manifestation of the inherent states of the system (e.g., physical, technical, organizational, social, cultural) that can be exploited by an adversary to adversely affect (cause harm or damage to) that system.

- *Intent* is the desire or motivation of an adversary to attack a target and cause adverse effects.
- *Capability* is the ability and capacity to attack and cause adverse effects.
- *Threat* is the *intent* and *capability* to adversely affect (cause harm or damage) the system by changing its states.
- *Risk* results from a threat adversely affecting a vulnerable system (Haimes, 2004).

AI is a significant advancement in cybersecurity, serving both as a tool for protection and a potential threat, depending on its use. We should never take the digital terrorists and their potential to destroy our critical infrastructure lightly. Hackers can imitate internal messages and exploit employee behavior to evade security measures. Case Study 18.1 illustrates the growing challenge of combating terrorism against critical infrastructure, including those driven by AI-driven threats, which are becoming increasingly difficult to detect and prevent.

18.5 THE DIGITAL NETWORK

Today's developing "information age" technology has intensified the importance of critical infrastructure protection, including energy-producing and storage systems, in which digital networks or cybersecurity have become as essential as physical security to protecting virtually all critical infrastructure sectors. The Department of Defense (DoD) determined that cyber threats to contractors' unclassified information systems represented an unacceptable risk of compromising DoD information and posed a significant risk to U.S. national security and economic security interests.

In the past few years, especially since 9/11, it has been somewhat routine for us to pick up a newspaper, magazine, or view a television news program where a significant topic of discussion is cybersecurity or the lack thereof. For example, recently, there has been discussion about Russian hackers trying to influence the U.S. 2016 elections. Many cyber intrusion incidents we read or hear about have added new terms or uses for old terms to our vocabulary. For example, old terms such as Botnets (short for robot networks, also balled bots, zombies, botnet fleets, and many others) are groups of computers that have been compromised with malware such as Trojan horses, worms, backdoors, remote control software, and viruses have taken on new connotations regarding cybersecurity issues. Relatively new terms such as scanners, Windows NT hacking tools, ICQ hacking tools, mail bombs, sniffer, logic bomb, nukers, dots, backdoor Trojan, key loggers, hackers' Swiss knife, password crackers, blended threats, Warhol worms, flash threats, targeted attacks, and BIOS crackers are now commonly read or heard about. New terms have evolved along with various control mechanisms. For example, because many control systems are vulnerable to attacks of varying degrees, these attack attempts range from telephone line sweeps (wardialing) to wireless network sniffing (wardriving), to physical network port scanning, and to physical monitoring and intrusion. When wireless network sniffing is performed at (or near) the target point by a pedestrian (warwalking), meaning that instead of a person being in an automotive vehicle, the potential intruder may be sniffing the network for weaknesses or vulnerabilities on foot, posing as a person walking. Still, they may have a handheld PDA device or laptop computer. Further, adversaries can leverage standard computer software programs, such as Adobe Acrobat and Microsoft Office, to deliver a threat by embedding exploits within software files that can be activated when a user opens a file within its corresponding program. Finally, the communications infrastructure and utilities depend on the IT Sector. This dependency is due to the reliance of the communications systems on the software that runs the control mechanism of the operating systems, the management software, the billing software, and any number of other software packages that is used by industry. Table 18.4 provides descriptions of common exploits or techniques, tactics, and practices used by digital/cyber adversaries.

Not all relatively new and universally recognizable IT, digital, or cyber terms have sinister connotations or meanings. Consider, for example, the following digital terms: backup, binary, bit, byte, CD-ROM, CPU, database, e-mail, HTML, icon, memory, cyberspace, modem, monitor, network,

TABLE 18.4
Common Methods of Digital/Cyber Exploits

Exploit	Description
Watering hole	A method by which threat actors exploit the vulnerabilities of carefully selected websites frequented by users of the targeted system. Malware is then injected to the targeted system via the compromised websites.
Phishing and spear phishing	A digital form of social engineering that uses authentic-looking e-mails, websites, or instant messages to get users to download malware, open malicious attachments, or open links that direct then to a website that requires information or executes malicious code.
Credentials-based	An exploit that takes advantage of a system's insufficient user authentication and/or any elements of cybersecurity supporting it, to include not limiting the number of failed login attempts, the use of hard-coded credentials, and the use of a broken or risky cryptographic algorithm.
Trusted third parties	An exploit that takes advantage of the security vulnerabilities of trusted third parties to gain access to an otherwise secure system.
Classic buffer overflow	An exploit that involves the intentional transmission of more data than a program's input buffer can hold, leading to the deletion of critical data and subsequent execution of malicious code.
Cryptographic weakness	An exploit that takes advantage of a network employing insufficient encryption when either storing or transmitting data, enabling adversaries to read and/or modify the data stream.
Structured query language (SQL) injection	An exploit that involves the alteration of a database search in a web-based application, which can be used to obtain unauthorized across to sensitive information in a database, resulting in data loss at corruption, denial or service, or complete host takeover.
Operating system command injection	An exploit that takes advantage of a system's inability to properly neutralize special elements used in operating system commands, allowing the adversaries to execute unexpected commands on the system by either modifying already evoked commands or evoking their own.
Cross-site scripting	An exploit that uses third-party web resources to run lines of programming code (referred to as scripts) within the victim's web browser or scriptable application. This occurs when a user, using a browser, visits a malicious website or clicks a malicious link. The most dangerous consequences can occur when this method is used to exploit additional vulnerabilities that may permit an adversary to steal cookers (data exchanged between a web server and a browser), log key stokes, capture screen shots, discover and collect network information, or remotely access and control the victim's machine.
Cross-site request forgery	An exploit takes advantage of an application that cannot, or does not, sufficiently verify whether a well-formed, valid, consistent request was intentionally provided by the user who submitted the request, tricking the victim into executing a falsified requires that results in the system or data being compromised.
Path traversal	An exploit that seeks to gain access to files outside of a restricted directory by modifying the directory pathname in an application that does not properly neutralize special elements (e.g., "…", "/", and "…/").
Integer overflow	An exploit where malicious code is inserted that leads to unexpected integer overflow, or wraparound, which can be used by adversaries to control looping or make security decisions in order to cause program crashes, memory corruption, or the execution of arbitrary code via buffer overflow.
Uncontrolled format string	Adversaries manipulate externally controlled format strings in print-style functions to gain access to information and/or execute unauthorized code or commands.
Open redirect	An exploit where the victim is tricked into selecting a URL (website location) that has been modified to direct them to an external, malicious site which may contain malware that can compromise the victim's machine.
Heap-based buffer overflow	Similar to classic buffer overflow, but the buffer that is overwritten is allocated in the heap portion of memory, generally meaning that the buffer was allocated using a memory allocation routine, such as "malloc ()."
Unrestricted upload of files	An exploit that takes advantage of insufficient upload restrictions, enabling adversaries to upload malware (e.g., .php) in place of the intended file type (e.g., .jpg).

(Continued)

TABLE 18.4 (*Continued*)
Common Methods of Digital/Cyber Exploits

Exploit	Description
Inclusion of functionality from un-trusted sphere	An exploit that uses trusted, third-party executable functionality (e.g., web widget or library) as a means of executing malicious code in software whose protection mechanism are unable to determent whether functionality a from a trusted sources, modified in transit, or being spoofed.
Certificate and certificate authority compromise	Exploits facilitated via the issuance of fraudulent digital certificates (e.g., Transport Layer Security and Secure Socket Layer). Adversaries use these certificates to establish secure connections with the target organization or individual by mimicking a trusted third party.
Hybrid of others	An exploit that combines elements of two or more of the aforementioned techniques.

Source: GAO (2015).

RAM, Wi-Fi (wireless fidelity), record, software, World Wide Web—none of these terms typically generate thoughts of terrorism in most of us.

18.6 ANALYZING LARGE DATABASES

One of AI's significant contributions to cybersecurity in energy storage operations is its ability to provide enhanced threat detection by analyzing vast amounts of data in real time. Typical threat detection methods struggle to keep up with cybercriminals' rapidly evolving tactics. This is where AI's benefits come into play, identifying patterns and irregularities in network traffic, e-mails, or application behavior that signal potential threats.

A good example of AI and how it detects what we used to call unknown security flaws—today this is known as detecting **zero-day vulnerabilities,** also known as a **0-day**, by its ability to spot or detect unusual activity that might escape conventional protocols and tools. This usually entails the detection of security holes in a computer system that developers and owners are unaware of. When terrorists exploit this methodology undetected, it is known as a zero-day exploit or attack (Zetter, 2014). When vulnerabilities in software are detected, patches or workaround instructions are created to mitigate them. Usually, a white-hat security hacker is hired to identify and reduce vulnerabilities. AI-driven platforms protect individuals and businesses against growing cyber threats to our energy storage systems.

18.7 NOTABLE ENERGY INFRASTRUCTURE CYBERATTACKS

Energy infrastructure and storage disruptions can affect multiple sectors, including government, industry, and security. The point is that attacks on energy production and storage can indirectly disrupt vulnerable targets with power interruptions. Energy storage systems are increasingly important and may be targets for physical and cyberattacks. Several significant attacks have targeted energy infrastructure:

- In 2005, a Progress Energy substation in Florida suffered a rifle attack on a transformer, which caused an oil tank leak, leading to an explosion and a local power outage (Peppard, 2005).
- In 2013, an individual in Arkansas carried out a series of attacks on an unprotected grid infrastructure. During the first attack in the series, a person tried to bring down a transmission tower by removing bolts from the tower's base and attempting to attach a cable to the moving train. Carter (2013) points out that the same individual set fire to a switching

station later that year and used a tractor to bring down two electric poles. The sabotage caused over $4.5 million in damage and disrupted power for about 9,000 people (FBI, 2015).

- In 2013, snipers shot 17 power transformers and cut communications cables for the Metcalf transmission substation near San Jose, C (Serrano and Halper, 2014). Pacific Grid and Electric operators quickly responded to this attack, which caused little service interruption to Silicon Valley. Presently, the perpetrators remain at large, and the reason for the attack remains unknown. On the positive or reactive side, the Metcalf incident has raised awareness for the physical security of power grids.

Cyberattacks on the power grid have also taken place:

- In December 2015, an attack on three Ukrainian regional power distribution utilities. Seven substations were remotely disconnected, causing a power outage that affected more than 200,000 customers (ICS-CERT, 2017). To ensure the operators were unaware of the situation, the perpetrators took action to hinder discovery and hamper system restoration.
- In 2016, another cyberattack outage happened in Ukraine (Greenberg, 2017). This incident was launched in a much more automated manner using sophisticated malware, *Crash Override*. This malware targeted power transmission control systems to trigger a sequence of events leading to a catastrophic outcome:
- Remotely opening circuit breakers to cause power outages. In addition, they exploited a vulnerability of protective relays to disable overload or fault protection devices.
- Following the manual restoration of service, the system remained unprotected. Uncleared powerline faults may permanently damage transmission devices.
- This result was a prelude to a sustained and large-scale power interruption.
- *The bottom line*: The attack was more ambitious than in previous years, but it failed to disrupt the system because the perpetrators had inadequate knowledge of it (Slowik, 2019).
- In 2018, cyber intrusions targeted the control of U.S. power utilities (Bade, 2018). The affected power company experienced intermittent service due to frequent firewall reboots, which ended after a patch was applied.

These episodes show the difficulty of adopting and maintaining cybersecurity defenses for grid operation, generator managers, and other power sector actors.

18.8 VULNERABILITY ASSESSMENT (VA)

Among many others, AI aims to improve automation and optimization. This has resulted in better performance across various sectors and applications and is subject to ongoing development. The benefits of AI in the science of energy storage are significant. However, along with acting as a shield, AI can also become a sword used by those who aim to harm, do damage, or put out of commission energy production and storage technologies and techniques.

This book focuses on AI applied to DERs. Although there are many DER definitions, this book defines DERs as small-scale power generation (1–10,000 kW), flexible load, or storage technologies that can provide an alternative to or enhance the traditional electric power system.

These can be located on an electric utility's distribution system, a subsystem of the utility's distribution system, or behind a customer's meter. In addition to electric storage, these may include variable generation, distributed generation, demand response, energy efficiency, thermal storage, or electric vehicles and their charging equipment.

The problem is that cybersecurity challenges for the electrical grid are growing with the high deployment of renewable energy sources and other DERs. Note that DERs already provide many automated features, and their deployment is coupled with transitioning to a digitally interconnected power grid. This is where swords can replace shields. Cyberattackers frequently evolve their techniques to attack information and operational technology systems.

The goal is to mitigate energy production and storage risk over the next decade. Systems must address cybersecurity challenges and be prepared for future threats and vulnerabilities. These mitigations form the base of a new framework for defining the defensive posture of the future grid.

The bottom line: Addressing cybersecurity challenges over the next decade must be a key priority for the DER industry, including utilities and DER owners, operators, developers, software and hardware vendors, and aggregators (aka data collectors) (US DOE, 2002).

18.8.1 Energy Security Assessment

In this day and age, when AI can be both shield and sword, as an energy storage manager or executive, it is essential to conduct an energy security assessment at your facility to identify threats to your energy supply and to ensure that a reliable energy supply is always available to carry out critical functions. To accomplish this, initiating an energy security process, conducting a vulnerability assessment, putting together an energy preparedness and operations plan, having a remedial action plan, and implementing them is vital.

Okay, all of this makes sense and is vitally important to recognize and to implement. To get started, an energy supply and storage operation should designate and assign an energy security manager to lead the energy security program at the site. Create an energy security planning team to provide site expertise and facility data and to make decisions regarding the site's mission and function. The team must grasp the energy security assessment program and its workflow. Put together a kickoff meeting to establish ground rules, orientation, analysis, or consulting and foster cooperation between all the organizations. The most important function the energy supply/storage team performs is vulnerability assessment (VA).

Vulnerability assessments have taken on increased importance due to 9/11, which prompted the DHS's directive to establish a Critical Infrastructure Protection Task Force to ensure that activities to protect and secure vital infrastructure like energy production, supply, and storage are comprehensive and carried out expeditiously. VA is one of the most critical tools for implementing security protection for energy supply and storage facilities. Another consequence is a heightened concern among citizens in the United States over the security of their energy infrastructure (i.e., the uninterrupted supply of electrical power and fuel to power vehicles, homes, and vital communications systems). As mentioned, along with other critical infrastructure, the AI and IT Services Sector is classified as "vulnerable" because inherent weaknesses in its operating environment could be exploited to cause harm to the system. There is also the possibility of a cascading effect—a chain of events—due to a terrorist act affecting IT Sector providers, which could cause corresponding damage (collateral damage) to other nearby users. In addition to significant damage to the nation's IT Services Sector, entities using and needing IT Services to function, like the energy supply and storage sector, can result in loss of life due to a lack of proper emergency response; shutdown of other industries; loss of electronic communication operational control networks could cause catastrophic environmental damage to rivers, lakes, and wetlands; and other long-term public health impacts.

Public and private members who utilize AI/IT Sector Services conduct risk assessments. These assessments look at issues and potential vulnerabilities within individual organizations and sector-wide. Since risk management is part of the business operations culture, regulators and private organizations have a long history of conducting regular risk assessments. In the private sector, some risk assessments are mandated through regulation and validated by the examination process. Furthermore, the private sector institutions conduct voluntary vulnerability/risk assessments to meet their business needs as part of their continuity planning and/or in conjunction with trade associations' recommendations and self-regulatory requirements (DHS, 2007).

18.8.2 Assessing Vulnerabilities

The AI/IT Services Sector conducts ongoing vulnerability assessments. What is a vulnerability assessment? For this text, and according to FEMA (2008), vulnerability is defined as any weakness an aggressor can exploit to make an asset susceptible to hazard damage. In addition, according to the Department of Homeland Security (DHS, 2009), vulnerabilities are physical features or operational attributes that render an entity open to exploitation or susceptible to a given hazard. Vulnerabilities may be associated with physical (e.g., a broken fence), cyber (e.g., lack of a firewall), or human (e.g., insider threats; untrained guards) factors.

Vulnerability assessments estimate the odds that a characteristic of, or flaw in, an infrastructure could make it susceptible to destruction, disruption, or exploitation based on its design, location, security posture, processes, or operations. Vulnerabilities are typically identified through internal assessments and information sharing with customers, vendors, and suppliers.

A vulnerability assessment methodology was developed as part of the AI/IT Services Sector-Specific Plan risk assessment methodology. The methodology examined physical, cyber, and human vulnerabilities and considered relevant national preparedness threat scenarios. The process varied depending on the architecture elements being studied and included subject matter expert interviews, site visits, and modeling and analysis.

The vulnerabilities of AI/IT Services architecture elements may vary depending on whether they are operational or implementation-specific. Operational vulnerabilities may include those that result from the inherent principles of network design, unanticipated network congestion caused by external factors, or collateral consequences from major disasters or events. Implementation-specific vulnerabilities may be very particular, from bugs in application software and protocol deficiencies to backdoors in vendor equipment firmware or software. The magnitude of the implementation vulnerabilities also varies depending on the exposure of the vulnerable equipment. While embedded firmware, for example, may have limited exposure to configuration and maintenance functions, systems such as the Domain Name Service (DNS) (i.e., a hierarchical and distributed name service that provides a naming system for computers, services, and other Internet services) require a high degree of exposure to provide service to customers (DHS, 2010).

Vulnerability assessments are conducted on many levels. These vulnerability assessments include examining the potential risks resulting from cross-sector dependency, sector-specific vulnerabilities, and dependencies on key assets, systems, technologies, and processes. Moreover, DHS has instituted a process to provide Awareness Training to asset owners/operators of financial services. The purpose of Awareness Training is to provide AI/IT Sector personnel with information about the place of their asset within the overall AI/IT Services mission requirements and acquisition process so they will understand their role and importance to the entities at the corporate and site levels (DHS, 2010). The training focuses on:

- Protection of AI/IT interests
- Protection of energy supply/storage sites
- Protection of the government, including federal interests
- It is vital that facilities foster relationships with local responders and federal, state, and local law enforcement/civil authorities for energy supply/storage recovery planning.

The Awareness Training also informs the asset owners/operators of the protection measures applied to their proprietary and business-sensitive information provided by and to the AI/IT Services Sector. Once critical AI/IT energy supply/storage assets are identified and prioritized, the next step is to conduct standardized assessments. DHS has established a standardized mission assurance assessment with various agencies for application to critical AI/IT energy supply/storage services assets. These assessments consider impact, vulnerability, and threat/hazard (whether from natural disaster, technological failure, human error, criminal activity, or terrorist attack). This approach to risk assessment ensures consideration of relevant factors for each AI/IT energy supply and storage services asset and a relative prioritization of risks to support military operations.

18.8.3 Insider Threat Vulnerability

The insider threat to an AI/IT energy supply and storage services organization and hostile and criminal cyber activities is not about rocket science; instead, it is about malicious software (malware) and the variety of forms of hostile or intrusive software, including computer viruses, worms, Trojan horses, ransomware, spyware, adware, scareware, and others. One does not need to be a super genius to understand an insider threat. The truth is the insider threat is … a human, a person, a mammal, *Homo sapiens*, a breathing organism, a heartbeat. The point is to protect an organization from insider threats, the energy supply/storage organization must "use tools to monitor the traffic in or out of the networks and be able to focus that monitoring on specific people who do something concerning or suspicious; moreover, nontechnical employee behavior must be monitored" (INSA, 2017).

"Behavior must be monitored" … yes, for sure. What monitoring comes down to, however, is awareness. An essential part of any successful vulnerability assessment process is awareness. Security and risk managers (and all employees in general) working in or with the critical infrastructure sectors must know the potential for insider threat vulnerability. Again, the keyword is AWARENESS. Awareness means that energy supply/storage sector personnel must know how to identify and act against insider threats. To achieve this critical goal, safety and security personnel must be provided with an overview of shared characteristics and indicators associated with malicious insiders and effective measures to counter insider threats.

18.8.3.1 Protecting against Insider Threats (FEMA, 2015)

As mentioned earlier, when analyzing threats to our nation's critical infrastructure, including energy supply and storage, we tend to focus on malicious actions from boat or plane loads of outside actors. Of equal concern (and even more so in the author's view) are threats from insiders, someone we have given legitimate access to information, systems, and resources. The measures we take to detect and protect against external threats may not be sufficient to address insider threats.

A malicious insider has access to inside knowledge of the organization and uses that knowledge to cause harm. The insider may be a current employee, a former employee, a service provider, or, especially in the current era, a planted person inside who has been brainwashed (radicalized) and turned into a terrorist waiting for the moment of maximum impact to people and property.

Given the importance of our nation's critical infrastructure, the actions taken by a malicious employee or service provider could have devastating consequences.

Let's look at some actual examples.

- A service provider employee at a nuclear facility stole two 5-gallon containers of low-enriched uranium dioxide and then attempted to extort $100,000 by threatening to disperse the material in an unnamed U.S. city.
- A power company field engineer, angry with his supervisor, disabled protection systems at a substation and forced the shutdown of the entire network. More than 800,000 customers lost power as a result.
- Two municipal employees used their access credentials to sabotage the system controlling the traffic lights of a major city, causing widespread traffic delays. The damage took four days to be repaired.
- A disgruntled supermarket meat packaging employee intentionally contaminated hamburger meat with a pesticide, causing various levels of illness in 92 consumers.

Insider threats endanger the integrity and security of our workplaces and our communities. This section helps you become aware of threat indicators and actions you can take. As stated earlier, and it can't be overstated, awareness is the first step to keeping our nation and workplaces safe.

18.8.3.2 Insider Threat Defined

The President's National Infrastructure Advisory Council (2008) defines the insider threat as follows:

> The insider threat to critical infrastructure is one or more individuals with access or inside knowledge of a company, organization, or enterprise that would allow them to exploit the vulnerabilities of the entity's security, systems, services, products, or facilities, intending to cause harm.

A person who takes advantage of access or inside knowledge in this manner is commonly called a "malicious insider."

18.8.3.3 Scope of Insider Threats

Insider threats can be accomplished through either physical or cyber means and may involve any of the following:

- *Physical or information-technology sabotage*: Involves modification or damage to an organization's facilities, property, assets, inventory, or systems to harm or threaten harm to an individual, the organization, or the organization's operations.
- *Theft of intellectual property*: Involves removing or transferring an organization's intellectual property outside the organization through physical or electronic means (also known as economic espionage).
- *Theft of economic fraud*: Involves the acquisition of an organization's financial or other assets through theft or fraud.
- *National security espionage*: Involves obtaining information or assets with a potential impact on national security through clandestine activities.

DID YOU KNOW?

The FBI testified in June 2012 that in the preceding year, economic espionage losses to the American economy totaled more than $13 billion. In the previous four years, the number of arrests the FBI made had doubled; indictments increased by a factor of five and convictions increased by a factor of eight (FBI, 2012). In another survey, security professionals found that 43.2% of respondents attributed some loss at their organization to insiders. Overall 46% of responds said the damage cause by insider attacks was more damaging than outside attacks (CSI, 2011).

18.8.3.4 Common Characteristics and Traits of Malicious Insiders

Based on research conducted by the Software Engineering Institute at Carnegie Mellon University and the U.S. Secret Service National Threat Assessment Center, malicious insiders are often perceived or known to be difficult or high-maintenance employees who are as follows:

- Obviously unhappy or extremely resentful.
- Having financial, performance, or behavioral problems.
- At risk (or perceived to be) for layoff or termination.

Keep in mind that not all malicious insiders fit this characterization. For example, an insider involved in national security espionage may exhibit few outward signs. In most cases, however, management and/or human resources personnel knew the employees and their issues before an incident.

18.8.3.5 Personal Factors Associated with Insiders

The following motives and personal situations are frequently linked with malicious insiders:

- **Personal or Behavioral Problems**
- Vulnerable to blackmail

- Experiencing family or financial problems
- Prone to compulsive or destructive behavior
- Subject to ego or self-image issues
- **Personal Desires**
- Seeking adventure or thrill
- Seeking approval and returned favors
- Professing allegiance
- **Workplace Issues**
- Experiencing problems at work
- Feeling anger or a need for revenge

18.8.3.6 Organizational Factors That Embolden Malicious Insiders

The following organizational factors have been known to encourage or present opportunities to potential malicious insiders:

- **Access and Availability**
- Ease of access to materials and information
- Ability to exit the facility or network with materials or information
- **Policies and Procedures**
- Undefined or inadequate policies and procedures
- Inadequate training
- Lack of training
- **Time Pressure and Consequences**
- Rushed employees
- Perception of lack of consequences

18.8.3.7 Inside Activities and Behavior You May See

Insider threats may be detected through particular activities and behaviors on the insider's part. This section of the presentation identifies those indicators. These activities and behaviors often appear unusual or suspicious. Remember that several explanations for a particular activity or behavior may be identified here. However, when combined with other factors, certain activities or behaviors point toward a possible insider threat. A combination or confluence of indicators should not be ignored.

18.8.3.8 Types of Insider Activities and Behavior

Unusual or suspicious insider activities and behavior can be described using the following categories:

- Inappropriate interest or acquisition
- Unauthorized or unusual computer use
- Unusual hours, contacts, or travel
- Secretive or peculiar behavior
- Personal or financial issues

Employer Actions

- Communicating and consistently enforcing security policies and controls.
- Ensuring that proprietary information and materials are protected adequately, if not robustly.
- Routinely monitoring computer networks for suspicious activity.
- Ensuring that security (including computer network security) personnel have the necessary tools.
- Consulting with legal and law enforcement experts as needed to ensure compliance with the law.

Employee Actions

Critical infrastructure organizations today employ several security measures to reduce the risk of insider threats. The measures involving employees include but are not limited to:

- Using appropriate screening processes to select new employees.
- Educating employees about security or other protocols.
- Encouraging and providing non-threatening, convenient ways for employees to report suspicious activity confidentially.
- Becoming familiar with behavior and activities associated with malicious insiders.
- Documenting and evaluating incidents of suspicious or disruptive behavior.
- Consulting with legal and law enforcement experts as needed to ensure compliance with the law.

18.9 THE VULNERABILITY ASSESSMENT (VA) IN ACTION[4]

A *vulnerability assessment* involves an in-depth analysis of the facility's functions, systems, and site characteristics to identify facility weaknesses and a lack of redundancy and determine mitigations or corrective actions that can be designed and implemented to reduce the vulnerabilities. A vulnerability assessment can be a stand-alone process or part of a full risk assessment. During this assessment, the analysis of site assets is based on (1) the identified threat, (2) the criticality of the assets, and (3) the chosen level of protection (i.e., based on willingness or unwillingness to accept risk).

It is important to point out that post 9/11, all sectors have taken great strides to protect their critical infrastructure. For instance, government and industry have developed vulnerability assessment methodologies for essential infrastructure systems and trained thousands of auditors and others to conduct them.

The complexity of vulnerability assessments will vary based on the design and operation of the financial services asset. The nature and extent of the VA will differ among systems based on several factors, including system size, potential population, and safety evaluations, which also vary based on knowledge and types of threats, available security technologies, and applicable local, state, and federal regulations. Preferably, a VA is "performance-based," meaning that it evaluates the risk to the financial services assets based on the effectiveness (performance) of existing and planned measures to counteract adversarial actions. According to US EPA (2002), the common elements of vulnerability assessments are as follows:

- Characterization of the AI/IT as applied in energy production, supply, and storage, including its mission and objectives.
- Identification and prioritization of adverse consequences to avoid.
- Determination of critical assets that might be subject to malevolent acts that could result in undesired consequences.
- Assessment of the likelihood (qualitative probability) of such malevolent acts from adversaries.
- Evaluation of existing countermeasures.
- Analyze current risks and develop a prioritized plan for risk reduction.

18.9.1 Benefits of Assessments

Energy supply and storage sector members should routinely perform vulnerability assessments to understand threats and vulnerabilities better, determine acceptable levels of risk, and stimulate action to mitigate identified vulnerabilities. These assessments are based upon the extensive knowledge of regulators and guidance issued and consider physical, cyber, and human vulnerabilities, available redundancy, the sector's reliance on sector-specific assets, systems, and processes, and cross-sector reliance on these factors. Consequence assessments include direct economic and national confidence impacts based on expert judgment and exercises. The direct benefits of performing a vulnerability assessment include the following:

- *Build and broaden awareness*: The assessment process directs senior management's attention to security. It brings security issues, risks, vulnerabilities, mitigation options, and best practices to the surface. Awareness is one of the least expensive and most effective methods for improving the organization's security posture.
- *Establish or evaluate against a baseline*: If a baseline has been established, an assessment is an opportunity for a checkup to gauge the improvement or deterioration of an organization's security posture. Suppose no previous baseline has been performed (or the work was not uniform or comprehensive). In that case, an assessment is an opportunity to integrate and unify previous efforts, define standard metrics, and establish a definitive baseline. The baseline can also be compared against best practices to provide perspective on an organization's security posture.
- *Identify vulnerabilities and develop responses*: Generating lists of vulnerabilities and potential responses is usually a core activity and outcome of an assessment. Sometimes, due to budget, time, complexity, and risk considerations, the response selected for many vulnerabilities may be non-action. Still, after completing the assessment process, these decisions will be conscious, with a documented decision process and item-by-item rationale available for revisiting issues at scheduled intervals. This information can help drive or motivate the development of a risk management process.
- *Categorize key assets and drive the risk management process*: An assessment can be a vehicle for reaching corporate-wide consensus on a hierarchy of key assets. This ranking, combined with threat, vulnerability, and risk analysis, is at the heart of any risk management process. For many organizations, Y2K was the first time a company-wide inventory and ranking of key assets was attempted. An assessment allows any organization to revisit that list from a broader perspective.
- *Develop and build internal skills and expertise*: When not implemented in an "audit" mode, a security assessment can be an excellent opportunity to build security skills and expertise within an organization. A well-structured assessment can have elements that serve as a forum for cross-cutting groups to come together and share issues, experiences, and expertise. External assessors can be instructed to emphasize "teaching and collaborating" rather than "evaluating" (the traditional role). Whatever the organization's current level of sophistication, a long-term goal should be to move the organization toward a capability for self-assessment.
- *Promote action:* Although disparate security efforts may be underway in an organization, an assessment can crystallize and focus management attention and resources on solving specific and systemic security problems. Often, the people in the trenches are well aware of security issues (and potential solutions) but cannot convert their awareness to action. An assessment provides an outlet for their concerns and the potential to surface these issues at appropriate levels (legal, financial, executive) and achieve action. A well-designed and executed assessment identifies vulnerabilities and makes recommendations; it also gains executive buy-in, identifies key players, and establishes a set of cross-cutting groups that can convert those recommendations into action.
- *Kick off an ongoing security effort*: An assessment can be used as a catalyst to involve people throughout the organization in security issues, build cross-cutting teams, establish permanent forums and councils, and harness the momentum generated by the assessment to make an ongoing institutional security effort. The assessment can lead to creating either an actual or a virtual (matrixed) security organization.

18.10 VULNERABILITY ASSESSMENT PROCESS

Table 18.5 provides an overview of the elements included in the assessment methodology. These elements are based on actual in-field experience and lessons learned.

TABLE 18.5
Basic Elements in Vulnerability Assessments

Element	Points to Consider

1. **Characterization of the communications entity, including its mission and objectives.**	• What are the essential missions of the system to be assessed? Define the highest priority services provided by the utility. Identify the industry's customers:
	  • General public
	  • Government
	  • Military
	  • Industrial
	  • Critical care
	  • Retail operations
	  • Firefighting
	• What are the system's most essential facilities, processes, and assets to achieve the mission objectives and avoid undesired consequences?
	Describe the:
	  • Industry facilities
	  • Operating procedures
	  • Management practices that are necessary to achieving mission objectives
	  • How the industry operates
	  • Treatment processes
	  • Storage methods and capacity
	  • Energy use and storage
	  • Distribution system
	In assessing critical assets, consider critical customers, dependence on other infrastructures (e.g., chemical, transportation, communications), contractual obligations, single points of failure, chemical hazards, and other aspects of the industry's operations, or availability of industry utilities that may increase or decrease the criticality of specific facilities, processes, and assets.

2. **Identification and prioritization consequences to avoid.**	• These consequences substantially disrupt the system's ability to provide a safe and reliable supply of materials. IT Services Sector systems should use the vulnerability assessment process to determine how to reduce the risk associated with the consequences of significant concern.
	• Ranges of consequences or impacts for each of these events should be identified and defined. Factors to be considered in assessing the consequences may include the following:
	  • Magnitude of service disruption
	  • Economic impact (such as replacement and installation costs for damaged critical assets or loss of revenue due to service outage)
	  • Number of illnesses or deaths resulting from an event
	  • Impact on public confidence in the material supply
	  • Chronic problems arising from specific events
	  • Other indicators of the impact of each event, as determined by the financial services sector.
	After the vulnerability assessment, risk reduction recommendations strive to prevent or reduce each consequence.

3. **Determination of critical assets that might be subject to some malevolent acts that could result in undesired consequences.**	• What are the malevolent acts that could reasonably **cause,** undesired?
	  • Electronic, computer, or other automated systems which are utilized by the financial sector entities (e.g., Supervisory Control and Data Acquisition (SCADA)
	  • The use, storage, or handling of various financial services supplies
	  • The operation and maintenance of such systems.

(Continued)

TABLE 18.5 (*Continued*)
Basic Elements in Vulnerability Assessments

Element	Points to Consider
4. **Assessment of the likelihood (qualitative probability) on malevolent acts from adversaries.**	• Determine the possible modes of attack that might result in consequences of significant concern based on critical assets of the financial services sector entity. The objective of this assessment step is to move beyond what is merely possible and determine the likelihood of a particular attack scenario. This is a challenging task as there is often insufficient information to assess the possibility of a specific event with any degree of certainty. The threats (the kind of adversary and the mode of attack) selected for consideration during a vulnerability assessment will dictate, to a great extent, the risk reduction measures that should be designed to counter the threat(s). Some vulnerability assessment methodologies refer to this as a "Design Basis Threat" (DBT) where the threat serves as the basis for the design of countermeasures and the benchmark against which vulnerabilities are assessed. It should be noted that there is no single DBT or threat profile for all financial systems in the United States. Differences in geographic location, size of the utility, previous attacks in the local area, and many other factors will influence the threat(s) that the energy supply/storage sector entity should consider in their assessments. IT Services Sector entities should consult with the local FBI and/or other law enforcement agencies, public officials, and others to determine the threats upon which their risk reduction measures should be based.

5. **Evaluation of existing countermeasures**
(Depending on countermeasures already in place, some critical assets may already be sufficiently and protection. This step will aid in identification of the areas of greatest risk concern and help to focus priorities for risk reduction.)
It is important to determine the performance characteristics. Poorly operated and maintained security technologies provide little or no protection.
	• What capabilities does the system currently employ for detection, delay, and response?
	  • Identify and evaluate current detection capabilities, such as intrusion detection systems, energy quality monitoring, operational, alarms, guard post orders, employee security awareness.
	  • Identify current delay mechanisms such as locks and key control, fencing, structure integrity of critical assets and vehicle access checkpoints.
	  • Identify existing policies and procedures for evaluation and response to intrusion and system malfunction alarms and cyber system intrusions.
	• What cyber protection system features does the facility have in place? Assess what protective measures are in-place for the SCADA and business-related computer information systems such as the following:
	  • Firewalls
	  • Modem access
	  • Internet and other external connections, including wireless data and voice communications
	  • Security policies and protocols
	It is important to identify whether vendors have access rights and/or "backdoors" to conduct system diagnostics remotely.
	• What security policies and procedures exist, and what is the compliance record for them? Identify existing policies and procedures concerning:
	  • Personal security
	  • Physical security
	  • Key and access badge control
	  • Control of system configuration and operational data
	  • Vendor deliveries
	  • Security training and exercise records

(*Continued*)

TABLE 18.5 (*Continued*)
Basic Elements in Vulnerability Assessments

Element	Points to Consider

6. **Analysis of current risk and development of a prioritized and plan for risk reduction.**	• Information gathered on threat, critical assets, IT Services Sector operations, consequences, and existing countermeasures should be analyzed to determine the current level of risk. The utility should then determine whether current risks are acceptable or risk reduction measures should be pursued.
	• Recommended actions should measurably reduce risks by reducing vulnerabilities and/or consequences through improved deterrence, delay, detection, and/or response capabilities or by improving operational policies or procedures. Selection of specific risk reduction actions should be completed before considering the cost of the recommended action(s). Facilities should carefully consider both short- and long-term solutions. An analysis of the cost of short- and long-term risk reduction actions may impact the utility's actions to achieve its security goals.
	• Facilities may also want to consider security improvements. Security and general infrastructure may provide multiple significant benefits. For example, improved treatment processes or system redundancies can reduce vulnerabilities and enhance day-to-day operation.
	• Generally, strategies for reducing vulnerabilities fall into three broad categories:
	  • Sound business practices—affect policies, procedures, and training to improve the overall security-related culture at the chemical facility. For example, it is essential to ensure rapid communication capabilities exist between public health authorities and local law enforcement and emergency responders.
	  • System upgrades—include changes in operations, equipment, processes, or infrastructure that make the system safer.
	  • Security upgrades—improve capabilities detection, delay, or response.

In Table 18.1, Step 3 deals with the identification of asset criticality. This is an essential step in any vulnerability assessment. Identifying asset criticality serves several functions:

- It enables more careful consideration of factors that affect risk, including threats, vulnerabilities, and consequences of loss or compromise of the asset.
- It enables more focused and thorough consideration of loss or compromise of the asset.
- It enables leaders to develop robust methods for managing the consequences of asset loss (restoration).
- It provides a means to increase the awareness of a broad range of employees about protecting truly critical assets and to differentiate in policies and procedures the heightened protection they require.

As previously indicated, identifying the criticality of assets is used primarily to focus the vulnerability analysis efforts. It also assists with the ranking of various recommendations for reducing vulnerabilities. As an example, let's take a look at the criticality of electric power assets and operations included in the regular operation of financial services sector assets:

Physical

- Generators
- Substations
- Transformers

- Transmission lines
- Distribution lines
- Control center
- Warehouses
- Office buildings
- Internal and external infrastructure dependencies

Cyber

- SCADA systems
- Networks
- Databases
- Business systems
- Telecommunications

Interdependencies

- Single-point nodes of failure
- Critical infrastructure components of high reliance

18.10.1 Vulnerability Assessment Methodology

The vulnerability assessment methodology consists of ten elements. Each element, along with a description, is listed below (US DOE, 2002).

1. Network architecture
2. Threat environment
3. Penetration testing
4. Physical security
5. Physical asset analysis
6. Operations security
7. Policies and procedures
8. Impact analysis
9. Infrastructure interdependence
10. Risk characterization

18.10.2 Network Architecture

This element analyzes the information assurance features of the information network(s) associated with the organization's critical information systems. Information examined should include network topology and connectivity (including subnets), principal information assets, interface and communication protocols, function and lineage of major software and hardware components (especially those associated with information security, such as intrusion detectors), and policies and procedures that govern the network's security features.

Procedures for information assurance in the system, including authentication and access authorization management, should be reviewed. The assessment should identify any obvious concerns about architectural vulnerabilities and operating procedures. Existing security plans should be evaluated, and the results of any prior testing should be analyzed. Results from the network architecture assessment should include potential recommendations for changes in the information architecture, functional areas, categories where testing is needed, and suggestions regarding system design that would enable more practical information and information system protection.

Three techniques are often used in conducting the network architecture assessment:

1. Analysis of network and system documentation during and after the site visit.
2. Interview with facility staff, managers, and the chief information officer.
3. Tours and physical inspections of key facilities.

18.10.3 Threat Environment

Development of a clear understanding of the threat environment is a fundamental element of risk management. When combined with an appreciation of the value of the information assets and systems, and the impact of unauthorized access and subsequent malicious activity, understanding threats provides a basis for better defining the level of investment needed to prevent such access.

The threat of a terrorist attack to the IT Services Sector infrastructure is real and could come from several areas, including physical, cyber, and interdependency. In addition, threats could come from individuals or organizations motivated by financial gain or people who derive pleasure from such penetration (e.g., recreational hackers, disgruntled employees). Other possible sources of threats are those who want to accomplish extremist goals (e.g., environmental terrorists, antinuclear advocates) or embarrass one or more organizations.

This element should include a characterization of these and other threats, identification of trends in these threats, and ways in which vulnerabilities are exploited. To the greatest extent possible, the characterization of the threat environment should be localized, that is, within the organization's service area.

18.10.4 Penetration Testing

Network penetration testing utilizes active scanning and penetration tools to identify vulnerabilities a determined adversary could exploit easily. Penetration testing can be customized to meet the specific needs and concerns of the financial services sector unit. In general, penetration testing should include a test plan and details on the rules of engagement (ROE). It should also include a general characterization of the critical information systems' access points and the access points to the essential information systems and communication interface connections, modem network connections, access points to principal network routers, and other external connections. Finally, penetration testing should include identified vulnerabilities and, in particular, whether access could be gained to the control network or specific subsystems or devices that have a critical role in assuring continuity of service.

Penetration testing consists of establishing the ground rules or ROE for the test; establishing a white cell for continuous communication; developing a format or methodology for the test; conducting the test; and generating a final report that details methods, findings, and recommendations.

Penetration testing methodology consists of three phases: reconnaissance, scenario development, and exploitation. A one-time penetration test can provide the utility with valuable feedback; however, it is far more effective if performed regularly. Repeated testing is recommended because new threats develop continuously, and the networks, computers, and architecture of the financial services sector unit or utility will likely change over time.

18.10.5 Physical Security

A critical dependency for the IT Services Sector and the other sectors is related to the physical security of the facilities. A physical security assessment aims to examine and evaluate the systems in place (or being planned) and identify potential improvements in this area for the sites assessed. Physical security systems include access controls, barriers locks and keys, badges and passes, intrusion detection devices and associated alarm reporting and display, closed-circuit television (assessment and surveillance), communications equipment (telephone, two-way radio, intercom, cellular), lighting (interior and exterior), power sources (line, battery, generator), inventory control, postings (signs), security system wiring, and protective force. Physical security systems are reviewed for design, installation, operation, maintenance, and testing.

The physical security assessment should focus on those sites directly related to the critical facilities, including information systems and assets required for operation. Typically included are facilities that house essential equipment, information assets, or networks dedicated to operating energy infrastructure: electric, oil, gas transmission, storage, or delivery systems. Other facilities can be included based on criteria specified by the assessed organization. Appropriate levels of physical security are contingent upon the value of company assets, the potential threats to these assets, and the cost associated with protecting the assets. Once the cost of implementing/maintaining physical security programs is known, it can be compared to the value of the company assets, thus providing the necessary information for risk management decisions. The focus of the physical security assessment task is determined by prioritizing the company assets; the most critical assets receive the majority of the assessment activity.

At the start of the assessment, survey personnel should develop a prioritized listing of company assets. This list should be discussed with company personnel to identify security strengths and weaknesses. During these initial interviews, the assessment area that would benefit the company most should be identified; once known, it should become the primary focus of the assessment activities.

The physical security assessment of each focus area usually consists of the following:

- Physical security program (general)
- Physical security program (planning)
- Barriers
- Access controls/badges
- Locks/keys
- Intrusion detection systems
- Communications equipment
- Protective force/local law enforcement agency

The key to reviewing the above topics is not to identify if they exist but to determine the appropriate level that is necessary and consistent with the value of the protected asset. The physical security assessment worksheets guide proper levels of protection.

Once the focus and content of the assessment task have been identified, the approach to conducting the assessment can be either at the "implementation level" or at the "organizational level." The approach taken depends on the maturity of the security program.

For example, a company with a solid security infrastructure (staffing plans/procedures, funding) should receive a cursory review of these items; however, facilities where the security programs are being implemented should receive a detailed review. Once reported, the security staff can act upon deficiencies found at the facilities.

For companies with an insufficient security organization, most of the time spent on the assessment should be spent at the organizational level to identify the appropriate staffing/funding necessary to implement security programs to protect company assets. Research into specific facility deficiencies should be limited to finding just enough examples to support any staffing/funding recommendations.

18.10.6 Physical Asset Analysis

The physical asset analysis examines the systems and physical operational assets to ascertain vulnerabilities. This element includes an examination of asset utilization, system redundancies, and emergency operating procedures. Consideration should also be given to the topology and operating practices for electric and gas transmission, processing, storage, and delivery, looking specifically for those elements that either singly or in concert with other factors provide a high potential for disrupting service. This assessment portion determines company and industry trends regarding these physical assets. Historic trends, such as asset utilization, maintenance, new infrastructure investments, spare parts, SCADA linkages, and field personnel, are part of the scoping element.

The proposed methodology for physical assets is based on a macro-level approach. The analysis can be performed with company, public, or both data. Some companies might not have readily available data or be reluctant to share that data.

Key outputs from the analysis should be graphs that show trends. The historic data analysis should be supplemented with on-site interviews and visits. Items to focus on during a site visit include the following:

- Trends in field testing
- Trends in maintenance expenditures
- Trends in infrastructure investments
- Historic infrastructure outages
- Critical system components and potential system bottlenecks
- Overall system operation controls
- Use and dependency of SCADA systems
- Linkages of operational staff with physical and IT security
- Adequate policies and procedures
- Communications with other regional financial assets
- Communications with external infrastructure providers
- Adequate organizational structure

18.10.7 Operations Security

Operations security (OPSEC) is the systematic process of denying potential adversaries (including competitors or their agents) information about the capabilities and intentions of the host organization. OPSEC involves identifying, controlling, and protecting generally nonsensitive activities concerning the planning and execution of sensitive activities. The OPSEC assessment reviews the processes and practices employed for denying adversary access to sensitive and nonsensitive information that might inappropriately aid or abet an individual's or organization's disproportionate influence over system operation. This assessment should include reviewing security training and awareness programs, discussions with key staff, and tours of appropriate principal facilities. Information that might be available through public access should also be reviewed.

18.10.8 Policies and Procedures

The policies and procedures by which security is administered (1) provide the basis for identifying and resolving issues; (2) establish the standards of reference for policy implementation; and (3) define and communicate roles, responsibilities, authorities, and accountabilities for all individuals and organizations that interface with critical systems. They are the backbone for decisions and day-to-day security operations. Security policies and procedures become particularly important when multiple parties must interact to affect the desired level of security and when substantial legal ramifications can result from policy violations. Policies and procedures should be reviewed to determine whether (1) address the key factors affecting security; (2) enable effective compliance, implementation, and enforcement; (3) reference or conform to established standards; (4) provide clear and comprehensive guidance; and (5) effectively address the roles, responsibilities, accountabilities, and authorities.

The policy and procedure assessment task aims to comprehensively understand how a facility protects its critical assets through developing and implementing policies and procedures. Understanding and assessing this area provides a means of identifying strengths and areas for improvement that can be achieved through:

- Modification of current policies and procedures
- Implementation of current policies and procedures

- Development and implementation of new policies and procedures
- Assurance of compliance with policies and procedures
- Cancellation of policies and procedures that are no longer relevant or are inappropriate for the facility's current strategy and operations

18.10.9 Impact Analysis

A detailed analysis should be conducted to determine the influence of the exploitation of unauthorized access to critical facilities or information systems might affect an organization's operations (e.g., market and/or physical operations). In general, such an analysis would require a thorough understanding of (1) the applications and their information processing, (2) decisions influenced by this information, (3) independent checks and balances that might exist regarding information upon which decisions are made, (4) factors that might mitigate the impact of unauthorized access, and (5) secondary impacts of such access. Similarly, the physical chain of events following disruption, including the primary, secondary, and tertiary impacts, should be examined.

The impact analysis aims to help estimate the detrimental impacts that could have on financial services sector units. It introduces risk characterization by providing quantitative estimates of these impacts so that the Energy Services Sector unit can implement a risk management program and weigh the risks and costs of various mitigation measures.

18.10.10 Infrastructure Interdependencies

The term "infrastructure interdependencies" refers to the physical and electronic (cyber) linkages within communications and among our nation's critical infrastructures—energy (electric power, oil, natural gas), telecommunications, transportation, water supply systems, banking and finance, emergency services, and government services. This task identifies the direct infrastructure linkages between and among the infrastructures that support critical facilities as recognized by the organization. The performance of this task requires a detailed understanding of an organization's functions and internal infrastructures and how these link to external infrastructures.

The purpose of the infrastructure interdependencies assessment is to examine and evaluate the infrastructures (internal and external) that support critical facility functions, along with their associated interdependencies and vulnerabilities.

18.10.11 Risk Characterization

Risk characterization provides a framework for prioritizing recommendations across all task areas. The recommendations for each task area are judged against a set of criteria to help prioritize the recommendations and assist the organization in determining the appropriate course of action. It provides a framework for assessing vulnerabilities, threats, and potential impacts (defined in the other tasks). In addition, the organization's existing risk analysis and management process should be reviewed and, if appropriate, utilized for prioritizing recommendations. The degree to which corporate risk management includes security factors is also evaluated.

18.11 VULNERABILITY ASSESSMENT PROCEDURES

Vulnerability assessment procedures can be conducted for Energy Services Sector assets using various methodologies. For example, checklist analysis is an effective technology. In addition, Pareto analysis (80/20 principle), relative ranking, pre-removal risk assessment (PRRA), change analysis, failure mode and effects analysis (FMEA), fault tree analysis, event tree analysis, what-if analysis, Hazard and Operability (HAZOP) can be used in conducting the assessment.

Okay, what is HAZOP?

HAZOP systematically identifies potential problems and issues within complex systems. This structured approach uncovers safety risks and processes inefficiencies by examining new designs and existing operations.

Based on personal experience, the what-if analysis and HAZOP seem to be the most user-friendly methodologies. Below is a sample procedural outline for a what-if analysis, followed by a brief explanation and outline for conducting HAZOP.

18.11.1 What-If Analysis Procedure/Sample What-If Questions

The steps in a what-if checklist analysis are as follows:

1. Select the team (personnel experienced in the process)
2. Assemble information (piping and instrumentation drawings (P&IDs), process flow diagrams (PFDs), operating procedures, equipment drawings, etc.)
3. Develop a list of what-if questions
4. Assemble your team in a room where each team member can view the information
5. Ask each what-if question in turn and determine the following:
   - What can cause the deviation from the design intent expressed by the question?
   - What adverse consequences might follow?
   - What are the existing design and procedural safeguards?
   - Are these safeguards adequate?
   - If these safeguards are inadequate, what additional ones does the team recommend?
6. Record the answers to these questions in tabular format as the discussion proceeds.
7. Do not restrict yourself to the list of questions that you developed before the project started. The team is free to ask additional questions at any time.
8. When you have finished the what-if questions, examine the checklist. This checklist ensures that the team has not forgotten anything. While you are reviewing the checklist, other what-if questions may arise.
9. Follow up on all recommendations and action items arising from the hazards evaluation.

18.11.2 HAZOP Analysis

The HAZOP analysis technique uses a systematic process to (1) identify possible deviations from normal operations and (2) ensure safeguards are in place to help prevent accidents. The HAZOP uses unique adjectives (such as speed, flow, and pressure) combined with process conditions (such as "more," "less," and "no,") to consider all credible deviations from normal conditions systematically. The adjectives, called guide words, are a unique feature of HAZOP analysis.

In this approach, each guide word is combined with relevant process parameters and applied at each point (study node, process section, or operating step) in the process that is being examined.

Guide Words	Meaning
No	Negation of the design intent
Less	Quantitative decrease
More	Quantitative increase
Part of	Other material present by intent
As well as	Other materials present unintentionally
Reverse	Logical opposite of the intent
Other than	Complete substitution

Common HAZOP Analysis Process Parameters

Flow	Time	Frequency	Mixing
Pressure	Composition	Viscosity	Addition
Temperature	pH	Voltage	Separation
Level	Speed	Information	Reaction

The following is an example of creating deviations using guide words and process parameters:

Guide Words		Parameter		Deviation
NO	+	FLOW	=	NO FLOW
MORE	+	PRESSURE	=	HIGH PRESSURE
AS WELL AS	+	ONE PHASE	=	TWO PHASE
OTHER THAN	+	OPERATION	=	MAINTENANCE
MORE	+	LEVEL	=	HIGH LEVEL

Guide words are applied to the more general parameters (e.g., react, mix) and the more specific parameters (e.g., pressure, temperature). With the general parameters, having more than one deviation from applying one guide word is not unusual. For example, "more reaction" could mean that a reaction occurs faster or that a greater quantity of product results. On the other hand, some combination of guide words and parameters will yield no sensible deviation (e.g., "as well as" with "pressure").

HAZOP Procedure

1. Select the team
2. Assemble information (P&IDs, PFDs, operating procedures, equipment drawings, etc.).
3. Assemble your team in a room where each team member can view P&IDs.
4. Divide the system you are reviewing into nodes (you can present the nodes, or the team can choose them as you go along).
5. Apply appropriate deviations to each node. For each deviation, address the following questions:
   - What can cause the deviation from design intent?
   - What adverse consequences might follow?
   - What are the existing design and procedural safeguards?
   - Are these safeguards adequate?
   - If these safeguards are not adequate, what does the team recommend?
6. Record the answers to these questions in tabular format as the discussion proceeds.

18.12 VULNERABILITY ASSESSMENT: CHECKLIST PROCEDURE

The checklist is one of the most straightforward methodologies when performing the vulnerability assessment of any Energy Services Sector unit or facility. The Building Vulnerability Assessment Checklist, developed by the Department of Veterans Affairs (VA) and part of EEMA 426, Reference Manual to Mitigate Potential Terrorist Attacks Against Buildings, is the reference manual that is recommended in this book and is available on the Internet at http://www.fema.gov/media-library-data/2013072. It is an excellent guide for conducting a viable Checklist-Type Vulnerability Assessment. This checklist will help you prepare your threat assessment because it allows for a consistent security evaluation of designs at various levels. The Checklist can be used as a screening tool for preliminary design vulnerability assessment and supports the preparation of all steps for use by the assessment teams during preparation for interviews with facility representatives to help assure that all relevant aspects of the financial services assets are considered in the survey.

The bottom line: Through the vulnerability assessments, the Energy Services Sector has determined that some of its most significant challenges are its dependence on the telecommunications network and the power grid.

18.13 THE BOTTOM LINE

In collaboration with many IT Sector stakeholders, the U.S. DHS identified cyber risk as significant to all critical infrastructure sectors. In the next chapter, as an example, we detail SCADA and its importance, significances, and vulnerabilities, all of which point to the need for planning, preparation, and mitigation procedures to ensure the resilience of Energy Sector systems and applications.

NOTES

1. Adapted from F.R. Spellman (2015). *IT and Homeland Security.* Lanham, MD: Bernan Press.
2. Information in this section is adapted from *America's Energy Infrastructure: A Comprehensive Delivery System.* Retrieved 04/16/22 @www.nell.doe.gov/publication/press/2001.
3. Adapted from *Terrorist Attacks* 2009. Retrieved 04/17/23 @http://www.infoplease.com/ipa/A0001454.html.
4. Much of the information in this section is from the U.S. Department of Energy (US DOE 2002) Vulnerability Assessment Methodology: Electric Power Infrastructure. Washington, DC; U.S. Army Research Laboratory (2000) Vulnerability Risk Assessment, ARL-TR-1045. Washington, DC: DOD; U.S. Department of Justice, (2002) A method to Assess the Vulnerability of U.S. Chemical Facilities. Washington, DC.

REFERENCES

Bade, G. (2018). *Russian Hackers Infiltrated Utility Control Rooms.* Washington, DC: US Department of Homeland Security.

Carter, C.J. (2013). Arkansas Man Charged with Power Grid Sabotage. *CNN*, Oct. 12, 2013.

CSI (2011). *2010/2011 Computer Crime and Security Survey.* Orlando, FL: Computer Security Institute.

DHS (2007). *The National Strategy for the Physical Protection of Critical Infrastructures and Key Assets.* Washington, DC: US Department of Homeland Security

DHS (2009). *National Infrastructure Protection Plan.* Retrieved 05/11/17 @ http://www.dhs.gov/library/assets/NIPP.Plan.pdf.

DHS (2010). *Homeland Security Directive 7: Critical Infrastructure Identification, Prioritization, and Protection.* Washington, DC: US Department of Homeland Security.

FBI (2012). *Insider Espionage. Report before the House Committee on Homeland Security Subcommittee on Counter Terrorism and Intelligence.* Washington, DC: Federal Bureau of Investigation.

FBI (2014). *Protecting Critical Infrastructure and the Importance of Partnerships.* Accessed @ https://www.fib.gov/news/speeches/protecitng-critical-infrastruce-andt-the-importantce-o.

FBI (2015). *Jason Woodring Pleads Guilty to Federal Charge Related to Attacks on Power Grid*, March 10, 2015. Washington, DC: Federal Bureau of Investigation.

FEMA (2008). *FEMA452: Risk Assessment A How to Guide.* Accessed 05/01/16 @ fema.gov/library/file?type=published/filetofile.

FEMA (2015). *Protecting Critical Infrastructure against Insider Threats.* Accessed 17 April 2015 @ http://emilms.fema.gov/IS0915/IABsummary.htm.

GAO (2015). *Critical Infrastructure Protection: Sector-Specific Agencies Need to Better Measure Cybersecurity Progress.* Washington, DC: United States Government Accountability Office.

Gellman, B. (2002). Cyber-Attacks by Al Qaeda Feared: Terrorists at Threshold of Using Internet as Tool of Bloodshed, Experts Say. *Washington Post*, June 27; p. A01.

Greenberg, A. (2017). Crash Override: The Malware that Took Down a Power Grid. *Wired*, June 2017.
ICS-CERT (2017). *Industrial Control Systems Cyber Emergency Response Team: Cyberattack against Ukrainian Critical Infrastructure.* IR-ALERT-H-056-01.
INSA (2017). *Building a Stronger Intelligence Community.* Arlington, VA: Intelligence and National Security Alliance. Accessed @ www.insooline.org.
National Infrastructure Advisory Council (2008). *First Report and Recommendations on the Insider Threat to Critical Infrastructure.* Washington, DC: National Infrastructure Advisory Council, U.S. Department of Justice.
NIPC (2002). *National Infrastructure Protection Center Report.* Washington, DC: National Infrastructure Protection Center.
Peppard, J. (2005). A Reward Is Offered for Shooting the Power Transformer. *WTSP News (Tampa).* Tampa, FL., Oct. 17, 2005.
Serrano, R.A. and Halper, E. (2014). Sophisticated But Low-Tech Power Grid Attack Baffles Authorities. *Los Angeles Times*, 2014.
Slowik, J. (2019). *CRASHOVERRIDED: Reassessing the 2016 Ukraine Electric Power Event as a Protection-Focused Attack.* Hanover, MD: Dragos, Inc.
US DOE (2002). *Vulnerability Assessment Methodology: Electric Power Infrastructure.* Washington, DC: US Department of Energy.
US EPA (2002). *Vulnerability Assessment Fact Sheet.* EPA 816-F-02-025. Accessed May 2006 @ www.epa.gov/ogwdw/security/index.html.
Zetter, K. (2014). *Hacker Lexicon: What is Zero Day*? Accessed 4/29/25 @ https://www.wired.com/2014/11/what-is-a-zero-day/).

19 SCADA and Energy

19.1 WHAT IS SCADA?

Many countries' critical infrastructure increasingly depends on Supervisory Control and Data Acquisition (SCADA) systems. One thing is sure: SCADA systems are vital in modern energy infrastructure management. They enable monitoring, controlling, and optimizing energy production, distribution, and storage processes. If we were to ask the specialist to define SCADA, the technical response could be outlined as follows:

- SCADA is a multi-tier system or interfaces with multi-tier systems.
- It is used for physical measurement and control endpoints via a remote terminal unit (RTU) and a programmable logic controller (PLC) to measure voltage, adjust a value, or flip a switch.
- It is an intermediate processor typically based on commercial third-party OSes—VMS, Unix, Windows, Linux.
- It involves human interfaces, for example, with graphical user interfaces (Windows GUIs).
- Its communication infrastructure consists of a variety of transport mediums such as analog, serial, Internet, radio, and Wi-Fi.
- A significant advantage of SCADA systems is their ability to integrate renewable energy sources into existing grids.

How about the non-specialist response?—for the rest of us who are non-specialists. Well, for those of us in this category, we could say that SCADA is a computer-based control system that remotely controls and monitors processes previously controlled and monitored manually. The philosophy behind SCADA control systems can be summed up by the phrase "If you can measure it, you can control it." SCADA allows an operator using a central computer to supervise (control and monitor) multiple networked computers at remote locations. Each remote computer can control mechanical processes (motors, generators, mixers, pumps, valves, etc.) and collect sensor data at its location. Thus, the phrase is formed: **S**upervisory **C**ontrol **a**nd **D**ata **A**cquisition, or **SCADA**.

The central computer is called the master terminal unit, or MTU. The MTU has two main functions: periodically obtaining data from RTUs/PLCs and controlling remote devices through the operator station. The operator interfaces with the MTU using software called human-machine interface, or HMI. The remote computer is called a PLC or a RTU. The RTU activates a relay (or switch) that turns mechanical equipment "on" and "off." The RTU also collects data from sensors. Sensors perform measurement, and actuators perform control.

In the initial stages, utilities ran wires, also known as hardwire or landlines, from the central computer (MTU) to the remote computers (RTUs). Because remote locations can be hundreds of miles from the central area, utilities began to use public phone lines and modems, leased telephone company lines, and radio and microwave communication. More recently, they have also started to use satellite links, the Internet, and newly developed wireless technologies.

DID YOU KNOW?

Modern RTUs typically use a ladder-logic approach to programming due to its similarity to standard electrical circuits. An RTU that employs this ladder logic programming is called a PLC.

DOI: 10.1201/9781003637851-22

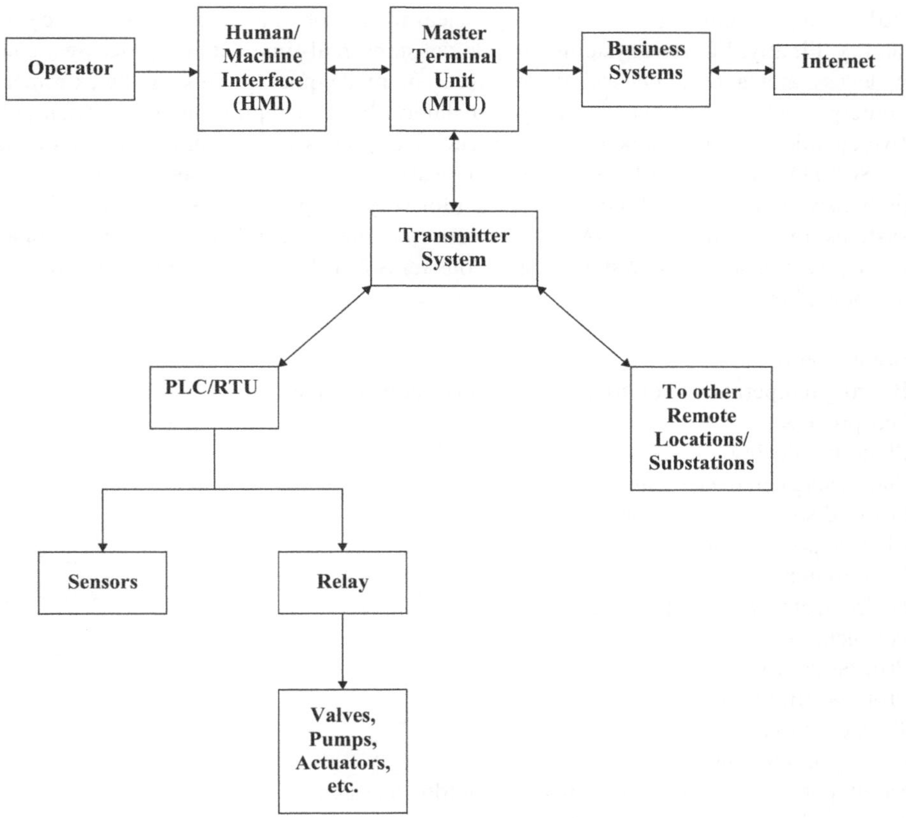

FIGURE 19.1 Representative SCADA network.

Because SCADA systems' sensors provided valuable information, many critical infrastructure entities, utilities, and other industries established "connections" between their SCADA systems and business systems. This allowed utility/industrial management and other staff access to valuable statistics, such as chemical usage. When utilities/industries later connected their systems to the Internet, they were able to provide stakeholders/stockholders with usage statistics on the IT segment and utility/industrial web pages. Figure 19.1 provides a basic illustration of a representative SCADA network. Note that firewall protection would typically be placed between the Internet and the business system and between the business system and the MTU.

19.2 SCADA APPLICATIONS

As stated above, SCADA systems can be designed to measure a variety of equipment operating conditions and parameters or volumes and flow rates of electricity, natural gas, and oil or oil and petrochemical mixture quality parameters, and to respond to change in those parameters either by alerting operators or by modifying system operation through a feedback loop system without having personnel physically visit each valve, process, or piece of other equipment daily to check it and/or ensure that it is functioning correctly. Automation and integration of large-scale, diverse assets require SCADA systems to provide the utmost flexibility, scalability, openness, and reliability. SCADA systems are used to automate certain energy production functions; these can be performed without initiation by an operator. In addition to processing equipment, SCADA systems can also integrate specific security alarms and equipment, such as cameras, motion sensors, lights, and data from card reading systems, thereby providing a clear picture of what is happening at areas throughout a facility. Finally, SCADA systems provide constant, real-time data on processes, equipment, location access, etc., and the necessary response to be made quickly. This can be extremely useful during emergencies, such as when energy distribution lines or piping break, or when potentially disruptive chemical reaction spikes appear in chemical processing operations. Currently, it can be said that SCADA has evolved from a simple indicating light and push-button control system into a comprehensive operation and handling system for very complex process control and safety shutdown systems. In a nutshell, SCADA results in an oversight system that requires fewer operators.

Today, many typical digital system applications for SCADA systems include, but are not limited to, those shown below:

- Boiler controls
- Bearing temperature monitors (electric generators and motors)
- Gas processing
- Plant monitoring
- Plant energy management
- Power distribution monitoring
- Electric power monitoring
- Fuel oil handling system
- Hydroelectric load management
- Petroleum pilot plants
- Process controls
- Process stimulators
- Tank controls
- Utility monitoring
- Safety parameters display systems and shutdown systems
- Tank level control and monitoring
- Turbine controls
- Turbine monitoring
- Virtual annunciator panels
- Alarm systems
- Security equipment
- Event logging

Because these systems can monitor multiple processes, equipment, and infrastructure and then provide quick notification of, or response to, problems or upsets, SCADA systems typically provide the first line of detection for atypical or abnormal conditions. For example, a SCADA system connected to sensors that measure specific machining quality parameters is measured outside a particular range. A real-time customized operator interface screen could display and control critical systems monitoring parameters.

The system could transmit warning signals back to the operators by initiating a call to a personal pager. This might allow the operators to initiate actions to prevent power outages, contamination, and energy supply disruption. Further automation of the system could ensure that the system initiates measures to rectify the problem. Preprogrammed control functions (e.g., shutting a valve, controlling flow, throwing a switch, or adding chemicals) can be triggered and operated based on the SCADA utility.

19.3 SCADA VULNERABILITIES

U.S. Electric Grid Gets Hacked

The Associated Press (AP) reported on April 9, 2009, that spies hacked into the U.S. energy grid and left behind computer programs (Trojan Horses) that would enable them to disrupt service, exposing potentially catastrophic vulnerabilities in key pieces of national infrastructure.

Even though terrorists, domestic and/or foreign, tend to aim their primary focus on the critical devices that control actual critical infrastructure production and delivery activities, according to USEPA (2005), SCADA networks were developed with little attention paid to security, making the security of these systems often weak. Studies have found that, while technological advancements introduced vulnerabilities, many critical infrastructures, sector plans/sites, and utilities have spent little time securing their SCADA networks. As a result, many SCADA networks may be susceptible to attacks and misuse. SCADA systems languished in obscurity, and this was the essence of their security; that is, until technological developments transformed SCADA from a backroom operation to a front-and-center visible control system.

Remote monitoring and supervisory control of processes developed in the early 1960s and adopted many technological advancements. The advent of minicomputers made it possible to automate many switches that were once manually operated. Advancements in radio technology have reduced the communication costs associated with installing and maintaining buried cables in remote areas. SCADA systems continued to adopt new communication methods, including satellite and cellular. As the price of computers and communications dropped, it became economically feasible to distribute operations and to expand SCADA networks to include even smaller facilities.

Advances in information technology and the necessity of improved efficiency have resulted in increasingly automated and interlinked infrastructures and created new vulnerabilities due to equipment failure, human error, weather and other natural causes, and physical and cyberattacks. Some areas and examples of possible SCADA vulnerabilities include (Wiles et al., 2007) the following:

- *Human*: People can be tricked or corrupted and may commit errors.
- *Communications*: Messages can be fabricated, intercepted, changed, deleted, or blocked.
- *Hardware*: Security features are not easily adapted to small self-contained units with limited power supplies.
- *Physical*: Intruders can break into a facility to steal or damage SCADA equipment.
- *Natural*: Tornadoes, floods, earthquakes, and other natural disasters can damage equipment and connections.
- *Software*: Programs can be poorly written.

Specific SCADA weaknesses and potential attack vectors include the following:

- Does not require any authorization
- Does not use encryption
- Does not properly recognize and properly handle errors and exceptions
- No authentication is required
- Data could be intercepted
- Manipulation of data
- Service denial
- *IP address spoofing*: Internet protocol packets with a false source IP address
- Session hijacking
- Unsolicited responses
- *Packet fuzzing*: It involves inputting false information, etc.
- Unauthorized control
- Log medication data

SCADA system computers and their connections are susceptible to different information system attacks and misuse, such as those mentioned above. The Computer Security Institute and the Federal Bureau of Investigation conduct an annual Computer Crime and Security Survey (FBI, 2007). The survey reported ten types of attacks or misuse and reported that viruses and denial of service had the most significant negative economic impact. The same study also found that 15% of the respondents reported abuse of wireless networks, which can be a SCADA component. On average, respondents from all sectors did not believe their organization invested enough in security awareness. For example, utilities reported a lower average computer security expenditure/investment per employee than other sectors, such as transportation, telecommunications, and financial.

Sandia National Laboratories' *Common Vulnerabilities in Critical Infrastructure Control Systems* described some of the common problems it has identified in the following five categories (Stamp et al., 2003):

1. *System data*: Important data attributes for security include availability, authenticity, integrity, and confidentiality. Data should be categorized according to its sensitivity, and ownership and responsibility must be assigned. However, SCADA data is often not classified, so identifying where security precautions are appropriate (for example, which communication links to secure databases require protection) is challenging.
2. *Security administration*: Vulnerabilities emerge because many systems lack a properly structured security policy (security administration is notoriously lax in the case of control systems), equipment and system implementation guides, configuration management, training, and enforcement and compliance auditing.
3. *Architecture*: Many common practices negatively affect SCADA security. For example, while it is convenient to use SCADA capabilities for other purposes, such as fire and security systems, these practices create single points of failure. Also, connecting SCADA networks to other automation and business networks introduces multiple entry points for potential adversaries.
4. *Network (including communication links)*: Legacy systems' hardware and software have minimal security capabilities, and the vulnerabilities of contemporary systems (based on modern information technology) are publicized. Wireless and shared links are susceptible to eavesdropping and data manipulation.
5. *Platforms*: Many platform vulnerabilities exist, including retained default configurations, poor password practices, shared accounts, inadequate protection for hardware, and nonexistent security monitoring controls. In most cases, important security patches are not installed, often due to concern about negatively impacting system operation; in some cases, technicians are contractually forbidden from updating systems by vendor agreements.

The following incident helps to illustrate some of the risks associated with SCADA vulnerabilities:

- During a vulnerability assessment, a contractor stated that personnel from his company penetrated the information system of a utility within minutes. Contractor personnel drove to a remote substation and noticed a wireless network antenna. Without leaving their vehicle, they plugged in their wireless radios and connected to the network within 5 minutes. Within 20 minutes, they had mapped the network, including SCADA equipment, and accessed the business network and data. This illustrates what a cybersecurity advisor from Sandia National Laboratories, specializing in SCADA, stated: utilities are moving to wireless communication without understanding the added risks.

19.3.1 The Increasing Risk

According to GAO (2015), security concerns about control systems (SCADA included) were historically related to protecting against physical attack and misusing refining and processing sites or distribution and holding facilities. However, more recently, there has been a growing recognition that control systems are now vulnerable to cyberattacks from numerous sources, including hostile governments, terrorist groups, disgruntled employees, and other malicious intruders.

In addition to control system vulnerabilities mentioned earlier, several factors have contributed to the escalation of risk to control systems, including (1) the adoption of standardized technologies with known vulnerabilities, (2) the connectivity of control systems to other networks, (3) constraints on the implementation of existing security technologies and practices, (4) insecure remote connections, and (5) the widespread availability of technical information about control systems.

19.3.2 Adoption of Technologies with Known Vulnerabilities

When a technology is not well known, widely used, understood, or publicized, it is difficult to penetrate and thus disable it. Historically, proprietary hardware, software, and network protocols made it difficult to know how control systems operate and how to hack into them. Today, however, to reduce costs and improve performance, organizations have been transitioning from proprietary systems to less expensive, standardized technologies such as Microsoft's Windows and Unix-like operating systems and the standard networking protocols used by the Internet. These widely used standardized technologies have commonly known vulnerabilities, and sophisticated and effective exploitation tools are widely available and relatively easy to use. Consequently, the number of people with the knowledge to wage attacks and the number of systems subject to attack have increased. Also, standard communication protocols and the emerging use of Extensible Markup Language (commonly referred to as XML) can make it easier for a hacker to interpret the content of communications among the components of a control system.

Control systems are often connected to other networks; enterprises usually integrate their control system with their enterprise networks. This increased connectivity has significant advantages, including providing decision-makers access to real-time information and allowing engineers to monitor and control the process control system from different points on the enterprise network. In addition, the enterprise networks are often connected to the networks of strategic partners and the Internet. Further, control systems increasingly use wide area networks and the Internet to transmit data to their remote or local stations and individual devices. This convergence of control networks with public and enterprise networks exposes the control systems to additional security vulnerabilities. Unless appropriate security controls are deployed in the enterprise and control system networks, enterprise security breaches can affect the control system's operation.

According to industry experts, control systems generally do not implement existing security technologies, strong user authentication, and patch management practices because they operate in real time, are not designed with cybersecurity in mind, and usually have limited processing capabilities.

Existing security technologies, such as authorization, authentication, encryption, intrusion detection, and network traffic and communication filtering, require more bandwidth, processing power, and memory than control system components typically have. Because controller stations are generally designed to do specific tasks, they use low-cost, resource-constrained microprocessors. Some devices in the electrical industry still use the Intel 8088 processor, introduced in 1978. Consequently, installing existing security technologies without seriously degrading the control system's performance is challenging.

Further, complex passwords and other strong password practices are not always used to prevent unauthorized access to control systems, in part because this could hinder a rapid response to safety procedures during an emergency. As a result, according to experts, weak passwords that are easy to guess, shared, and infrequently changed are reportedly common in control systems, including the use of default passwords or even no password at all.

In addition, although modern control systems are based on standard operating systems, they are typically customized to support control system applications. Consequently, vendor-provided software patches are generally incompatible or cannot be implemented without compromising service, shutting down "always-on" systems, or affecting interdependent operations.

Insecure connections exacerbate potential vulnerabilities in control systems. Organizations often leave access links—such as dial-up modems to equipment and control information—open for remote diagnostics, maintenance, and examination of system status. Such links may not be protected with authentication or encryption, which increases the risk that hackers could use these insecure connections to break into remotely controlled systems. Also, control systems often use wireless communications systems, which are especially vulnerable to attack, or leased lines that pass through commercial telecommunications facilities. Without encryption to protect data as it flows through these insecure connections or authentication mechanisms to limit access, there is limited protection for the integrity of the information being transmitted.

Potential hackers and intruders can access public information about infrastructures and control systems. A university graduate student whose dissertation reportedly mapped every business and industrial sector in the American economy to the fiber-optic network that connects them, using material available publicly on the Internet, none of which was classified, demonstrated the availability of this infrastructure and vulnerability data.

19.3.3 Cyber Threats to Control Systems

There is a consensus—and increasing concern—among government officials and experts on control systems about potential cyber threats to the control systems that govern our critical infrastructures. As components of control systems increasingly make crucial decisions that humans once made, the possible effect of a cyber threat becomes more devastating. Such cyber threats could come from numerous sources, ranging from hostile governments and terrorist groups to disgruntled employees and malicious intruders.

In July 2002, the National Infrastructure Protection Center (NIPC) reported that the potential for compound cyber and physical attacks, referred to as "swarming attacks," is an emerging threat to the U.S. critical infrastructure. As NIPC reports, the effects of a swarming attack include slowing or complicating the response to a physical attack. For instance, a cyberattack that disabled the water supply or the electrical system in conjunction with a physical attack could deny emergency services the necessary resources to manage the consequences, such as controlling fires, coordinating actions, and generating light.

Control systems, such as SCADA, can be vulnerable to cyberattacks. Entities or individuals with malicious intent might take one or more of the following actions to attack control systems successfully:

- Disrupt the operation of control systems by delaying or blocking the flow of information through control networks, thereby denying the networks' availability for control system operations.
- Make unauthorized changes to programmed instructions in PLCs, RTUs, or Distributed Control System controllers, change alarm thresholds, or issue unauthorized commands to control equipment, which could potentially result in damage to equipment (if tolerances are exceeded), premature shutdown of processes (such as prematurely shutting down transmission lines), or even disabling of control equipment.
- Send false information to control system operators either to disguise unauthorized changes or to initiate inappropriate actions by system operators.
- Modify the control system software, producing unpredictable results.
- Interfere with the operation of safety systems.

In addition, remote sites are often unstaffed and may not be physically monitored in control systems that cover a wide geographic area. If such remote systems are physically breached, the attackers could establish a cyber connection to the control network.

19.3.4 Securing Control Systems

Several challenges must be addressed to secure control systems against cyber threats effectively. These challenges include (1) the limitations of current security technologies in securing control systems; (2) the perception that securing control systems may not be economically justifiable; and (3) the conflicting priorities within organizations regarding the security of control systems.

A significant challenge in effectively securing control systems is the lack of specialized security technologies. The computing resources in control systems needed to perform security functions tend to be limited, making it very difficult to use security technologies within control system networks without severely hindering performance.

Securing control systems may not be perceived as economically justifiable. Experts and industry representatives have indicated that organizations may be reluctant to spend more money to secure control systems. Hardening the security of control systems would require industries to expend more resources, including acquiring more personnel, providing training for personnel, and potentially prematurely replacing current systems that typically have a lifespan of about 20 years.

Finally, several experts and industry representatives indicated that the responsibility for securing control systems typically includes two separate groups: IT security personnel and control system engineers and operators. IT security personnel tend to focus on securing enterprise systems, while control system engineers and operators tend to be more concerned with the reliable performance of their control systems. Further, they indicate that, as a result, those two groups do not always fully understand each other's requirements and collaborate to implement secure control systems.

19.4 STEPS TO IMPROVE SCADA SECURITY

The President's Critical Infrastructure Protection Board and the Department of Energy (DOE) have developed the steps outlined below to help organizations improve the security of their SCADA networks. DOE (2001) points out that these steps are not meant to be prescriptive or all-inclusive. However, they do address essential actions to improve the protection of SCADA networks. The steps are divided into two categories: specific actions to improve implementation and actions to establish essential underlying management processes and policies.

19.4.1 21 Steps to Increase SCADA Security (DOE, 2001)

The following steps focus on specific actions to be taken to increase the security of SCADA networks:

1. *Identify all connections to SCADA networks*
   Conduct a thorough risk analysis to assess the risk and necessity of each connection to the SCADA network. Develop a comprehensive understanding of all connections to the SCADA network and how well those connections are protected. Identify and evaluate the following types of connections:
   - Internal local area and wide area networks, including business networks
   - The Internet
   - Wireless network devices, including satellite uplinks
   - Modem or dial-up connections
   - Connections to business partners, vendors, or regulatory agencies
2. *Disconnect unnecessary connections to the SCADA network*
   To ensure the highest degree of security of SCADA systems, isolate the SCADA network from other network connections to the greatest possible degree. Any connection to another network introduces security risks, especially if the connection creates a pathway from or to the Internet. Although direct connections with other networks may allow vital information to be passed efficiently and conveniently, insecure connections are not worth the risk; isolation of the SCADA network must be a primary goal to provide the needed protection. Strategies such as using "demilitarized zones" (DMZs) and data warehousing can facilitate the secure data transfer from the SCADA network to business networks. However, they

must be designed and implemented correctly to avoid introducing additional risk through improper configuration.

3. *Evaluate and strengthen the security of any remaining connections to the SCADA networks*
   Conduct penetration testing or vulnerability analysis of any remaining connections to the SCADA network to evaluate the protection posture associated with these pathways. Use this information with risk management processes to develop a robust protection strategy for any paths to the SCADA network. Since the SCADA network is only as secure as its weakest connecting point, it is essential to implement firewalls, intrusion detection systems (IDSs), and other appropriate security measures at each entry point. Configure firewall rules to prohibit access from and to the SCADA network and be specific when permitting approved connections. For example, an independent system operator (ISO) should not be granted "blanket" network access simply because there is a need for a connection to specific components of the SCADA system. Strategically place IDSs at each entry point to alert security personnel of potential breaches of network security. Organization management must understand and accept the responsibility and risks associated with any connection to the SCADA network.
4. *Harden SCADA networks by removing or disabling unnecessary services*
   SCADA control servers built on commercial or open-source operating systems can be exposed to attack through default network services. Remove or disable unused services and network daemons to reduce the risk of direct attack to the greatest extent possible. This is particularly important when SCADA networks are interconnected with other networks. Do not permit a service or feature on a SCADA network unless a thorough risk assessment of the consequences of allowing the service/feature shows that the benefits of the service/feature far outweigh the potential for vulnerability exploitation. Examples of services to remove from SCADA networks include automated meter reading/remote billing systems, e-mail services, and Internet access. An example of a feature to disable is remote maintenance—numerous secure configurations, such as the National Security Agency's series of security guides. Additionally, work closely with SCADA vendors to identify secure configurations and coordinate any changes to operational systems to ensure that removing or disabling services does not cause downtime, interruption of service, or loss of support.
5. *Do not rely on proprietary protocols to protect your system*
   Some SCADA systems use unique, proprietary protocols for communication between field devices and servers. The security of SCADA systems is often based solely on the secrecy of these protocols. Unfortunately, obscure protocols provide very little "real" security. Do not rely on proprietary protocols or the default configuration settings to protect your system. Additionally, demand that vendors disclose any backdoors or vendor interfaces to your SCADA systems, and expect them to provide systems that can be secured.
6. *Implement the security features provided by device and system vendors*
   Older SCADA systems (most systems in use) have no security features. SCADA system owners must insist that their system vendor implement security features through product patches or upgrades. Some newer SCADA devices are shipped with basic security features, which are usually disabled to ensure ease of installation.

   Analyze each SCADA device to determine whether security features are present. Additionally, factory default security settings (such as in computer network firewalls) are often set to provide maximum usability, but minimal security. Set all security features to provide the maximum security only after a thorough risk assessment of the consequences of reducing the security level.
7. *Establish strong controls over any medium used as a backdoor into the SCADA network*
   Strong authentication must be implemented to ensure secure communications where backdoors or vendor connections exist in SCADA systems. Modems, wireless, and wired networks used for communications and maintenance represent a significant vulnerability to the SCADA

network and remote sites. Successful "war dialing" or "war driving" attacks could allow attackers to bypass all other controls and directly access the SCADA network or resources. Disable inbound access and replace it with a callback system to minimize the risk of such attacks.

8. *Implement internal and external intrusion detection systems and establish 24-hour-a-day incident monitoring*
To effectively respond to cyberattacks, establish an intrusion detection strategy that includes alerting network administrators of malicious network activity originating from internal or external sources. IDS monitoring is essential 24 hours a day; this capability can easily be set up through a pager. Additionally, incident response procedures must be in place to allow an effective response to any attack. To complement network monitoring, enable daily logging on all systems and audit system logs to detect suspicious activity as soon as possible.
9. *To identify security concerns, perform technical audits of SCADA devices, networks, and any other connected networks*
Technical audits of SCADA devices and networks are critical to ongoing security effectiveness. Many commercial and open-source security tools allow system administrators to conduct audits of their systems/networks to identify active services, patch level, and common vulnerabilities. Using these tools will not solve systemic problems but eliminate the "paths of least resistance" that an attacker could exploit. Analyze identified vulnerabilities to determine their significance and take corrective actions as appropriate. Track corrective actions and analyze this information to identify trends. Additionally, retest systems after corrective actions have been taken to ensure that vulnerabilities are eliminated. Scan non-production environments actively to identify and address potential problems.
10. *Conduct physical security surveys and assess the security of all remote sites connected to the SCADA network*
Any location connected to the SCADA network is a target, especially unmanned or unguarded remote sites. Conduct a physical security survey and inventory access points at each facility connected to the SCADA system. Identify and assess any source of information, including remote telephone/computer networks/fiber-optic cables that could be tapped, radio and microwave links that are exploitable computer terminals that could be accessed, and wireless local area network access points. Identify and eliminate single points of failure. The site's security must be adequate to detect or prevent unauthorized access. Do not allow "live" network access points at remote, unguarded sites simply for convenience.
11. *Establish SCADA "Red Teams" to identify and evaluate possible attack scenarios*
Establish a "Red Team" to identify potential attack scenarios and evaluate system vulnerabilities. Use a variety of people who can provide insight into weaknesses of the overall network, SCADA system, physical systems, and security controls. People who work on the system daily have great insight into the vulnerabilities of your SCADA network and should be consulted when identifying potential attack scenarios and possible consequences. Also, ensure that the risk from a malicious insider is thoroughly evaluated, given that this represents one of the greatest threats to an organization. Feed information from the "Red Team" evaluation into risk management processes to assess the information and establish appropriate protection strategies.

The following steps focus on management actions to establish an effective cybersecurity program:

12. *Clearly define cybersecurity roles, responsibilities, and authorities for managers, system administrators, and users*
Organization personnel must understand the specific expectations of protecting information technology resources by defining clear and logical roles and responsibilities. In

addition, key personnel must be given sufficient authority to carry out their duties. Too often, good cybersecurity is left up to the individual's initiative, leading to inconsistent implementations and ineffective security. Establish a cybersecurity organizational structure that defines roles and responsibilities and identifies how cybersecurity issues are escalated and who is notified in an emergency.

13. *Document network architecture and identify systems that serve critical functions or contain sensitive information requiring additional protection*
Develop and document a robust information security architecture as part of a process to establish an effective protection strategy. Organizations must design their network with security in mind and continue to have a strong understanding of their network architecture throughout its life cycle. Of particular importance, an in-depth knowledge of the functions that the systems perform and the sensitivity of the stored information is required. Risk cannot be adequately assessed without this understanding, and protection strategies may not be sufficient. Documenting the information security architecture and its components is critical to understanding the overall protection strategy and identifying single points of failure.
14. *Establish a rigorous, ongoing risk management process*
A thorough understanding of the risks to network computing resources from denial-of-service attacks and the vulnerability of sensitive information to compromise is essential to an effective cybersecurity program. Risk assessments from the technical basis of this understanding are critical to formulating effective strategies to mitigate vulnerabilities and preserve the integrity of computing resources. Initially, perform a baseline risk analysis based on the current threat assessment for developing a network protection strategy. Due to rapidly changing technology and the emergence of new threats daily, an ongoing risk assessment process is also needed so that routine changes can be made to the protection strategy to ensure it remains effective. Fundamental to risk management is the identification of residual risk with a network protection strategy in place and acceptance of that risk by management.
15. *Establish a network protection strategy based on the principle of defense-in-depth*
Defense-in-depth is a fundamental principle that must be part of any network protection strategy. Defense-in-depth must be considered early in the design phase of the development process and must be an integral consideration in all technical decision-making associated with the network. Utilize technical and administrative controls to mitigate threats from identified risks to the most significant degree possible at all levels of the network. Single points of failure must be avoided, and cybersecurity defense must be layered to limit and contain the impact of any security incidents. Additionally, each layer must be protected against other systems at the same level. For example, to protect against the inside threat, restrict users' access to only those resources necessary to perform their job functions.
16. *Identify cybersecurity requirements*
Organizations and companies need structured security programs with mandated requirements to establish expectations and allow personnel to be held accountable. Formalized policies and procedures are typically used to establish and institutionalize a cybersecurity program. A formal program is essential for establishing a consistent, standards-based approach to cybersecurity throughout an organization and eliminates sole dependence on individual initiative. Policies and procedures also inform employees of their cybersecurity responsibilities and the consequences of failing to meet them. They also provide guidance regarding actions to be taken during a cybersecurity incident and promote efficient and effective actions during a crisis. To identify cybersecurity requirements, include user agreements, notifications, and warning banners. Establish requirements to minimize the

threat from malicious insiders, including the need for conducting background checks and limiting network privileges to those necessary.

17. *Establish effective configuration management processes*
Configuration management is a fundamental management process needed to maintain a secure network. It needs to cover both hardware and software configurations. Changes to hardware or software can easily introduce vulnerabilities that undermine network security. Processes must evaluate and control any change to ensure the network remains secure. Configuration management begins with well-tested and documented security baselines for your various systems.
18. *Conduct routine self-assessments*
Robust performance evaluation processes are needed to provide organizations with feedback on the effectiveness of cybersecurity policy and technical implementation. A mature organization can self-identify issues, conduct root cause analyses, and implement practical corrective actions that address individual and systemic problems. Self-assessment processes, usually part of an effective cybersecurity program, include routine scanning for vulnerabilities, automated network auditing, and self-assessments of organizational and individual performance.
19. *Establish system backups and disaster recovery plans*
Establish a disaster recovery plan for rapid recovery from any emergency (including a cyberattack). System backups are essential to any plan and allow rapid network reconstruction. Routinely exercise disaster recovery plans to ensure that they work and that personnel are familiar with them. Based on lessons learned from exercises, make appropriate changes to disaster recovery plans.
20. *Senior organizational leadership should establish expectations for cybersecurity performance and hold individuals accountable for their performance*
Effective cybersecurity performance requires commitment and leadership from senior managers in the organization. Senior management must establish an expectation for strong cybersecurity and communicate this to organizational subordinate managers. It is also essential that senior organizational leadership establish a structure for implementing a cybersecurity program. This structure will promote consistent implementation and the ability to sustain a strong cybersecurity program. It is then essential for individuals to be held accountable for their cybersecurity performance. This includes managers, system administrators, technicians, and users/operators.
21. *Establish policies and conduct training to minimize the likelihood that organizational personnel will inadvertently disclose sensitive information regarding SCADA system design, operations, or security controls*
Release data related to the SCADA network only on a strict, need-to-know basis, and only to people authorized to receive such information. "Social engineering," gathering information about a computer or computer network via questions to naïve users, is often the first step in a malicious attack on computer networks. The more information revealed about a laptop or computer network, the more vulnerable the computer/network is. Never divulge data revealed to a SCADA network, including the names and contact information about the system operators/administrators, computer operating systems, and/or physical and logical locations of computers and network systems over telephones or to personnel unless they are explicitly authorized to receive such information. Any requests for information by unknown people need to be sent to a central network security location for verification and fulfillment. People can be a weak link in an otherwise secure network. Conduct training and information awareness campaigns to ensure personnel remain diligent in guarding sensitive network information, notably their passwords.

19.5 THE BOTTOM LINE

Because AI is both a shield and a sword, we must think far and continuously ask: what if? We must determine whether to use consultants to consider the advice, incorporate it into the vulnerability analysis process, and create compelling and synergistic remediation projects. Energy production, supply, and storage managers must carefully develop remediation plans to support the site's mission and function. These managers must maintain ongoing reviews and prepare updates of the energy security plan.

The real bottom line: do not underestimate the crucial importance of information security throughout the assessment process.

REFERENCES

DOE (2001). *21 Steps to Improve Cyber Security of SCADA Networks.* Washington, DC: Department of Energy.

FBI (2007). *Ninth Annual Computer Crime and Security Survey.* Washington, DC: FBI: Computer Crime Institute and Federal Bureau of Investigations, U.S. Department of Justice.

GAO (2015). *Critical Infrastructure Protection: Sector-Specific Agencies Need to Better Measure Cybersecurity Progress.* Washington, DC: United States Government Accountability Office.

Stamp, J. et al. (2003). *Common Vulnerabilities in Critical Infrastructure Control Systems*, 2nd ed. Sandia National Laboratories, Sandia, White Sands, NM.

USEPA (2005). *EPA Needs to Determine What Barriers Prevent Water Systems from Securing Known SCADA Vulnerabilities.* In: Harris, J. Final Briefing Report. Washington, DC: USEPA.

Wiles, J. et al. (2007). *Techno Security's™ Guide to Securing SCADA.* Burlington, MA: Elsevier, Inc.

20 Security Techniques and Hardware

Protecting employees and members of the public who visit any facility is a complex and challenging responsibility. It is also one of the company's top priorities. If your facility still uses key locks, does anyone control those keys? Does anyone know who has the keys? Does anyone know how many keys have been issued? Is there a record of each key that has been issued?

The bottom line: I know your password…therefore, you belong to me.

20.1 CYBER-PHYSICAL SECURITY PROVISIONS

When walking into any business or other pedestrian-entry energy production, supply, or storage system, we usually pay little attention to our surroundings. Why? Well, we become used to the surroundings and the ambience through our continued usage of such a place or from the perusal of such services. More than likely, we are in a hurry, as usual, and the surroundings are not that important to us; instead, the business at hand is much more critical.

Let's say on the other hand, we are part of that group of people who take it all in, no matter where or when. If this is the case, a few things stand out to us when we walk into the business building. First, it is usually busy, crowded, and hectic. The standard cashier greets us as we stand to pay for whatever goods we have purchased. While the cashier totals our purchases, we might glance to our right or left or above the place we are standing; there we would probably notice the CCTV security cameras here and there, some pointed right at us. We presume that the teller behind the teller window is nearby and easily reachable by an alarm or panic button, which will sound the alarm and hopefully send security and the police scrambling to respond.

These security devices are apparent to us, or at least we assume they are present. However, what we do not see, anticipate, expect, comprehend, or think is present (or should be) is really what the emphasis is on herein. Note that many devices and security systems are available to protect energy sector assets that are not open to pedestrian traffic, and fewer are available to energy facilities or facilities that use Artificial Intelligence (AI)/Information Technology (IT) services that are open to pedestrian traffic (malls, stores, music halls, sports stadiums, businesses, etc.). Many are not readily visible to most people in locations where various security devices are viable. However, remember that some devices and security techniques discussed in the following are only feasible in IT assets that are not open to pedestrian traffic.

The bottom line: The physical security of any facility depends on several security decisions that can be identified through a comprehensive risk-management process. Risk management aims to identify an achievable level of protection for a company that is as comprehensive as possible to the level of risk without exceeding it. The achievable level of protection that should be sought and pursued when dealing with minimizing and safeguarding printed materials with sensitive information, ensuring mail security, disposing of trash securely, disposing of electronic equipment securely, and training employees in physical facility security procedures. In this chapter, we focus on the latter, facility security procedures. We can't expect AI/IT as applied to energy sector equipment and operations to be secure unless the physical structure housing such items and functions is secure.

DOI: 10.1201/9781003637851-23

20.2 THE MULTIPLE-BARRIER APPROACH

Ideally, in a perfect world, all energy sector physical sites/facilities/assets would be secured in a layered fashion (aka the multiple barrier approach). Layer security systems are vital. Using the protection "in depth" principle, requiring an adversary to defeat several protective barriers or security layers to accomplish its goal, the financial services sector's physical infrastructure can be more secure. Protection in depth is a term commonly used by the military to describe security measures that reinforce one another, masking the defense mechanisms from the view of intruders, and allowing the defender time to respond to intrusion or attack.

A prime example of using the multi-barrier approach to ensure the practices of the bottled water industry demonstrate security and safety. In the aftermath of 9/11 and the increased emphasis on Homeland Security, a shifted paradigm of national security and vulnerability awareness has emerged. Recall that in the immediate aftermath of the 9/11 tragedies, emergency responders and others responded quickly and worked to exhaustion. In addition to the emergency responders, bottled water companies responded immediately by donating several million bottles of water to the crews at the crash sites in New York, at the Pentagon, and in Pennsylvania. International Bottled Water Association (IBWA, 2004) reports that "within hours of the first attack, bottled water was delivered where it mattered most; to emergency personnel on the scene who required ample water to stay hydrated as they worked to rescue victims and clean up debris" (p. 2).

Bottled water companies continued to provide bottled water to responders and rescuers at the 9/11 sites throughout the post-event process. However, these patriotic actions by the bottled water companies raise the question: How do we ensure the safety and security of the bottled water provided to anyone? IBWA (2004) have the answer: Using a multi-barrier approach, along with other principles, will enhance the safety and security of bottled water. IBWA (2004) describe their multi-barrier approach as follows:

> **A multi-barrier approach**—Bottled water products are produced utilizing a multi-barrier approach, from source to finished product, that helps prevent possible harmful contaminants (physical, chemical, or microbiological) from adulterating the finished product as well as storage, production, and transportation equipment. Measures in a multi-barrier approach may include source protection, source monitoring, reverse osmosis, distillation, filtration, ozonation, or ultraviolet (UV) light. Many steps in a multi-barrier system may effectively safeguard bottled water from microbiological and other contamination. Piping in and out of plants, as well as storage silos and water tankers are also protected and maintained through sanitation procedures. In addition, bottled water products are bottled in a controlled, sanitary environment to prevent contamination during the filling operation.
>
> **p. 3**

In-depth protection describes a layered security approach for areas not open to pedestrian traffic in the energy sector security infrastructure. An in-depth protection strategy uses several security techniques and/or devices against an intruder and does not rely on a single defensive mechanism to protect infrastructure. By implementing multiple layers of security, a hole or flaw in one layer is covered by the other layers. An intruder will have to intrude through each layer without being detected in the process—the layered approach implies that no matter how an intruder attempts to accomplish his goal, he will encounter practical elements of the physical protection system.

For example, an adequate security layering approach requires that an adversary penetrate multiple, separate barriers to gain entry to a critical TARGET at an energy sector facility. Protection in depth (multiple layers of security) helps to ensure that the security system remains effective in the event of a failure or an intruder bypassing a single layer of security.

Again, layered security starts with the outer perimeter (the fence—the first line of physical security) of the facility and goes inward to the facility, the buildings, structures, other individual assets, and finally to the contents of those buildings—the TARGETs.

The area between the outer perimeter and structures or buildings is known as the site. This open site area provides an incomparable opportunity for early identification of an unauthorized intruder and initiation of early warning/response. This open space area is commonly used to calculate the standoff distance; that is, it is the distance between the outside perimeter (public areas to the fence) to TARGET or critical assets (buildings/structures) inside the perimeter (inside the fence line—the restricted access area).

The open area, between the perimeter fence and Target (e.g., operations center), if adequately outfitted with various security devices, can also provide layered protection against intruders. For example, lighting is a deterrent. Based on personal experience, an open area within the plant site that is almost as well-lit at night as would be expected during daylight hours is the rule of thumb. In addition, strategically placed motion detectors and crash barriers at perimeter gate openings and in front of vital structures are recommended. Armed mobile guards who roam the interior of the plant site regularly provide the ultimate security in the site area.

The next layer of physical security is the outside wall of the target structure itself. Notwithstanding the door, window, and/or skylight entry, walls prevent most intruders from entering quickly. If doors can only be entered using card reader access, security is shored up or enhanced to an extent. The same can be said for windows and skylights that are small enough to prohibit regular human entry. Buildings' same "weak" spots can be bastioned with break-proof or reinforced security glass.

Properly designed interior features of buildings provide the final layer of security. Examples include internal doors and walls, equipment cages, and backup or redundant equipment.

In the preceding discussion, the conditions described referred to perfect world conditions; that is, to those conditions that we "would want" (i.e., the security manager's proverbial wish list) to be incorporated into the design and installation of new financial services sector infrastructure. Post 9/11, in a not-so-perfect world, however, many of the peripheral (fence line) measures described above are more difficult to incorporate into the financial services sector infrastructure. This is not to say that energy sector sites and facilities do not have fence lines or fences; many of them do. These fences are designed to keep vandals, thieves, and trespassers out. The problem is not only the fact that many of these facilities were constructed several years ago, before urban encroachment encircled the many sites, allowing, at present, little room for security stand backs or setbacks to be incorporated into electrical power stations, plants, and/or critical equipment locations. Based on personal observation, many of these fences face busy city streets or closely abut structures outside the fence line. The point is that when one sits down to plan a security upgrade, these factors must be considered.

The fly in the ointment for the energy sector assets is accessibility. People who use Energy Services in person must use the facilities to conduct business. Thus, the multiple barrier approach to security in a mall, for example, is unrealistic.

Managers of the energy sector infrastructure have four primary security areas to manage. These security areas are listed and described below.

- *Physical security*: In the energy sector, where practicable, physical security techniques and practices have the most effect at "fenced" locations. A systems approach is best at such places, where resources, procedures, and policies plan and support detection, assessment, communication, and response.
- *Cyber/information technology security*: Only using key operating systems that have been appropriately vetted and scrubbed of alleged Chinese and/or Russian Trojan Horses hacked into the North American electrical grid is essential. The only positive way to ensure the security of the North American Grid is to disconnect its cyber and other digital systems from the Internet. This step is impractical, but it points to the need to conduct frequent system audits and install firewall protection in digital components and other systems to prevent hacking. Frequent third-party penetration testing is advised.

- *Employment screening*: Mitigates the threat from the enemy at the water cooler (inside the organization). We are always amazed whenever we conduct security audits for various companies. Often, a simple check, such as reviewing an employee's driving record, reveals that the employee has no license, is driving on a suspended license, or has a horrific driving record. Hiring standards and pre-employment background investigations may help ensure the trustworthiness and reliability of personnel with unescorted access to critical facilities.
- *Protecting potentially sensitive information*: The old saying that goes—a secret can best be held between three people so long as two of the three are dead—makes the point that reducing the likelihood that information could be used by those intent on disrupting operations or causing death and destruction in Energy Services Sector sites is crucial. Information should only be shared within an organization on a need-to-know basis.

For existing facilities, security upgrades should be based on the results generated from the vulnerability assessment, which characterizes and prioritizes the targeted assets. Those vulnerabilities identified must be protected.

The following sections describe various security hardware and/or devices. These devices serve the primary purpose of providing security against physical and/or digital intrusion. They are designed to delay and deny intrusion and are generally coupled with detection and assessment technology. Possible additional security measures, based on the vulnerability assessment that may be recommended (covered in this text), include the following (NAERC, 2002):

- Electronic security
- Closing non-essential perimeter and internal portals
- Physical barriers such as bollards or Jersey walls
- Fencing
- Lighting
- Security surveys
- Vulnerability assessments
- Availability of security resources
- General personnel and security officer training
- Law Enforcement Liaison
- Ensure availability of essential spare parts (machines, repair parts, wire, pipe, valves, transformers, etc.) for critical facilities

However, remember that no matter the type of security device or system employed, financial services sector systems cannot be immune to all possible intrusions or attacks. Whenever a facility safety/security manager tells us that he or she has secured their site 100%, we are reminded of Schneier's (2000) view of security: "You can't defend. You can't prevent it. The only thing you can do is detect and respond." When it comes to making "anything" absolutely secure from intrusion or attack, there is inherently, or otherwise, no silver bullet.

In the next section, security hardware devices are discussed in detail. Before we describe these hardware devices, remember that in addition to security hardware devices to help protect and monitor Energy Services Sector assets, a few employee practices and actions can be taken to preserve Energy Services assets. For example, how is it disposed of when a commercial asset computer system fails and must be disposed of? Is there a procedure or practice to prevent valuable information on the system's hard drive from being pulled from a trash heap or a dumpster dive and used by potential enemies? Are shredders used? Are they state-of-the-art shredders that prevent scraps from being reassembled by enemies? Are building cleaning crews properly vetted and supervised? Do you have a team that routinely inspects suspended ceilings for bugs, cameras, and listening recorders? Do you have key stroke reader capability; that is, can you record what messages are being sent by employees? Do you routinely check hard-wired phone lines? Have you removed all door signs that tell anyone what is on the other side of the room? Have you trained your employees to be slightly suspicious of just about anything and everything? Have you opened your manholes lately to see what is inside?

All the practices just mentioned do not require security hardware such as barriers, motion detectors, fences, locks, biometric systems, video cameras, armed guards, and electrified fences. Instead, they require common sense, awareness, and alert and engaged supervisors and employees. The point is that what is heard at the water cooler is sometimes more significant than any security alarm apparatus can provide.

Again, remember that many of the security hardware devices described in the following only apply to non-pedestrian facilities; we cannot barricade or fence in sports arenas and shopping malls, for example.

20.3 SECURITY HARDWARE DEVICES[1]

USEPA (2005) groups infrastructure security devices or products described below into four general categories:

- Physical asset monitoring and control devices
- Cyber protection devices
- Communication/integration
- Environmental monitoring devices

20.3.1 Physical Asset Monitoring and Control Devices

20.3.1.1 Aboveground, Outdoor Equipment Enclosures

Energy Services Sector facilities and sites can have multiple structural components spread over a wide area. They typically include a centralized production, storage, and distribution center and component supply facilities distributed throughout the region. One of the primary reasons for constructing structural components is that house operational equipment aboveground eliminates the safety risks associated with confined space entry, which is often required to maintain equipment located below ground. In addition, space restrictions usually limit the amount of equipment that can be located inside. There are concerns that some types of equipment (such as backflow prevention devices—to prevent chemicals and fuel waste from entering plant and offsite potable water systems) can, under certain circumstances, discharge fuel slurry or waste mixtures that could flood pits, vaults, or equipment rooms. Regarding electrical power, electrical substations are not usually suited for underground installation. Therefore, many critical electrical equipment pieces are outdoors and aboveground in properly fenced, insulated, or isolated configurations to prevent accidental electrical shock or short circuits/fires in equipment.

Experience demonstrates that many different system components can be and are often installed outdoors and aboveground, many controlled by wireless communication devices. Examples of the types of components that could be included in commercial facilities or their ancillaries are:

- Backflow prevention devices
- Air release and control valves
- Pressure vacuum breakers
- Oil and gas pumps and motors
- Petrochemical storage and feed equipment
- Meters
- Sampling equipment
- Instrumentation
- Electrical substations
- Oil and natural gas pipelines

One of the most effective security measures for protecting aboveground equipment, where feasible, is to place it inside a building or an exterior fenced structure. When/where this is not possible, enclosing the equipment or parts of the equipment using some commercial or homemade add-on structure may help to prevent tampering with the equipment. These types of add-on structures or enclosures, which are designed to protect people and animals from electrocution and to protect equipment both from the elements and unauthorized access or tampering, typically consist of a box-like fenced structure that is placed over or around the entire component, or over/around critical parts of the element (i.e., valves), and is then secured to delay or prevent intruders from tampering with the equipment. The enclosures are typically locked or otherwise anchored to a solid foundation, which makes it difficult for unauthorized personnel to remove the enclosures and access the equipment.

Standardized aboveground enclosures are available in various materials, sizes, and configurations. Many options and security features are also available for each type of enclosure, allowing system operators to customize an enclosure for a specific application and/or price range. In addition, most manufacturers can custom-design enclosures if standard, off-the-shelf enclosures do not meet users' needs.

Many of these enclosures are designed to meet specific standards. For example, the American Society of Sanitary Engineers (ASSE) has developed Standard #1060, *Performance Requirements for Outdoor Enclosures for Backflow Prevention Assemblies*. Suppose an enclosure will be used to house a backflow preventer. In that case, this standard specifies the acceptable construction materials for the enclosure and the performance requirements that the enclosure should meet, including specifications for freeze protection, drainage, air inlets, access for maintenance, and hinge requirements. ASSE #1060 also states that the enclosure should be lockable to enhance security.

Electrical substations and electrical equipment enclosures must meet the requirements and recommendations of various OSHA Standards, the National Fire Protection Association (NFPA), the National Electrical Codes (NEC), the Institute of Electrical and Electronics Engineers (IEEE), and local code requirements.

Equipment enclosures can generally be categorized into one of four main configurations, which include:

- One piece, drop over enclosures
- Hinged or removable top enclosures
- Sectional enclosures
- Shelters with access locks

All enclosures, including those with integral floors, must be secured to a foundation to prevent them from being moved or removed. Un- or poorly anchored enclosures may be blown off the protected equipment or defeated by intruders. In either case, this may result in the equipment beneath the enclosure becoming exposed and damaged. Therefore, ensuring that the enclosure is securely anchored will increase the security of the protected equipment.

The three basic foundation types that can anchor the aboveground equipment enclosure are concrete footers, concrete slabs-on-grade, or manufactured fiberglass pads. The most common foundations utilized for equipment enclosures are standard or slab-on-grade footers; however, local climate and soil conditions may dictate whether either can be used. These foundations can be either precast or poured in place at the installation site. Once the foundation is installed and appropriately cured, the equipment enclosure is bolted or anchored to secure it.

An alternative foundation, specifically for use with smaller Hot Box® enclosures, is a manufactured fiberglass pad known as the Glass Pad™. The Glass Pad™ has the center cut out so it can be dropped directly over the enclosed equipment. Once the pad is set level on the ground, it is backfilled over a two-inch flange around its base. Depending on the enclosure configuration, the enclosure is placed on top of the foundation and is locked in place with either a staple or a slotted anchor.

One of the primary attributes of a security enclosure is its strength and resistance to breaking and penetration. Accordingly, the materials from which the enclosure is constructed will be important in determining the strength of the enclosure, and thus, its usefulness for security applications. Enclosures are typically manufactured for either fiberglass or aluminum. Except for the one-piece, drop-over enclosure, which is usually fabricated from fiberglass, each configuration described above can be constructed from either material. In addition, enclosures can be custom-manufactured from polyurethane, galvanized steel, or stainless steel. Galvanized or stainless steel is often offered as an exterior layer, or "skin," for an aluminum enclosure. Although they are typically utilized in underground applications, precast concrete structures can also be used as aboveground equipment enclosures. However, precast structures are much heavier and more challenging to maneuver than their fiberglass and aluminum counterparts. Concrete is also brittle, which can be a security concern. However, products can be applied to concrete structures to strengthen and minimize security risks (i.e., epoxy coating). Because precast concrete structures can be purchased from any concrete producer, this document does not identify specific vendors for these types of products.

In addition to the construction materials, enclosure walls can be configured or reinforced to give them added strength. Adding insulation is one option that can strengthen the structural characteristics of an enclosure; however, some manufacturers offer additional features to add strength to exterior walls. For example, while most enclosures are fabricated with flat wall construction, some vendors manufacture fiberglass shelters with ribbed exterior walls. These ribs increase the wall's structural integrity and allow the fabrication of standard shelters up to 20 feet in length. Another vendor has developed a proprietary process that uses a series of integrated fiberglass beams placed throughout a foam inner core to tie together the interior and exterior walls and roof. Yet another vendor constructs aluminum enclosures with horizontal and vertical redwood beams for structural support.

Other security features that can be implemented aboveground, such as outdoor equipment enclosures, include locks, mounting brackets, tamper-resistant doors, and exterior lighting.

20.3.1.2 Active Security Barriers (Crash Barriers)

Terrorist vehicle attacks are on the rise. Consider the Halloween 2017 truck attack in New York City that killed eight and injured many. We can take steps to lessen the possibility of many of these attacks. For example, active security barriers (also known as crash barriers) are large structures placed in roadways at entrance and exit points to protect facilities and control vehicle access to these areas. These barriers are placed perpendicular to traffic to block the roadway, so the only way traffic can pass is to move the barrier out of the roadway. These barriers are typically constructed from sturdy materials, such as concrete or steel, so vehicles cannot penetrate them. They are also designed at a certain height off the roadway so cars cannot cross them.

The key difference between active security barriers, which include wedges, crash beams, gates, retractable bollards, and portable barricades; and passive security barriers, which include non-moveable bollards, jersey barriers, and planters, is that active security barriers are designed so that they can be raised and lowered or moved out of the roadway easily to allow authorized vehicles to pass them. Many of these barriers are designed to be opened and closed automatically (i.e., mechanized gates, hydraulic wedge barriers). In contrast, others easily open and close manually (swing crash beams, manual gates). In contrast to active barriers, passive barriers are permanent, non-movable barriers. Thus, they are typically used to protect the perimeter of a protected facility, such as sidewalks and other areas that do not require vehicular traffic to pass through. Several major active security barriers, such as wedge barriers, crash beams, gates, bollards, and portable/removable barricades, are described below.

Wedge barriers are plated, rectangular steel buttresses approximately 2–3 feet high that can be raised and lowered from the roadway. When in the open position, they are flush with the roadway, and vehicles can pass over them. However, in the closed (armed) position, they project up from the road at a 45-degree angle, with the upper end pointing toward the oncoming vehicle and the base of the barrier away from it. Generally, wedge barriers are constructed from heavy-gauge steel or concrete containing an impact-dampening iron rebar core that is strong and resistant to breaking or cracking, allowing them to withstand the impact from a vehicle attempting to crash through them. In addition, both of these materials help transfer the impact's energy over the barrier's entire volume, thus helping to prevent the barrier from being sheared off its base. In addition, because the barrier is angled away from traffic, the force of any vehicle impacting the barrier is distributed over the entire surface of the barrier and is not concentrated at the base, which helps prevent the barrier from breaking off at the base. Finally, the angle of the barrier helps hang up any vehicles attempting to drive over it.

Wedge barriers can be fixed or portable. Fixed wedge barriers can be mounted on the roadway's surface ("surface-mounted wedges"), in a shallow mount on the road's surface, or installed entirely below the road surface. Surface-mounted wedge barricades operate by rising from a flat position on the roadway's surface. In contrast, shallow-mount wedge barriers rise from their resting position below the road surface. In comparison, below-surface wedge barriers operate by rising from beneath the road surface. The shallow-mounted and surface-mounted barriers require little or no excavation and thus do not interfere with buried utilities. All three barrier mounting types project above the road surface and block traffic when raised into the armed position. Once they are disarmed and lowered, they are flush with the road, allowing traffic to pass. Portable wedge barriers that are moved into place on wheels are removed after the barrier has been set into place.

Installing rising wedge barriers requires the preparation of the road surface. Installing surface-mounted wedges does not require excavating the road; however, the road surface must be intact and strong enough to allow the bolts anchoring the wedge to the road surface to attach correctly. Shallow-mount and below-surface wedge barricades require excavation of a large pit to accommodate the wedge structure and any arming/disarming mechanisms. Generally, the bottom of the excavation pit is lined with gravel to allow for drainage. Areas not sheltered from rain or surface runoff can install a gravity drain or a self-priming pump. Table 20.1 lists the pros and cons of wedge barriers.

TABLE 20.1
Pros and Cons of Wedge Barriers

Pros	Cons
Can be surface-mounted or completely installed below the roadway surface.	Installations below the surface of the roadway will require construction that may interfere with buried utilities.
Wedge barriers have a quick response time (usually 3.5–10.5 seconds, but the barrier can be 1–3 seconds in emergency situations. Because emergency activation of the barrier causes more wear and tear on the system than does normal activation, it is recommended for use only in true emergencies.	Regular maintenance is needed to keep the wedge fully operational.
Surface or shallow-mount wedge authorized barricades are being utilized in locations damaged. With a high-water table and/or corrosive soils.	Improper use of the system may result in authorized vehicles being hung up by the barrier and damaged. Guards must be trained to use the system properly to ensure that this does not happen. Safety technologies may also be installed to reduce the risk of the wedge activating under an authorized vehicle.
All three wedge barrier designs have a high crash rating, thereby allowing them to be employed for higher security applications.	
These types of barriers are extremely visible, which may deter potential intruders.	

Source: USEPA (2005).

Crash beam barriers consist of aluminum beams that can be opened or closed across the roadway. While there are several different crash beam designs, every crash beam system consists of an aluminum beam supported on each side by a solid footing or buttress, which is typically constructed from concrete, steel, or some other strong material. Beams usually contain an interior steel cable (generally at least one inch in diameter) to give the beam added strength and rigidity. A heavy-duty hinge or other mechanism connects the beam to one of the footings to swing or rotate out of the roadway when it is open. It can swing back across the road when it is in a closed (armed) position, blocking the road and inhibiting access by unauthorized vehicles. The non-hinged end of the beam can be locked into its footing, thus providing anchoring for the beam on both sides of the road and increasing its resistance to any vehicle attempting to penetrate through it. In addition, if the crash beam is hit by a car, the aluminum beam transfers the impact energy to the interior cable, which in turn transfers the impact energy through the footings and into their foundation, thereby minimizing the chance that the impact will snap the beam and allow the intruding vehicle to pass through.

Crash beam barriers can employ drop-arm, cantilever, or swing beam designs. Drop-arm crash beams operate vertically, raising and lowering the beam across the road. Cantilever crash beams are projecting structures that are opened and closed by extending the beam from the hinge buttress to the receiving buttress on the opposite side of the road. In the swing beam design, the beam is hinged to the buttress so it swings horizontally across the road. Generally, swing beam and cantilever designs are used at locations where a vertical lift beam is impractical. For example, the swing beam or cantilever designs are utilized at entrances and exits with overhangs, trees, or buildings that would physically block the operation of the drop-arm beam design.

Installing any of these crash beam barriers involves excavating a pit approximately 48 inches deep for both the hinge and the receiver footings. Due to the depth of the excavation, the site should be inspected for underground utilities before digging begins. Table 20.2 lists the pros and cons of crash beams.

In contrast to wedge barriers and crash beams, which are typically installed separately from a fence line, *gates* are often integrated units of a perimeter fence or wall around a facility.

Gates are movable pieces of fencing that can be opened and closed across a road. When the gate is closed (armed), the leaves lock into steel buttresses embedded in the concrete foundation on both sides of the roadway, thereby blocking access to the roadway. Generally, gate barricades are constructed from a combination of heavy-gauge steel and aluminum that can absorb an impact from vehicles attempting to ram through them. Any remaining impact energy not absorbed by the gate material is transferred to the steel buttresses and their concrete foundation.

TABLE 20.2
Pros and Cons of Crash Beams

Pros	Cons
Requires little maintenance, while providing long-term durability.	Crash beams have a slower response time (normally 9.5–15.3 seconds but can be reduced to 7–10 seconds in emergency situations) than do other types of active security barriers, such as wedge barriers. Because emergency activation of the barrier causes more wear and tear on the system than does normal activation, it is recommended for use only in true emergency situations.
No excavation is required in the roadway itself to install crash beams.	All three crash beam designs possess a low crash rating relative to other types of barriers, such as wedge barriers, such as wedge barriers, and thus they typically are used for lower security applications.
	Certain crash barriers may not be visible to oncoming traffic and therefore may require additional lighting and/or other warning markings to reduce the potential for traffic to accidentally run into the beam.

Source: USEPA (2005).

Gates can utilize cantilever, linear, or swing designs. Cantilever gates are projecting structures that operate by extending the gate from the hinge footing across the roadway to the receiver footing. A linear gate is designed to slide across the road on tracks via a rack-and-pinion drive mechanism. Swing gates are hinged so that they can swing horizontally across the road.

Installation of the cantilever, linear, or swing gate designs described above involves excavating a pit approximately 48 inches deep for both the hinge and receiver footings to which the gates are attached. Due to the depth of the excavation, the site should be inspected for underground utilities before digging begins. Table 20.3 lists the pros and cons of gates.

Bollards are vertical barriers at least 3 feet tall and 0.4–2 feet in diameter that are typically set 4–5 feet apart to block vehicles from passing between them. Smaller bollards, usually 4 inches diameter pipes filled with concrete, are installed in parking areas to prevent vehicles from striking walls or windows or to protect walkway areas. Bollards can be fixed in place, removable, or retractable. Fixed and removable bollards are passive barriers that are typically used along building perimeters or on sidewalks to prevent vehicles from entering, while allowing pedestrians to pass. In contrast to passive bollards, retractable bollards are active security barriers that can easily be raised and lowered to allow vehicles to pass between them. Thus, they can be used in driveways or on roads to control vehicular access. When the bollards are raised, they protect above the road surface and block the roadway; when lowered, they sit flush with the road surface and thus allow traffic to pass over them. Retractable bollards are typically constructed from steel or other materials with a low weight-to-volume ratio, requiring low power to raise and lower. Steel is also more resistant to breaking than a more brittle material, such as concrete, and can better withstand direct vehicular impact without breaking apart.

Retractable bollards are installed in a trench dug across a roadway, typically at an entrance or gate. Installing retractable bollards requires preparing the road surface. Depending on the vendor, bollards can be installed in a continuous slab of concrete or in individual excavations with concrete poured in place. The required excavation for a bollard is typically slightly broader and slightly more profound than the bollard height when extended above ground. The bottom of the excavation

TABLE 20.3
Pros and Cons of Gates

Pros	Cons
All three gate designs possess an intermediate crash rating, thereby allowing them to be utilized for medium, to higher security applications.	Gates have a slower response time (normally 10–15 seconds, but can be reduced to 7–10 seconds in emergency situations) than do other types of active security barriers, such as wedge barriers. Because emergency activation of the barrier causes more wear and tear on the system than does normal activation, it is recommended for use only in true emergency situations.
Requires very little maintenance.	
Can be tailored to blend in with perimeter fencing.	
Gate construction requires no roadway excavation.	
Cantilever gates are useful for roads with high crowns or drainage gutters.	
These types of barriers are extremely visible, which may deter intruders.	
Gates can also be used to control pedestrian traffic.	

Source: USEPA (2005).

TABLE 20.4
Pros and Cons of Retractable Bollards

Pros	Cons
Bollards have a quick response time (normally 3–10 seconds, but can be reduced to 1–3 seconds in emergency situations).	Bollard installations will require construction below the surface of the roadway, which may interfere with buried utilities.
Bollards have an intermediate crash rating, which allows them to be utilized for medium to higher security applications.	Some maintenance is needed to ensure the barrier is free to move up and down.
	The distance between bollards must be decreased (i.e., more bollards must be installed along the same perimeter) to make these systems effective against small vehicles (i.e., motorcycles).

Source: USEPA (2005).

is usually lined with gravel to allow drainage. The bollards are then connected to a control panel, which controls the raising and lowering of the bollards. Installation typically requires mechanical, electrical, and concrete work; if utility personnel with these skills are available, the utility can install the bollards themselves. Table 20.4 lists the pros and cons of retractable bollards.

Portable/removable barriers, including removable crash beams and wedge barriers, are mobile obstacles that can be moved in and out of position on a roadway. For example, a crash beam may be removed entirely and stored off-site when unnecessary. Wedge barriers equipped with wheels that can be removed after the barricade is towed into place are another example.

When portable barricades are needed, they can be moved into position rapidly. To provide them with added strength and stability, they are typically anchored to buttress boxes that are located on either side of the road. These buttress boxes, which may or may not be permanent, are usually filled with sand, water, cement, gravel, or concrete to make them heavy and aid in stabilizing the portable barrier. In addition, these buttresses can help dissipate any impact energy from vehicles crashing into the barrier itself.

Because these barriers are not anchored into the roadway, they do not require excavation or other related construction for installation. In contrast, they can be assembled and made operational in a short period of time. The primary shortcoming of this type of design is that these barriers may move if vehicles hit them. Therefore, it is essential to carefully assess the placement and anchoring of these types of obstacles to ensure they can withstand the kinds of impacts that may be anticipated at that location. Table 20.5 lists the pros and cons of portable/removable barricades.

TABLE 20.5
Pros and Cons of Portable/Removable Barricades

Pros	Cons
Installing portable barricades requires no foundation or roadway excavation.	Portable barriers may move slightly when hit by a vehicle, resulting in lower crash resistance.
Can be moved in and out of position in a short period of time.	Portable barricades typically require 7.75–16.25 seconds to move into place, and thus, they are considered to have a medium response time when compared with other active barriers.
Wedge barriers equipped with wheels can be easily towed into place.	
Minimal maintenance is needed to keep barriers fully operational.	

Source: USEPA (2005).

Because the primary threat to active security barriers is vehicles attempting to crash through them, their most important attributes are their size, strength, and crash resistance. Other essential features of an active security barrier are the mechanisms by which the barrier is raised and lowered to allow authorized vehicle entry, and other factors such as weather resistance and safety features.

20.3.1.3 Alarms

An *alarm system* is an electronic monitoring system used to detect and respond to specific events, such as unauthorized access to an asset or a possible fire. In chemical processing systems, alarms are also used to alert operators when process operating or monitoring conditions go out of preset parameters (i.e., process alarms). These alarms are primarily integrated with process monitoring and reporting systems (i.e., SCADA systems). Note that this discussion does not focus on alarm systems unrelated to a facility's processes.

Alarm systems can be integrated with fire detection systems, intrusion detection systems (IDSs), access control systems, or closed-circuit television (CCTV) systems, such that these systems automatically respond when the alarm is triggered. For example, a smoke detector alarm can be set up to notify the fire department when smoke is detected automatically, or an intrusion alarm can automatically trigger cameras to turn on in a remote location so that personnel can monitor that location.

An alarm system consists of sensors that detect different types of events; an arming station that is used to turn the system on and off; a control panel that receives information, processes it, and transmits the alarm; and an annunciator that generates a visual and/or audible response to the alarm. When a sensor is tripped, it sends a signal to a control panel, which triggers a visual or audible alarm and/or notifies a central monitoring station. A more complete description of each of the components of an alarm system is provided below.

Detection devices (also called *sensors*) are designed to detect a specific type of event (such as smoke and intrusion). Depending on the type of event they are designed to detect, sensors can be located inside or outside the facility or other asset. When an event is detected, the sensors use a communication method (such as wireless radio transmitters, conductors, or cables) to send signals to the control panel to generate the alarm. For example, a smoke detector sends a signal to a control panel when it detects smoke.

Alarms use either normally closed (NC) or normally open (NO) electric loops, or "circuits," to generate alarm signals. These two types of circuits are discussed separately below.

In NC loops or circuits, all of the system's sensors and switches are connected in series. The contacts are "at rest" in the closed (on) position, and current continually passes through the system. However, when an event triggers the sensor, the loop is opened, breaking the flow of current through the system and triggering the alarm. NC switches are used more often than are NO switches because the alarm will be activated if the loop or circuit is broken or cut, thereby reducing the potential for circumventing the alarm. This is known as a "supervised" system.

In NO loops or circuits, all of the system's sensors and switches are connected in parallel. The contacts are "at rest" in the open (off) position, and no current passes through the system. However, when an event triggers the sensor, the loop is closed. This allows current to flow through the loop, powering the alarm. NO systems are not "supervised" because the alarm will not be activated if the loop or circuit is broken or cut. However, adding an end-of-line resistor to an NO loop will cause the system to alarm if tampering is detected.

An *arming station*, which is the main user interface with the security system, allows the user to arm (turn on), disarm (turn off), and communicate with the system. How a specific system is armed will depend on how it is used. For example, while IDSs can be armed for continuous operation (24 hours/day), they are usually armed and disarmed according to the work schedule at a specific location so that personnel going about their daily activities do not set off the alarms. In contrast, fire protection systems are typically armed 24 hours/day.

A *control panel* receives information from the sensors and sends it to an appropriate location, such as a central operations station or a 24-hour monitoring facility. Once the alarm signal is received at the central monitoring location, personnel monitoring for alarms can respond (such as sending security teams to investigate or dispatching the fire department).

An *annunciator* responds to the detection of an event by emitting a signal. This signal may be visual, audible, electronic, or a combination. For example, fire alarm signals will always be connected to audible annunciators, whereas intrusion alarms may not be.

Alarms can be reported locally, remotely, or both locally and remotely. Below, we discuss local and remotely (centrally) reported alarms in more detail.

A *local alarm* emits a signal at the event's location (typically using a bell or siren). A "local only" alarm emits a signal at the event's location. Still, it does not transmit the alarm signal to any other location (i.e., it does not transmit the alarm to a central monitoring location). Typically, the purpose of a "local only" alarm is to frighten away intruders, and possibly to attract the attention of someone who might notify the proper authorities. Because no signal is sent to a central monitoring location, personnel can only respond to a local alarm if they are in the area and can hear and/or see the alarm signal.

Fire alarm systems must have local alarms, including audible and visual signals. Most fire alarm signal and response requirements are codified in the National Fire Alarm Code, NFPA 72. NFPA 72 discusses the application, installation, performance, and maintenance of protective signaling systems and their components. In contrast to fire alarms, which require a local signal when fire is detected, many IDSs do not have a local alert device, because monitoring personnel do not wish to inform potential intruders that they have been detected. Instead, these systems silently alert monitoring personnel that an intrusion has been detected, thus allowing monitoring personnel to respond.

In contrast to systems that are set up to transmit "local only" alarms when the sensors are triggered, systems can also be set up to transmit signals to a *central location*, such as to a control room or guard post at the utility, or to a police or fire station. Most fire/smoke alarms are set up to signal both at the location of the event and at a fire station or central monitoring station. Many insurance companies require that facilities install certified systems that include alarm communication to a central station. For example, systems certified by the Underwriters Laboratory (UL) require that the alarm be reported to a central monitoring station.

The main differences between alarm systems lie in the types of event detection devices used in different systems. *Intrusion sensors*, for example, consist of two main categories: perimeter sensors and interior (space) sensors. *Perimeter intrusion sensors* are typically applied on fences, doors, walls, windows, etc., and are designed to detect an intruder before he/she accesses a protected asset (i.e., perimeter intrusion sensors are used to detect intruders attempting to enter through a door, and window). In contrast, *interior intrusion sensors* are designed to detect an intruder who has already accessed the protected asset (i.e., interior intrusion sensors are used to detect intruders once they are already within a protected room or building). These two types of detection devices can be complementary, and they are often used together to enhance security for an asset. For example, a typical intrusion alarm system might employ a perimeter glass-break detector that protects against intruders accessing a room through a window and an ultrasonic interior sensor that detects intruders who have entered the room without using the window. Table 20.6 lists and describes types of perimeter and interior sensors.

Fire Detection/Fire Alarm Systems consist of different fire detection devices and fire alarm systems available. These systems may detect fire, heat, smoke, or a combination of any of these. For example, a typical fire alarm system might consist of heat sensors located throughout a facility that detect high temperatures or a certain change in temperature over a fixed period. A different system

TABLE 20.6
Perimeter and Interior Sensors

Type of Perimeter Sensor	Description
Foil	Foil is a thin, fragile, lead-based metallic tape that is applied to glass windows and doors. The tape is applied to the window or door, and electric wiring connects this tape to a control panel. The tape functions as a conductor and completes the electric circuit with the control panel. When an intruder breaks the door or window, the fragile foil breaks, opening the circuit and triggering an alarm condition.
Magnetic switches (reed switches)	The most widely used perimeter sensor. They are typically used to protect doors, as well as windows that can be opened (windows that cannot be opened are more typically protected by foil alarms).
Glass break detectors	Placed on glass and sense vibrations in the glass when it is disturbed. The two most common types of glass-break detectors are shock sensors and audio discriminators.
Type of Interior Sensor	**Description**
Passive infrared (PIR)	Presently the most popular and cost-effective interior sensor. PIR detectors monitor infrared radiation (energy in the form of heat) and detect rapid changes in temperature within a protected area. Because infrared radiation is emitted by all living things, these types of sensors can be very effective.
Quad PIRs	Consist of two dual-element sensors combined in one housing. Each sensor has a separate lens and a separate processing circuitry, which allows each lens to be set up to generate a different protection pattern
Ultrasonic detectors	Emit high-frequency sound waves, and sense movement in a protected area by sensing changes in these waves. The sensor emits sound waves that stabilize and set a baseline condition in the area to be protected. Any subsequent movement within the protected area by a would-be intruder will cause a change in these waves, thus creating an alarm condition.
Microwave detectors	Emit ultra-high-frequency radio waves, and the detector senses any changes in these waves as they are reflected throughout the protected space. Microwaves can penetrate through walls, and thus a unit placed in one location may be able to protect multiple rooms.
Dual technology devices	Incorporate two types of sensor technology (such as PIR and microwave technology) together in one housing. When both technologies sense an intrusion, an alarm is triggered.

Source: USEPA (2005).

might be outfitted with both smoke and heat detection devices. A summary of several different types of fire/smoke/heat detection sensors is provided in Table 20.7.

Once an alarm system sensor detects an event, it must communicate an alarm signal. The two basic types of alarm communication systems are hardwired and wireless. Hardwired systems rely on wires from the control panel to each detection device and annunciator. Wireless systems transmit signals from a transmitter to a receiver through the air, primarily using radio or other waves. Hardwired systems are usually lower-cost, more reliable (not affected by terrain or environmental factors), and significantly easier to troubleshoot than wireless systems. However, a significant disadvantage of hardwired systems is that hardwiring all locations may not be possible (for example, it may be challenging to hardwire remote locations). In addition, running wires to their required locations can be time-consuming and costly. The significant advantage of using wireless systems is that they can often be installed in areas where hardwired systems are not feasible. However, wireless components can be much more expensive when compared to hardwired systems. In addition, in the past, it has not been easy to perform self-diagnostics on wireless systems to confirm that they are communicating properly with the controller. Presently, most wireless systems incorporate supervising circuitry, which allows the subscriber to know immediately if there is a problem with

TABLE 20.7
Fire/Smoke/Heat Detection Sensors

Detector Type	Description
Thermal detector	Sense when temperatures exceed a set threshold (fixed temperature detectors) or when the rate of change of temperature increases over a fixed time period (rate-of-rise detectors).
Duct detector	Is located within the haring and ventilation ducts of the facility. This sensor detects the presence of smoke within the system's return or supply ducts. A sampling tube can be added to the detector to help span the width of the duct.
Smoke detectors	Sense invisible and/or visible products of combustion. The two principal types of smoke detectors are photoelectric and ionization detectors. The major differences between these devices are described below: • Photoelectric smoke detectors react to visible particles of smoke. These detectors are more sensitive to the cooler smoke with large smoke particles that is typical of smoldering fires. • Ionization smoke detectors are sensitive to the presence of ions produced by the chemical reactions that take place with a few smoke particles, such as those typically produced by fast-burning/flaming fires.
Multi-sensor detectors	Are a combination of photoelectric and thermal detectors. The photoelectric sensor serves to detect smoldering fires, while the thermal detector senses the heat given off from fast-burning/flaming fires.
Carbon monoxide (CO) detectors	Are used to indicate the outbreak of fire by sensing the level of carbon monoxide in the air. The detector has an electrochemical cell which senses carbon monoxide, but not some or other products of combustion.
Beam detectors	Are designed to protect large, open spaces such as industrial warehouses. These detectors consist of three parts: the transmitter, which projects a beam of infrared light; the receiver, which registers the light and produces an electrical signal; and the interface, which processes the signal and generates alarm or fault signals. In the event of a fire, smoke particles obstruct the beam of light. Once a preset threshold is exceeded, the detector will go into alarm.
Flame detectors	Sense either ultraviolet (UV) or infrared (IR) radiation emitted by a fire.
Air-sampling detectors	Actively and continuously sample the air from a protected space and are able to sense the pre-combustion stages of incipient fire.

Source: USEPA (2005).

the system (such as a broken detection device or a low battery), or if a protected door or window has been left open.

20.3.1.4 Backflow Prevention Devices

Backflow prevention devices are designed to prevent backflow, which is the reversal of the normal and intended direction of water flow in a water system. Backflow is a potential problem in a petrochemical processing system because if incorrectly cross-connected to potable water, it can spread contaminated water back through a distribution system. For example, backflow at uncontrolled cross connections (cross-connections are any actual or potential connection between the public water supply and a source of chemical contamination) or pollution can allow pollutants or contaminants to enter the potable water system. More specifically, the backflow from private plumbing systems, industrial areas, hospitals, and other hazardous contaminant-containing systems into public water mains and wells poses serious public health risks and security problems. Cross-contamination from private plumbing systems can contain biological hazards (such as bacteria or viruses) or toxic substances that can contaminate and sicken an entire population in the event of backflow. The majority of historical incidence of backflow have been accidental, but growing concern that contaminants could be intentionally backfed into a system is prompting increased awareness for private homes, businesses, industries, and areas most vulnerable to intentional strikes. Therefore, backflow prevention is a significant tool for protecting water systems.

Backflow may occur under two types of conditions: backpressure and backsiphonage. *Backpressure* is the reverse of normal flow direction within a piping system, resulting from higher downstream pressure than the supply pressure. These reductions in the supply pressure occur whenever the amount of water being used exceeds the amount of water supplied, such as during water line flushing, firefighting, or breaks in water mains. *Backsiphonage* is the reverse of normal flow direction within a piping system caused by negative pressure in the supply piping (i.e., the reversal of normal flow in a system caused by a vacuum or partial vacuum within the water supply piping). Backsiphonage can occur where there is a high velocity in a pipeline; when there is a line repair or break that is lower than a service point; or when there is lowered primary pressure due to high water withdrawal rate, such as during firefighting or water main flushing.

Various types of backflow preventers are appropriate for use to prevent backflow. The primary types of backflow preventers are:

- Air gap drains
- Double check valves
- Reduced pressure principle assemblies
- Pressure vacuum breakers

20.3.1.5 Biometric Security Systems

Biometrics involves measuring the unique physical characteristics or traits of the human body. In ancient times, biometrics involved judging one's accent, body hair, or face to determine friend or foe. It is well known that any aspect of the body that is measurably different from person to person, for example, fingerprints or eye characteristics, can serve as a unique biometric identifier for that individual. Biometric systems recognize fingerprints, palm shape, eyes, face, voice, and signature, which comprise the bulk of the current biometric systems, although other biological features exist.

Biometric security systems use biometric technology and locking mechanisms to control access to specific assets. To access an asset controlled by a biometric security system, an individual's biometric trait must be matched with an existing profile stored in a database. Suppose there is a match between the two. In that case, the locking mechanisms (a physical lock at a doorway, an electronic lock at a computer terminal, or some other type of lock) are disengaged, and the individual is given access to the asset.

A biometric security system typically comprises the following components:

- A sensor, which measures/records a biometric characteristic or trait.
- A control panel serves as the connection point between various system components. The control panel communicates information back and forth between the sensor and the host computer, and controls access to the asset by engaging or disengaging the system lock based on internal logic and information from the host computer.
- A host computer, which processes and stores the biometric trait in a database.
- Specialized software, which compares an individual image taken by the sensor with a stored profile or profiles.
- A locking mechanism that is controlled by the biometric system.
- A power source to power the system.

20.3.1.6 Biometric Hand and Finger Geometry Recognition

Hand and finger geometry recognition identifies an individual through the unique "geometry" (shape, thickness, length, width, etc.) of that individual's hand or fingers. Hand geometry recognition has been employed since the early 1980s and is among the most widely used biometric technologies for controlling access to essential assets. It is easy to install and use and is appropriate for use in any location requiring two-finger, highly accurate, non-intrusive biometric security. For example, it is used in workplaces, day care facilities, hospitals, universities, airports, refineries, and power plants.

A newer option within hand geometry recognition technology is finger geometry recognition (not to be confused with fingerprint recognition). Finger geometry recognition relies on the same scanning methods and technologies as hand geometry recognition. Still, the scanner only scans two of the user's fingers instead of their entire hand. Finger geometry recognition has been in commercial use since the mid-1990s and is mainly used in time and attendance applications (i.e., to track when individuals have entered and exited a location). To date, the only large-scale commercial use of two-finger geometry for controlling access is at Disney World, where season pass holders use the geometry of their index and middle fingers to gain access to the facilities.

To use a hand or finger geometry unit, an individual presents his or her hand or fingers to the biometric unit for "scanning." The scanner consists of a Charged Coupled Device (CCD), a high-resolution digital camera, a reflective platen on which the hand is placed, and a mirror or mirrors that help capture different angles of the hand or fingers. The camera "scans" individual geometric characteristics of the hand or fingers by taking multiple images while the user's hand rests on the reflective platen. The camera also captures "depth," or three-dimensional information, through light reflected from the mirrors and the reflective platen. This live image is then compared to a "template" previously established for that individual when they were "enrolled" in the system. If the live scan of the individual matches the stored template, the individual is "verified" and is given access to that asset. Typically, verification takes about 2 seconds. In access control applications, the scanner is usually connected to an electronic lock, which unlocks the door, turnstile, or other entry barrier when the user is verified. The user can then proceed through the entrance. In time and attendance applications, the time that an individual checks in and out of a location is stored for later use.

As discussed above, hand and finger geometry recognition systems can be used in several applications, including access control and time and attendance tracking. While time and attendance tracking can be used for security, it is primarily used for operations and payroll purposes (i.e., clocking in and out). In contrast, access control applications are more likely to be security-related. Biometric systems are widely used for access control and can be used on various types of assets, including entryways, computers, and vehicles. However, because of their size, hand/finger recognition systems are used mainly in entryway access control applications.

20.3.1.7 Biometric Overview-Iris Recognition

The iris, the colored or pigmented area of the eye surrounded by the sclera (the white portion), is a muscular membrane that controls the amount of light entering the eye by contracting or expanding the pupil (the dark center of the eye). The dense, unique connective tissue patterns in the human iris were first noted in 1936. Still, it was not until 1994, when algorithms for iris recognition were created and patented, that commercial applications using biometric iris recognition began to be used extensively. There are now two vendors producing iris recognition technology: the original developer of these algorithms and a second company, which has developed and patented a different set of algorithms for iris recognition.

The iris is ideal for identifying individuals because it is formed *in utero*, and its unique patterns stabilize around eight months after birth. No two irises are alike: neither an individual's right nor left irises, nor the irises of identical twins. The cornea (the transparent cover over the eye) protects the iris. Therefore, it is not subject to aging or physical changes (and potential variation) common to other biometric measures, such as the hand, fingerprints, and the face. Although some limited changes can occur naturally over time, these changes generally occur in the iris' melanin and therefore affect only the eye's color, and not its unique patterns (in addition, because iris scanning uses only black and white images, color changes would not affect the scan anyway). Thus, barring specific injuries or certain rate surgeries directly affecting the iris, the iris' unique patterns remain relatively unchanged over an individual's lifetime.

Iris recognition systems employ a monochromatic or black and white video camera that uses visible and near-infrared light to take video of an individual's iris. Video is used rather than still photography as an extra security procedure. The video is used to confirm the normal continuous fluctuations of the pupil as the eye focuses, which ensures that the scan is of a living human being, and not a photograph or some other attempted hoax. A high-resolution image of the iris is then captured or extracted from the video, using a device often referred to as a "frame grabber." The unique characteristics identified in this image are then converted into a numeric code stored as a "template" for that user.

20.3.1.8 Integration of AI and Biometrics

Biometric systems have become more adaptive, accurate, and scalable as a result of AI integration. Algorithms used in AI-driven machine learning and deep learning have enabled biometric technologies to analyze large datasets, recognize complex patterns, and verify identities in real time.

The bottom line: AI-powered systems have augmented traditional biometric methods by providing continuous improvement.

20.3.1.9 Card Identification/Access/Tracking Systems

A card reader system is an electronic identification system used to identify a card and then perform an action associated with that card. Depending on the system, the card may determine where a person is or where they were at a particular time, or authorize another action, such as disengaging a lock. For example, a security guard may use his card at card readers located throughout a facility to indicate that he has checked a specific location at a particular time. The reader will store he information and/or send it to a central location, where it can be checked later to ensure that the guard has patrolled the area. Other card reader systems can be associated with a lock, so the cardholder must have their card read and accepted by the reader before the lock disengages.

A complete card reader system typically consists of the following components:

- Access cards that the user carries
- Card readers, which read the card signals and send the information to control units
- Control units, which control the response of the card reader to the card
- A power source

A "card" may be a typical card or device, such as a key fob or wand. These cards store electronic information, ranging from a simple code (i.e., the alphanumeric code on a Proximity card) to individualized personal data (i.e., biometric data on a Smartcard). The card reader reads the information stored on the card and sends it to the control unit, determining the appropriate action to take when a card is presented. For example, in a card access system, the control unit compares the information on the card versus stored access authorization information to determine if the cardholder is authorized to proceed through the door. If the data stored in the card reader system indicates that the key is permitted to allow entrance through the doorway, the system disengages the lock, and the key holder can proceed through the door.

There are many different types of card reader systems on the market. The primary differences between them are how data is encoded on the cards, how it is transferred between the card and the card reader, and the types of applications for which they are best suited. However, all card systems are similar in how the card reader and control unit interact to respond to the card.

While card readers are similar in how the card reader and control unit interact to control access, they are different in the way data is encoded on the cards and how this data is transferred between the card and the card reader. There are several types of technologies available for card reader systems. These include:

- Proximity
- Wiegand
- Smartcard

- Magnetic Stripe
- Bar Code
- Infrared
- Barium Ferrite
- Hollerith
- Mixed Technologies

Table 20.8 summarizes various aspects of card reader technologies. The determination of the security rate (low, moderate, or high) is based on the level of technology a given card reader system has and how simple it is to duplicate that technology, thus bypassing security. Vulnerability ratings were based on whether the card reader can be damaged easily due to frequent use or difficult working conditions (i.e., weather conditions if the reader is outside). Often, this is influenced by the number of moving parts in the system; the more moving parts, the greater the system's potential susceptibility to damage. The life cycle rating is based on the durability of a given card reader system over its entire operational period. Systems requiring frequent physical contact between the reader and the card often have a shorter life cycle due to the wear and tear to which the equipment is exposed. The vulnerability and life cycle ratings are reciprocal for many card reader systems. For instance, a given system with a high vulnerability rating will almost always have a shorter life cycle.

Card reader technology can be implemented for facilities of any size and with any number of users. However, because individual systems vary in the complexity of their technology and the level of security they can provide to a facility, individual users must determine the

TABLE 20.8
Card Reader Technology

Types of Card Readers	Technology	Life Cycle	Vulnerability	Level of Security
Proximity	Embedded radio frequency circuits encoded with unique information	Long	Virtually none	Moderate-high
Wiegand	Short lengths of small-diameter, special alloy wire with unique magnetic properties	Long	Low susceptibility to damage. high durability due to embedded wires.	Moderate-expensive
Magnetic Stripe	Electromagnetic charges to encode information on a piece of tape attached to the back of the card.	Moderate	Moderately susceptible to damage due to frequency of use.	Low-moderate
Bar Code	Series of narrow and wide bars and spaces.	Short	High; easily damaged.	Low
Hollerith	Holes punched in a plastic or paper card and read optically.	Short	High; easily damaged from frequent use.	Low
Infrared	An encoded shadow pattern within the card, read using an infrared scanner.	Moderate	IR scanners are optical and thus, vulnerable to contamination.	High
Barium Ferrite	Uses small bits of magnetized barium ferrite, placed inside a plastic card. The polarity and location of the "spots" determines the coding.	Moderate	Low susceptibility to damage; durable since spots are embedded in the material.	Moderate-high
Smartcards	Patterns or series of narrow and wide bars and spaces.	Short	High susceptibility to damage, low durability.	Highest

Source: USEPA (2005).

appropriate system for their needs. Some essential features to consider when selecting a card reader system include:

- The technological sophistication and security level of the card system
- The size and security needs of the facility
- The frequency with which the card system will be used. For systems that will experience a high frequency of use, it is important to consider a system with a longer life cycle and lower vulnerability rating, thus making it more cost-effective to implement.
- The conditions in which the system will be used (i.e., will it be used on the interior or exterior of buildings and does it require light or humidity controls). Most card reader systems can operate under normal environmental conditions, and therefore, this would be a mitigating factor only in extreme conditions.
- System costs

20.3.1.10 Exterior Intrusion-Buried Sensors

Buried sensors are electronic devices designed to detect potential intruders. They are buried along the perimeters of sensitive assets and can detect intruder activity both above and below ground. Some systems comprise individual, stand-alone sensor units, while others consist of buried cables.

Four types of buried sensors rely on different triggers: pressure or seismic, magnetic field, ported coaxial cable, and fiber-optic cables. These four sensors are all covert and terrain-following, meaning they are hidden from view and follow the contour of the terrain. The four types of sensors are described in more detail below. Table 20.9 presents the distinctions between the four types of buried sensors.

20.3.1.11 Exterior Intrusion Sensors

An exterior intrusion sensor is a detection device that is used in an outdoor environment to detect intrusions into a protected area. These devices are designed to detect an intruder and then communicate an alarm signal to an alarm system. The alarm system can respond to the intrusion in many different ways, such as by triggering an audible or visual alarm signal or sending an electronic signal to a central monitoring location that notifies security personnel of the intrusion.

An intrusion sensor can be used to protect many kinds of assets. Intrusion sensors that protect physical space are classified according to whether they protect indoor, or "interior" space (i.e., an entire building or room within a building), or outdoor, or "exterior" space (i.e., a fence line or perimeter). Interior intrusion sensors are designed to protect the interior space of a facility by detecting an intruder who is attempting to enter, or who has already entered, a room or building. In contrast, exterior intrusion sensors are designed to detect an intrusion into a protected outdoor/exterior area. Exterior protected areas are typically arranged as zones or exclusion areas placed so that the intruder is detected early in the intrusion attempt before the intruder can access more valuable

TABLE 20.9
Types of Buried Sensors

Type	Description
Pressure or seismic	Responds to disturbances in the soil.
Magnetic field	Responds to a change in the local magnetic field caused by the movement of nearby metallic material.
Ported coaxial cables	Respond to the motion of a material with a high dielectric constant or high conductivity near the cables.
Fiber-optic cables	Respond to a change in the shape of the fiber, which can be sensed using sophisticated sensors and computer signal processing.

Source: Adapted from Garcia (2001).

assets (e.g., into a building located within the protected area). Early detection creates additional time for security forces to respond to the alarm.

Exterior intrusion sensors are classified according to how the sensor detects the intrusion within the protected area. The three classes of exterior sensor technology include:

- Buried line sensors
- Fence-associated sensors
- Freestanding sensors

1. *Buried-line sensors*: As the name suggests, buried line sensors are sensors that are buried underground, and are designed to detect disturbances within the ground, such as disturbances caused by an intruder digging, crawling, walking, or running on the monitored ground. Because they sense ground disturbances, these sensors can detect intruder activity both on the surface and below ground. Individual types of exterior buried line sensors function differently, including by detecting motion, pressure, or vibrations within the protected ground, or by detecting changes in some kind of field (e.g., magnetic field) that the sensors generate within the protected ground. Specific buried line sensors include pressure or seismic sensors, magnetic field sensors, ported coaxial cables, and fiber-optic cables. Details on each of these sensor types are provided below.
   - *Buried-line pressure* or *seismic sensors* detect physical disturbances to the ground, such as vibrations or soil compression, caused by intruders walking, driving, digging, or physically contacting the protected ground. These sensors detect disturbances from all directions and, therefore, can protect an area radially outward from their location; however, because detection may weaken as a function of distance from the disturbance, choosing the correct burial depth from the design area will be crucial. In general, sensors buried at a shallow depth protect a relatively small area but have a high probability of detecting intrusion within that area. In contrast, sensors buried at a deeper depth protect a more expansive area but have a lower probability of detecting intrusion into that area.
   - *Buried line magnetic field sensors* detect changes in a local magnetic field caused by the moment of metallic objects within that field. This type of sensor can detect ferrous metal objects worn or carried by an intruder entering a protected area on foot and driving vehicles into the protected area.
   - *Buried line coaxial cable sensors* detect the motion of any object (i.e., human body, and metal) possessing high conductivity and located near the cables. An intruder entering the protected space creates an active electric field disturbance, triggering an alarm.
   - *Buried line fiber-optic cable sensors* detect changes in the attenuation of light signals transmitted within the cable. When the soil around the cable is compressed, the cable is distorted, and the light signal transmitted through the cable changes, initiating an alarm. This type of sensor is easy to install because it can be buried at a shallow burial depth (only a few centimeters) and still be effective.
2. *Fence-associated sensors*: Fence-associated sensors are either attached to an existing fence or are installed in such a way as to create a wall. These sensors detect disturbances to the wall, such as those caused by an intruder attempting to climb the fence or by an intruder trying to cut or lift the fence fabric. Exterior fence-associated sensors include fence-disturbance sensors, taut-wire sensor fences, and electric field or capacitance sensors. Details on each of these sensor types are provided below.
   - *Fence-disturbance sensors* detect the motion or vibration of a fence, such as that which can be caused by an intruder attempting to climb or cut through the fence. In general, fence disturbance sensors are used on chain link fences or on other fence types where a moveable fence fabric is hung between fence posts.
   - *Taut-wire sensor fences* are similar to fence-disturbance sensors except that instead of attaching the sensors to a loose fence fabric, the sensors are attached to a wire stretched

tightly across the fence. These types of systems are designed to detect changes in the tension of the wire rather than vibrations in the fence fabric. Taut-wire sensor fences can be installed over existing fences or as stand-alone fence systems.

- *Electric field or capacitance sensors* detect changes in capacitive coupling between wires attached to, but electrically isolated from, the fence. As opposed to other fence-associated intrusion sensors, both electric field and capacitance sensors generate an electric field that radiates out from the fence line, resulting in an expanded zone of protection relative to other fence-associated sensors, and allowing the sensor to detect an intruder's presence before they arrive at the fence line. Note: proper spacing is necessary during installation of the electric field sensor to detect a would-be intruder from slipping between widely spaced wires.

3. *Free-standing sensors*: These sensors, which include active infrared, passive infrared, bistatic microwave, monostatic microwave, dual-technology, and video motion detection (VMD) sensors, consist of individual sensor units or components that can be set up in a variety of configurations to meet a user's needs. They are installed aboveground, and depending on how they are oriented relative to each other, they can be used to establish a protected perimeter or space. More details on each of these sensor types are provided below.
   - *Active infrared sensors* transmit infrared energy into the protected space and monitor for changes in this energy caused by intruders. In a typical application, an infrared light beam is transmitted from a transmitter to a receiver unit. If an intruder crosses the beam, the beam is blocked, and the receiver unit detects a change in the amount of light received, triggering an alarm. Different sensors can see single- and multiple-beam arrays. Single-beam infrared sensors transmit a single infrared beam. In contrast, multiple-beam infrared sensors transmit two or more beams parallel. This multiple-beam sensor arrangement creates an infrared "fence."
   - *Passive infrared (PIR) sensors* monitor the ambient infrared energy in a protected area and evaluate changes in that ambient energy that may be caused by intruders moving through the protected area. Detection ranges can exceed 100 yards on cold days, with size and distance limitations dependent upon the background temperature. PIR sensors generate a non-uniform detection pattern (or "curtain") that has areas (or "zones") of more sensitivity and regions of less sensitivity. The detector's lenses determine the specific shape of the protected area. The general shape familiar to many detection patterns is a series of long "fingers" emanating from the PIR and spreading in various directions. When intruders enter the detection area, the PIR sensor detects differences in temperature due to the intruder's body heat and triggers an alarm. While the PIR leaves unprotected areas between its fingers, an intruder would be detected if he passed from a non-protected area to a protected area.
   - *Microwave sensors* detect changes in received energy generated by an intruder's motion entering a protected area. Monostatic microwave sensors incorporate a transmitter and a receiver in one unit, while bistatic sensors separate the transmitter and the receiver into different units. Monostatic sensors are limited to a coverage area of 400 feet, while bistatic sensors can cover up to 1,500 feet. For bistatic sensors, a zone of no detection exists in the first few feet in front of the antennas. This distance from the antennas to the point at which the intruder is first detected is known as the offset distance. Due to this offset distance, antennas must be configured to overlap one another (as opposed to being adjacent), thereby creating long perimeters with a continuous line of detection.
   - *Dual-technology sensors* consist of two different sensor technologies incorporated into one sensor unit. For example, a dual-technology sensor could consist of a passive infrared detector and a monostatic microwave sensor integrated into the same sensor unit.

- *VMD sensors* monitor video images from a protected area for image changes. Video cameras detect unauthorized intrusion into the protected area by comparing the most recent image against a previously established one. Cameras can be installed on towers or other tall structures to monitor a large area.

20.3.1.12 Fences

A fence is a physical barrier that can be set up around the perimeter of an asset. Fences often consist of pieces (such as individual pickets in a wooden wall, or particular sections of a wrought iron fence) that are fastened together. Separate wall sections are fastened together using posts, which are sunk into the ground to provide stability and strength for the fence sections hung between them. Gates are installed between individual fence sections to allow access inside the fenced area.

Many fences are used as decorative architectural features to separate physical spaces from each other. They may also be used to physically mark the location of a boundary (such as a fence installed along a property line). However, a fence can also serve as an effective means for physically delaying intruders from gaining access to a financial service sector asset. For example, many utilities install fences around their primary facilities, remote pump stations, hazardous petrochemical materials storage areas, or sensitive areas. Access to the area can be controlled through security at gates or doors through the fence (for example, by posting a guard at the gate or locking it). To access the asset, unauthorized persons could go around or through the fence.

Fences are often compared with walls when determining the appropriate system for perimeter security. While both fences and walls can provide adequate perimeter security, fences are frequently easier and less expensive to install than walls. However, they do not usually offer the same physical strength as walls. In addition, many fences have gaps between the pieces that make up the fence (i.e., the spaces between chain links in a chain link fence or the space between pickets in a picket fence). Thus, many types of fences allow the interior of the fenced area to be seen. This may allow intruders to gather important information about the locations or defenses of vulnerable areas within the facility.

Numerous materials are used to construct fences, including chain link iron, aluminum, wood, or wire. Some fences, such as split rails or pickets, may not be appropriate for security because they are traditionally low and not physically strong. Potential intruders may be able to easily defeat these fences by jumping or climbing over them or by breaking through them. For example, the rails in a split fence may be damaged easily.

Important security attributes of a fence include the height at which it can be constructed, the strength of the material comprising the fence, the method and strength of attaching the individual sections of the fence to the posts, and the fence's ability to restrict the view of the assets inside the fence. Additional considerations should include the ease of installing the wall and removing and reusing fence sections. Table 20.10 compares the critical security and usability features of various fence types.

TABLE 20.10
Comparison of Different Fence Types

Specifications	Chain Link	Iron	Wire (Wirewall)	Wood
Height limitations (inch)	12	12	12	8
Strength	Medium	High	High	Low
Installation requirements	Low	High	High	Low
Ability to remove/reuse	Low	High	Low	High
Ability to replace/repair	Medium	High	Low	High

Source: USEPA (2005).

Some fences can include additional measures to delay or detect potential intruders. Such measures may include the addition of barbed wire, razor wire, or other deterrents at the top of the fence. Barbed wire is sometimes employed at the base of fences as well. This can impede a would-be intruder's progress, even reaching the wall. Fences may also be fitted with security cameras to provide visual surveillance of the perimeter. Finally, some facilities have installed motion sensors along their fences to detect movement on the wall. Several manufacturers have combined these multiple perimeter security features into one product, offering alarms and other security features.

Correctly implementing a fence can make it a much more effective security measure. Security experts recommend the following when a facility constructs a fence:

- The fence should be at least 7–9 feet high.
- Any outriggers, such as barbed wire, affixed on top of the fence should be angled out and away from the facility, not in toward the facility. This will make climbing the fence more difficult and prevent ladders from being placed against it.
- Other types of hardware can increase the security of the fence. This can include installing concertina wire along the fence (this can be done in front of the fence or at the top), or adding intrusion sensors, a camera, or other hardware to the fence.
- All undergrowth should be cleared for several feet (typically 6 feet) on both sides of the fence. This will allow any patrols in the area to see the fence clearly.
- Any trees with limbs or branches hanging over the fence should be trimmed so that intruders cannot use them to go over the fence. Also, it should be noted that fallen trees can damage fences, and so management of trees around the fence can be important. This can be especially important in areas where the fence goes through a remote area.
- Fences that do not block the view from outside the fence to inside the fence allow patrols to see inside the fence without having to enter the facility.
- "No Trespassing" signs posted along the fence can be a valuable tool in prosecuting any intruders who claim that the fence was broken and that they did not enter through the fence illegally. Adding signs that highlight the local ordinances against trespassing can further persuade simple troublemakers from illegally jumping/climbing the fence. Electrical substations and other electrical component installations should have visible signage warning of HIGH VOLTAGE and the dangers of Electrical Shock.

20.3.1.13 Films for Glass Shatter Protection

Many financial services sector entities have numerous windows on the outside of buildings, indoors, and in interior offices. In addition, many facilities have glass doors or other glass structures, such as glass walls or display cases. These glass objects are potentially vulnerable to shattering when heavy objects are thrown or launched at them, when explosions occur near them, or when there are high winds (for exterior glass). If the glass is shattered, intruders may potentially enter an area. In addition, broken glass projected into a room from an explosion or an object being thrown through a door or window can injure and potentially incapacitate personnel in the room. Materials that prevent glass from shattering can help to maintain the integrity of the door, window, or other glass object, and can delay an intruder from gaining access. These materials can also prevent flying glass and thus reduce potential injuries.

Materials designed to prevent glass from shattering include specialized films and coatings. These materials can be applied to existing glass objects to improve their strength and ability to resist breaking. The films have been tested against many scenarios that could result in glass breakage, including penetration by blunt objects, bullets, high winds, and simulated explosions. Thus, the films are tested against both simulated weather scenarios (which could include both the high winds and the force of objects blown into the glass), as well as more criminal/terrorist scenarios where the glass is subject to explosives or bullets. Many vendors provide information on the results of these tests; thus, potential users can compare different product lines to determine which products best suit their needs.

The primary attributes of films for shatter protection are:

- The materials from which the film is made
- The adhesive that bonds the film to the glass surface
- The thickness of the film

Standard glass safety films are made from high-strength polyester. Polyester provides both strength and elasticity, which is important in absorbing an object's impact, spreading the force of the impact over the entire film, and resisting tearing. Polyester is also designed to resist scratching, which can result when films are cleaned with abrasives or other industrial cleaners.

The bonding adhesive is important in ensuring that the film does not tear away from the glass surface. This can be especially important when the glass is broken, so that the film does not peel off the glass and allow it to shatter. In addition, films applied to exterior windows can be subject to high concentrations of UV light, which can break down bonding materials.

Film thickness is measured in gauge or mils. According to test results reported by several manufacturers, film thickness appears to affect resistance to penetration/tearing, with thicker films being more resistant to penetration and tearing. However, the appreciation of a thicker film did not decrease glass fragmentation.

Many manufacturers offer films in different thicknesses. The "standard" film is usually one 4 mil layer; thicker films are typically composed of several layers of the standard 4 mil sheet. However, newer technologies have allowed the polyester to be "micro layered" to produce a more substantial film without significantly increasing its thickness. In this micro-layering process, each laminate film is composed of multiple micro-thin layers of polyester woven together at alternating angles. This provides increased strength for the film while maintaining the flexibility and thin profile of one film layer.

As described above, many vendors test their products in various scenarios, leading to glass shattering, including simulated bomb blasts and the simulation of the glass being struck by wind-blown debris. Some manufacturers refer to the Government Services Administration standard for bomb blasts, which requires resistance to tearing for a 4-PSI blast. Other manufacturers use other measures and tests for resistance to tearing. Many of these tests are not "standard," in that none of the accepted standards-setting institutions has adopted any standard testing or reporting methods. However, many of the vendors publish the procedure and the results of these tests on their websites, which may allow users to evaluate the protective properties of these films. For example, several vendors evaluate the "protectiveness" of their movies and the "hazard" resulting from blasts near windows with and without protective films. Protectiveness is usually assessed based on the percentage of glass ejected from the window, and the height at which that ejected glass travels during the blast (for example, if the blasted glass tends to project upward into a room—potentially toward people's faces—it is a higher hazard than if it is blown downward into the room toward people's feet). There are some standard measures of glass breakage. For example, several vendors indicated that their products exceed the American Society for Testing and Materials (ASTM) standard 64Z-95 "Standard Test Method for glazing and Glazing Systems Subject to Air Blast Loadings." Vendors often compare the results of penetration or force tests, ballistics tests, or simulated explosions with unprotected glass versus glass onto which their films have been applied. Results generally show that applying films to the glass surfaces reduces breakage/penetration of the glass and can reduce the amount and direction of glass ejected from the frame. This, in turn, reduces the hazard from flying glass.

In addition to these types of tests, many vendors conduct standard physical tests on their products, such as tensile and peel strength tests. Tensile strength indicates the strength per material area, while the peel strength indicates the force it would take to peel the product from the glass surface. Several vendors suggest that their products exceed the American National Standards Institute (ANSI) standard Z97.1 for tensile strength and adhesion.

Vendors typically have a warranty against peeling or other forms of deterioration of their products. However, the warranty requires that manufacturer-certified technicians install the films to ensure that they are applied correctly and that the warranty is in effect. Warranties from different manufacturers may vary. Some may only cover the cost of replacing the material, while others include material plus installation. Because installation costs are significantly greater than material costs, different warranties may represent a large difference in potential costs.

20.3.1.14 Fire Hydrant Locks

Fire hydrants are installed at strategic locations throughout a community's water distribution system to supply water for firefighting. However, because there are many hydrants in a system and they are often located in residential neighborhoods, industrial districts, and other areas where they cannot be easily observed and/or guarded, they are potentially vulnerable to unauthorized access. Many municipalities, states, and EPA Regions have recognized this potential vulnerability and have instituted programs to lock hydrants. For example, EPA Region 1 has included locking hydrants as number 7 on its "Drinking Water Security and Emergency Preparedness" Top Ten List for small groundwater suppliers.

A "hydrant lock" is a physical security device designed to prevent unauthorized access to the water supply through a hydrant. These locks can also ensure water and water pressure availability to firefighters and avoid water theft and associated lost water revenue. They have been successfully used in numerous municipalities and in various climates and weather conditions.

Fire hydrant locks are steel covers or caps locked in place over the operating nut of a fire hydrant. The lock prevents unauthorized people from accessing the operating nut and opening the fire hydrant valve. The lock also makes removing the bolts from the hydrant and accessing the system more difficult. Finally, hydrant locks shield the valve from being broken off. Should a vandal attempt to breach the hydrant lock by force and succeed in breaking the lock, the vandal will only succeed in bending the operating valve. If the hydrant's operating valve is bent, the hydrant will not be operational, but the water asset remains protected and inaccessible to vandals. However, the entire hydrant will need to be replaced.

Hydrant locks are designed so that special "key wrenches" can operate the hydrants without removing the lock. These specialized wrenches are generally distributed to the fire department, public works department, and other authorized personnel so that they can access the hydrants as needed. A municipality generally keeps an inventory of wrenches and their serial numbers so that the location of all wrenches is known. These operating key wrenches may only be purchased by registered lock owners.

The most essential features of hydrants are their strength and the security of their locking systems. The locks must be strong so that they cannot be broken. Hydrant locks are constructed from stainless or alloyed steel. Stainless steel locks are stronger and are ideal for all climates; however, they are more expensive than alloy locks. The locking mechanisms for each fire hydrant lock system ensure that the hydrant can only be operated by authorized personnel with the specialized key to work the hydrant.

20.3.1.15 Hatch Security

A hatch is a door installed on a horizontal plane (such as in a floor, a paved lot, or a ceiling) instead of on a vertical plane (such as in a building wall). Hatches are usually used to access assets located underground (such as hatches to basements or underground vaults and storage areas), or to assets located above ceilings (such as emergency roof exits). At chemical industrial facilities, hatches are typically used to access underground vaults containing pumps, meter chambers, valves, or piping, or to access the interior of chemical tanks or covered water reservoirs. Securing a hatch by locking it or upgrading materials to give the hatch added strength can help delay unauthorized access to any asset behind the hatch.

Like all doors, a hatch consists of a frame anchored to the horizontal structure, a door or doors, hinges connecting the doors to the frame, and a latching or locking mechanism that keeps the hatch door or doors closed.

It should be noted that improving hatch security is straightforward. Hatches with upgraded security features can be installed new or retrofitted for existing applications. Depending on the application, the primary security-related attributes of a hatch are the strength of the door and frame, its resistance to the elements and corrosion, its ability to be sealed against water or gas, and its locking features.

Hatches must be strong and lightweight to withstand typical static loads (such as people or vehicles walking or driving over them) while still being easy to open. In addition, because hatches are typically installed at outdoor locations, they are usually designed from corrosion-resistant metal that can withstand the elements. Therefore, hatches are generally constructed from high-gauge steel or lightweight aluminum.

Aluminum is typically the material of choice for hatches because it is lightweight and more corrosion-resistant than steel. Aluminum is not as rigid as steel, so that aluminum hatch doors may be reinforced with aluminum stiffeners for extra strength and rigidity. The doors are usually constructed from single or double material layers (or "leaves"). Single-leaf designs are standard for smaller hatches, while double-leaf designs are required for larger hatches. In addition, aluminum products do not require painting. This is reflected in the warranties available for different products. Product warranties range from ten years to a lifetime.

Steel is heavier per square foot than aluminum, and thus, steel hatches will be heavier and more difficult to open than aluminum hatches of the same size. However, heavy-steel hatch doors may have spring-loaded, hydraulic, or gas openers or other specialized features that help open and keep the hatch open.

Many hatches are installed in outdoor areas, roadways, or pedestrian areas. Therefore, the hatch installed for any application must be designed to withstand the expected load at that location. Hatches are typically solid to withstand either pedestrian or vehicle loading. Pedestrian loading hatches are generally designed to withstand either 150 or 300 pounds per square foot (psf) of loading. The vehicle loading standard is the American Association of State Highway and Transportation Officials (AASHTO) H-20-wheel loading standard of 16,000 pounds over an 8-inch by 20-inch area. It should be noted that these design parameters are for static loads and not dynamic loads; thus, the loading capabilities may not reflect potential resistance to other types of loads that may be more typical of an intentional threat, such as repeated blows from a sledgehammer or pressure generated by bomb blasts or bullets.

The typical design for a watertight hatch includes a channel frame that directs water away from the hatch. This can be especially important in a storage tank hatch because it prevents liquid contaminants from being dumped on the hatch and leaking through into the interior. Hatches can also be constructed with gasket seals that are air-, odor-, and gas-tight.

Typically, hatches for pedestrian loading applications have hinges on the hatch's exterior, while hatches designed for H-20 loads have hinges in the hatch's interior. Hinges on the hatch's exterior may be removed, allowing intruders to remove the hatch door and access the asset behind the hatch. Therefore, installing H-20 hatches even for applications that do not require H-20 loading levels may increase security, because intruders cannot tamper with the hinges and circumvent the hatch this way.

In addition to the location of the hinges, stock hinges can be replaced with heavy-duty or security hinges that are more resistant to tampering.

The hatch locking mechanism is perhaps the most critical part of hatch security. There are several locks that can be implemented for hatches, including:

- Slam locks (internal locks that are located within the hatch frame
- Recessed cylinder locks
- Bolt locks
- Padlocks

20.3.1.16 Ladder Access Control

Financial services sector facilities have several assets raised aboveground level, including electrical substations, transmitting stations, raised conduit systems, and roof access points into buildings. In addition, communications equipment, antennae, or other electronic devices may be located on the top of these raised assets. Typically, these assets are reached by ladders permanently anchored to the asset. For example, raised petrochemical/water tanks are usually accessed by ladders bolted to one of the legs of the tank. Controlling access to these raised assets by controlling access to the ladder can increase security at a financial services sector facility.

A typical ladder access control system consists of a cover that is locked or secured over the ladder. The cover can be a casing that surrounds most of the ladder, or a door or shield that covers only part of the ladder. In either case, several rungs of the ladder (the number of rungs depends on the size of the cover) are made inaccessible by the cover, and these rungs can only be accessed by opening or removing the cover. The cover is locked, so only authorized personnel can open or remove it and use the ladder. Ladder access controls are usually installed at several feet aboveground level, and they typically extend several feet up the ladder so that they cannot be circumvented by someone accessing the ladder above the control system.

The essential features of ladder access control are the size and strength of the cover and its ability to lock or otherwise be secured from unauthorized access.

The covers are constructed from aluminum or some type of steel. This should provide adequate protection from being pierced or cut through. The metals are corrosion-resistant, so they will not corrode or become fragile from extreme weather conditions in outdoor applications. The bolts used to install each of these systems are galvanized steel. In addition, the bolts for each cover are installed on the inside of the unit, so they cannot be removed from the outside.

20.3.1.17 Locks

A lock is a type of physical security device that can delay or prevent a door, a gate, a window, a manhole, a filing cabinet drawer, or some other physical feature from being opened, moved, or operated. Locks typically operate by connecting two pieces, such as connecting a door to a door jamb or a manhole to its casing. Every lock has two modes—engaged (or "locked") and disengaged (or "opened"). When a lock is disengaged, the asset on which the lock is installed can be accessed by anyone, but when the lock is engaged, access to the locked asset.

Before discussing locks and their applicability, it is important to discuss key control. Based on our experience, many financial services sector facilities (and others) have no idea how many keys for various site/equipment locks have been issued to employees over the years. Many facilities issue keys to employees hired with no accountability for the keys upon the employee's departure. Needless to say, this is not a good security policy. You can have the best-made locks installed throughout your facilities, but if you do not have proper key control, you do not have proper security.

Locks are excellent security features because they have been designed to function in many ways and to work on many different types of assets. Locks can also provide various levels of security depending on how they are designed and implemented. The security provided by a lock is dependent on several factors, including its ability to withstand physical damage (i.e., can it be cut off, broken, or otherwise physically disabled) as well as its requirements for supervision or operation (i.e., combinations may need to be changed frequently so that they are not compromised and the locks remain secure). While there is no single definition of the "security" of a lock, locks are often described as minimum, medium, or maximum security. Minimum security locks are those that can be easily disengaged (or "picked") without the correct key or code, or those that can be disabled easily (such as small padlocks that can be cut with bolt cutters). Higher security locks are more complex and thus are more challenging to pick or are sturdier and more resistant to physical damage.

Many locks, such as door locks, only need to be unlocked from one side. For example, most door locks require a key to be unlocked only from the outside. A person opens such devices, called single-cylinder locks, from the inside by pushing a button or by turning a knob or handle. Double-cylinder locks require a key to be locked or unlocked from both sides.

20.3.1.18 Manhole Intrusion Sensors

Manholes are found at some financial services sector sites. Manholes are designed to provide access to underground utilities, meter vaults, petrochemical pumping rooms, etc.; therefore, they are potential entry points to a system. Because many utilities run under other infrastructure (roads, buildings), manholes also provide possible access points to critical infrastructure as well as petrochemical process assets. In addition, because the portion of the system to which manholes provide entry is primarily located underground, access to a system through a manhole increases the chance that an intruder will not be seen. Therefore, protecting manholes can be critical to guarding an entire plant site and the surrounding community.

Multiple methods exist for protecting manholes, including preventing unauthorized personnel from physically accessing them and detecting attempts at unauthorized access.

A manhole intrusion sensor is a physical security device designed to detect unauthorized access to a facility through a manhole. Monitoring a manhole that provides access to a chemical plant or processing system can mitigate two distinct types of threats. First, monitoring a manhole may detect unauthorized personnel's access to chemical systems or assets through the manhole. Second, monitoring manholes may also allow the detection of intruders attempting to place explosive or other destructive weapons of mass destruction (WMD) devices into the energy storage system.

Several different technologies have been used to develop manhole intrusion sensors, including mechanical systems, magnetic systems, and fiber optic and infrared sensors. Some of these intrusion sensors have been specifically designed for manholes, while others consist of standard, off-the-shelf intrusion sensors implemented in a system specifically designed for manhole applications.

20.3.1.19 Manhole Locks

A "manhole lock" is a physical security device designed to delay unauthorized access to the financial services sector facility or system through a manhole.

20.3.1.20 Passive Security Barriers

One of the most basic threats facing any facility is intruders accessing it with the intention of causing damage to its assets. These threats may include intruders actually entering the facility and intruders attacking the facility from outside without actually entering it (i.e., detonating bombs near enough to the facility to cause damage within its boundaries).

Security barriers are one of the most effective ways to counter the threat of intruders accessing a facility or the perimeter. Security barriers are large, heavy structures that control access through a perimeter by either vehicles or personnel. They can be used in many ways, depending on how/where they are located in the facility. For example, security barriers can be used on or along driveways or roads to direct traffic to a checkpoint (i.e., a facility may install jersey barriers on a road to direct traffic in a specific direction). Security barriers (crash beams, gates) can be installed at the checkpoint so that guards can regulate which vehicles can access the facility. Finally, other security barriers (i.e., bollards or security planters) can be used along the perimeter to establish a protective buffer between the facility and approaching vehicles. Establishing such a protective buffer can help in mitigating the effects of the type of bomb blast described above, both by potentially absorbing some of the explosion, and also by increasing the "stand-off" distance between the blast and the facility (the force of an explosion is reduced as the shock wave travels further from the source, and thus the further the explosion is from the target, the less effective it will be in damaging the target).

Security barriers can be either "active" or "passive." "Active" barriers, which include gates, retractable bollards, wedge barriers, and crash barriers, are readily movable. Thus, they are typically used in areas where they must be moved often to allow vehicles to pass—such as in roadways at entrances and exits to a facility. In contrast to active security barriers, "passive" security barriers, which include jersey barriers, bollards, and security planters, are not designed to be moved on a regular basis, and thus, they are typically used in areas where access is not required or allowed—such as along building perimeters or in traffic control areas. Passive security barriers are typically large, heavy structures that are usually several feet high, and they are designed so that even heavy-duty vehicles cannot go over or through them. Therefore, they can be placed in a roadway parallel to the flow of traffic so that they direct traffic in a certain direction (such as to a guardhouse, a gate, or some other sort of checkpoint), or perpendicular to traffic such that they prevent a vehicle from using a road or approaching a building or area.

20.3.1.21 Security for Doorways-Side Hinged Doors

Doorways are the main access points to a facility or rooms within a building. They are used on the exterior or in the interior of buildings to provide privacy and security for the areas behind them. Different types of doorway security systems may be installed in different doorways, depending on the needs or requirements of the buildings or rooms. For example, exterior doorways tend to have heavier doors to withstand the elements and to provide some security at the entrance of the building. Interior doorways in office areas may have lighter doors that may be primarily designed to provide privacy rather than protection. Therefore, these doors may be made of glass or lightweight wood. Doorways in industrial areas may have sturdier doors than other interior doors and may be designed to provide protection or security for areas behind the doorway. For example, fireproof doors may be installed in chemical storage areas or other areas where fire is dangerous.

Because they are the main entries into a facility or a room, doorways are often prime targets for unauthorized entry into a facility or an asset. Therefore, securing doorways may be a significant step in providing security at a facility.

A doorway includes four main components:

- The door, which blocks the entrance. The primary threat to the actual door is breaking or piercing through the door. Therefore, the primary security features of doors are their strength and resistance to various physical threats, such as fire or explosions.
- The door frame, which connects the door to the wall. The primary threat to a door frame is that the door can be pried away from the frame. Therefore, the primary security feature of a door frame is its resistance to prying.
- The hinges, which connect the door to the door frame. The primary threat to door hinges is that they can be removed or broken, allowing intruders to remove the entire door. Therefore, security hinges are designed to be resistant to breaking. They may also be designed to minimize the threat of removal from the door.
- The lock, which connects the door to the door frame. Use of the lock is controlled through various security features, such as keys and combinations, so only authorized personnel can open the lock and go through the door. Locks may also incorporate other security features, such as software or other systems to track overall use of the door or to track individuals using the door, etc.

Each of these components is integral to providing security for a doorway. Upgrading the security of only one of these components while leaving the other components unprotected may not increase the overall security of the doorway. For example, many facilities upgrade door locks as an essential step in improving their security. However, suppose the facilities do not also focus on increasing security for the door hinges or frame. In that case, the door may remain vulnerable to being removed from its frame, defeating the door lock's increased security.

The primary attribute for the security of a door is its strength. Many security doors are 4–20-gauge hollow metal doors consisting of steel plates over a hollow cavity reinforced with steel stiffeners to give the door extra stiffness and rigidity. This increases resistance to blunt force used to try to penetrate through the door. The space between the stiffeners may be filled with specialized materials to provide fire-, blast-, or bullet resistance to the door.

The Windows and Doors Manufacturers Association has developed a series of performance attributes for doors. These include:

- Structural Resistance
- Forced Entry Resistance
- Hinge Style Screw Resistance
- Split Resistance
- Hinge Resistance
- Security Rating
- Fire Resistance
- Bullet Resistance
- Blast Resistance

The first five bullets explain a door's resistance to standard physical breaking and prying attacks. These tests are used to evaluate the strength of the door, the resistance of the hinges, and the frame in a standardized way. For example, the Rack Load Test simulates a prying attack on a corner of the door. A test panel is restrained at one end, and a third corner is supported. Loads are applied and measured at the fourth corner. The Door Impact Test simulates a battering attack on a door and frame using impacts of 200-foot pounds by a steel pendulum. The door must remain fully operable after the test. It should be noted that door glazing is also rated for resistance to shattering etc. Manufacturers will be able to provide security ratings for these door features as well.

Door frames are integral to doorway security because they anchor the door to the wall. Door frames are typically constructed from wood or steel and are installed to extend several inches over the doorway cut into the wall. For added security, frames can have varying degrees of overlapping with, or wrapping over, the underlying wall. This can make prying the frame from the wall more difficult. A frame formed from a continuous piece of metal (as opposed to a frame constructed from individual metal pieces) will prevent prying between pieces of the frame.

Many security doors can be retrofitted into existing frames; however, many security door installations include replacing the door frame as well as the door itself. For example, bullet-resistance per Underwriter's Laboratory (UL) 752 requires resistance of the door and frame assembly, and thus, replacing the door alone would not meet UL 752 requirements.

20.3.1.22 Valve Lockout Devices

Valves are utilized as control elements in fuel oil/natural gas and petrochemical process piping networks. They regulate the flow of liquids and gases by opening, closing, or obstructing a flow passageway. Valves are typically located where flow control is necessary. They can be located in-line or at pipeline, tank entrance, and exit points. They can serve multiple purposes in a process pipe network, including:

- Redirecting and throttling the flow
- Preventing backflow

- Shutting off the flow to a pipeline or tank (for isolation purposes)
- Releasing pressure
- Draining extraneous liquid from pipelines or tanks
- Introducing chemicals into the process network
- As access points for the sampling process of water

Valves may be located either aboveground or belowground. It is critical to protect against valve tampering. For example, tampering with a pressure relief valve could result in a pressure buildup and a potential explosion in the piping network. On a larger scale, adding a contaminant or non-compatible chemical substance to the chemical processing system through an unprotected valve could result in the catastrophic release of that contaminant to the general population.

Different security products are available to protect aboveground versus ground valves. For example, valve lockout devices can be purchased to protect valves and valve controls located aboveground. Vaults containing underground valves can be locked to prevent access to these valves.

As described above, a lockout device can be used as a security measure to prevent unauthorized access to aboveground valves located within petrochemical processing systems. Valve lockout devices are locks specially designed to fit over valves and valve handles to control their ability to be turned or seated. These devices can be used to lock the valve into the desired position. Once the valve is closed, it cannot be turned unless an authorized individual has unlocked or removed the locking device.

Various valve lockout options are available for industrial use, including:

- Cable lockouts
- Padlocked chains/cables
- Valve-specific lockouts

Many of these lockout devices are not specifically designed for use in the financial services sector industry (i.e., chains, padlocks), but are available from a local hardware store or manufacturer specializing in safety equipment. Other lockout devices (for example, valve-specific lockouts or valve box-locks) are more specialized and must be purchased from safety or valve-related equipment vendors.

The three most common types of valves for which lockout devices are available are gate, ball, and butterfly valves. Each is described in more detail below.

- *Gate valve lockouts*: Gate valve lockouts are designed to fit over the operating hand wheel of the gate valve to prevent it from being turned. The lockout is secured with a padlock. Two types of gate valve lockouts are available: diameter-specific and adjustable. Diameter-specific lockouts are available for handles ranging from 1 to 13 inches in diameter. Adjustable gate valve lockouts can fit any handle ranging from 1 to 6+ inches in diameter.
- *Ball valve lockouts*: Several configurations are available to lock out ball valves, all of which are designed to prevent rotation of the valve handle. The three major configurations available are a wedge shape for 1- to 3-inch valves, a lockout that covers 3/8- to 8-inch ball valve handles, and a universal lockout that can be applied to quarter-turn valves of varying sizes and geometric handle dimensions. All three ball valve lockouts can be installed by sliding the lockout device over the handle and securing it with a padlock.
- *Butterfly valve lockouts*: The butterfly valve lockout functions similarly to the ball valve lockout. The polypropylene lockout device is placed over the valve handle and secured with a padlock. This type of lockout has been commonly used in the bottling industry.

A significant difference between valve-specific lockout devices and the padlocked chain or cable lockouts discussed earlier is that they do not need to be secured to an anchoring device in the floor or the piping system. In addition, valve-specific lockouts eliminate potential tripping or access hazards caused by chains or cable lockouts applied to valves near walkways or frequently maintained equipment.

Valve-specific lockout devices are available in various colors, which can be useful in distinguishing different valves. For example, different colored lockouts can be used to distinguish the type of liquid passing through the valve (i.e., treated, untreated, potable, petrochemical) or to identify the party responsible for maintaining the lockout. Implementing a system of different-colored locks on operating valves can increase system security by reducing the likelihood of an operator inadvertently opening the wrong valve and causing a problem in the system.

20.3.1.23 Security for Vents

Vents are installed in some aboveground financial services sector storage areas to allow safe venting of off-gases. The specific vent design for any given application will vary depending on the design of the chemical storage vessel. However, every vent has an open-air connection between the storage container and the outside environment. Although these air exchange vents are integral to covered or underground chemical storage containers, they also represent a potential security threat. Improving vent security by making the vents tamper-resistant or by adding other security features, such as security screens or security covers, can enhance the security of the entire petrochemical processing system.

Many municipalities already have specifications for vent security at their local chemical industrial assets. These specifications typically include the following requirements:

- Vent openings are to be angled down or shielded to minimize the entrance of surface and/or rainwater into the vent through the opening
- Vent designs are to include features to exclude insects, birds, animals, and dust
- Corrosion-resistant materials are to be used to construct the vents.

20.3.1.24 Visual Surveillance Monitoring

Visual surveillance is used to detect threats through continuous observation of essential or vulnerable areas of an asset. The observations can also be recorded for later review or use (for example, in court proceedings). A visual surveillance system can monitor various parts of production, distribution, or pumping/compressing systems, including the facility's perimeter, outlying pumping stations, or entry or access points into specific buildings. These systems are also helpful in recording individuals who enter or leave a facility, thereby helping to identify unauthorized access. Images can be transmitted live to a monitoring station, monitored in real time, recorded, and reviewed later. Many financial services sector facilities have found that a combination of electronic surveillance and security guards provides an effective means of facility security.

Visual surveillance is provided through a CCTV system, in which an image's capture, transmission, and reception are localized within a closed "circuit." This differs from other broadcast images, such as over-the-air television, which is broadcast over the air to any receiver within range.

At a minimum, a CCTV system consists of:

- One or more cameras
- A monitor for viewing images
- A system for transmitting images from the camera to the monitor.

Specific attributes and features of camera systems, lenses, and lighting systems are presented in Table 20.11.

TABLE 20.11
Attributes of Camera, Lenses, and Lighting Systems

Attribute	Discussion
	Camera Systems
Camera Type	Major factors in choosing the correct camera are the resolution of the image required and the lighting of the area to be viewed. • *Solid State* (including charge-coupled devices, charge priming devices, charge injection devices, and metal oxide substrates): These cameras are becoming predominant in the marketplace because of their high resolution and their elimination of problems inherent in tube cameras. • *Thermal*: These cameras are designed for night vision. They require no light and use differences in temperature between objects in the field of view to produce a video image. Resolution is low compared to other cameras, and the technology is currently expensive relative to other technologies. • *Tube*: These cameras can provide high resolution burnout and must be replaced after 1–2 years. In addition, tube performance can degrade over time. Finally, tube cameras are prone to burn images in the tube replacement.
Resolution (the ability to see fine details)	The user must determine the amount of resolution required, depending on the level of detail required for threat determination. A high definition focus with a wide field of vision will give an optimal viewing area.
Field of vision width	Cameras are designed to cover a defined field of vision, which is usually defined in degrees. The wider the field of vision, the more area a camera will be able to monitor.
Type of image produced (color, black and white, thermal)	Color images may allow the identification of distinctive markings, while black and white images may provide sharper contrast. Thermal imaging allows the identification of heat sources (such as human beings or other living creatures) from low-light environments; however, thermal images are not effective in identifying specific individuals (i.e., for subsequent legal processes).
Pan/Tilt/Zoom (PTZ)	Panning (moving the camera in a horizontal plane), tilting (moving the camera in a vertical plane), and zooming (moving the lens to focus on objects that are at different distances from the camera) allow the camera to follow a moving object. Different systems allow these functions to be controlled manually or automatically. Factors to be considered in PTZ cameras are the degree of coverage for pan and tilt functions and the power of the zoom lens.
	Lenses
Format	Lens format determines the maximum image size to be transmitted.
Focal Length	This is the distance from the lens to the center of the focus. The greater the focal length, the higher the magnification, but the narrower the field of vision.
F Number	F number is the ability to gather light. Smaller F numbers may be required for outdoor applications where light cannot be controlled as easily.
Distance and width approximation	The distance and width approximations are used to determine the geometry of the space that can be monitored at the best resolution.
	Lighting Systems
Intensity	Light intensity must be great enough for the camera type to produce sharp images. Light can be generated from natural or artificial sources. Artificial sources can be controlled to produce the amount and distribution of light required for a given camera and lens.
Evenness	Light must be distributed evenly over the field of view so that there are no darker or shadowy areas. If there are lighter versus darker areas, brighter areas may appear washed out (i.e., details cannot be distinguished) while no specific objects can be viewed from darker areas.
Location	Light sources must be located above the camera so that light does not shine directly into the camera.

Source: USEPA (2005).

20.4 COMMUNICATION INTEGRATION

This section discusses devices necessary for communication and integration of IT/AI Services Sector industrial processing operations, such as electronic controllers, two-way radios, and wireless data communications. Regarding security applications, electronic controllers automatically activate equipment (such as lights, surveillance cameras, audible alarms, or locks) when triggered. Triggering could be in response to various scenarios, including tripping an alarm or a motion sensor, breaking a window or a glass door, a variation in vibration sensor readings, or simply through input from a timer.

Two-way wireless radios allow two or more users with radios tuned to the same frequency to communicate instantaneously without the radios being physically linked together with wires or cables.

Wireless data communications devices enable data transmission between computer systems without physically linking individual components together via wires or cables. In financial processing systems, these devices are often used to link remote monitoring stations or portable computers (e.g., laptops) to computer networks without using physical wiring connections.

20.4.1 Electronic Controllers

An electronic controller is a piece of electronic equipment that receives incoming electrical signals and uses preprogrammed logic to generate electronic output signals based on the incoming signals. While electronic controllers can be implemented for any application that involves inputs and outputs (for example, control of a piece of machinery in a factory), in a security application, these controllers essentially act as the system's "brain," and can respond to specific security-related inputs with preprogrammed output responses. These systems combine the control of electronic circuitry with a logic function such that circuits are opened and closed (and thus, equipment is turned on and off) through some preprogrammed logic. The basic principle behind the operation of an electrical controller is that it receives electronic input from sensors or any device generating an electrical signal (for example, electrical signals from motion sensors), and then uses it preprogrammed logic to produce electrical outputs (for example, these outputs could turn on power to a surveillance camera or an audible alarm). Thus, these systems automatically generate a preprogrammed, logical response to a preprogrammed input scenario.

The three major types of electronic controllers are timers, electromechanical relays, and programmable logic controllers (PLCs), which are often called "digital relays." Each of these types of controllers is discussed in more detail below.

Timers use internal signals/inputs (in contrast to externally generated inputs) and generate electronic output signals at certain times. More specifically, timers control electric current flow to any application to which they are connected. They can turn the current on or off on a pre-specified schedule by the user. A typical timer range (a time that can be programmed to elapse before the timer activates linked equipment) is from 0.2 seconds to 10 hours, although some of the more advanced timers have ranges of up to 60 hours. Timers are helpful in fixed applications that don't require frequent schedule changes. For example, a timer can be used to turn on the lights in a room or building at a certain time every day. Timers are usually connected to their own power supply (usually 120–240 V).

In contrast to timers, which have internal triggers based on a regular schedule, electromechanical relays and PLCs have both external inputs and external outputs. However, PLCs are more flexible and more powerful than electromechanical relays, and thus, this section focuses primarily on PLCs as the predominant technology for security-related electronic control applications.

Electromechanical relays are simple devices that use a magnetic field to control a switch. Voltage applied to the relay's input coil creates a magnetic field, which attracts an internal metal switch. This causes the relay's contacts to touch, closing the switch and completing the electrical circuit. This activates any linked equipment. These types of systems are often used for high-voltage applications, such as in some automotive and other manufacturing processes.

20.4.1.1 Two-Way Radios

As discussed here, two-way radios are limited to direct unit-to-unit radio communication via single unit-to-unit transmission and reception or multiple hand-held units to a base station radio contact and distribution system. Radio frequency spectrum limitations apply to all handheld units and are directed by the FCC. This also distinguishes a handheld unit from a base station or base station unit (such as those used by an amateur [ham] radio operator), which operate under different wavelength parameters.

Two-way radios allow a user to instantly contact another user or group of users on the same frequency and to transmit voice or data without wires. They use "half-duplex" communications, or communication that can be transmitted or received, but not transmitted and received simultaneously. In other words, only one person can talk, while other personnel with a radio(s) can only listen. To talk, the user depresses the talk button and speaks into the radio. The audio then transmits the voice wirelessly to the receiving radios. When the speaker has finished speaking and the channel has cleared, users on any receiving radios can communicate to answer the first transmission or to begin a new conversation. In addition to carrying voice data, many types of wireless radios also allow the transmission of digital data, and these radios may be interfaced with computer networks that can use or track this data. For example, some two-way radios can send information such as global positioning system (GPS) data or the ID of the radio. Some two-way radios can also send data through a SCADA system.

Wireless radios broadcast these voice or data communications over the airwaves from the transmitter to the receiver. While this can be an advantage in that the signal emanates in all directions and does not need a direct physical connection to be received at the receiver, it can also make the communications vulnerable to being blocked, intercepted, or otherwise altered. However, security features are available to ensure the communications are not tampered with.

20.4.1.2 Wireless Data Communications

A wireless data communication system consists of two components: a "Wireless Access Point" (WAP), and a "Wireless Network Interface Card" (sometimes also referred to as a "Client"), which work together to complete the communications link. These wireless systems can link electronic devices, computers, and computer systems together using radio waves, thus eliminating the need for these components to be directly connected through physical wires. While wireless data communications have widespread application in energy production, distribution, and storage systems, they also have limitations. First, wireless data connections are limited by the distance between components (radio waves scatter over a long distance and cannot be received efficiently, unless special directional antennas are used). Second, these devices only function if the individual components are in a direct line of sight with each other, since radio waves are affected by interference from physical obstructions. However, in some cases, repeater units can amplify and retransmit wireless signals to circumvent these problems. The two components of wireless devices are discussed in more detail below.

1. *WAP*: The WAP provides a wireless data communication service. It usually consists of a housing (which is constructed from plastic or metal, depending on the environment it will be used in) containing a circuit board; flash memory that holds software; one of two external ports to connect to existing wired networks; a wireless radio transmitter/receiver; and one or more antenna connections. Typically, the WAP requires a one-time user configuration to allow the device to interact with the local area network (LAN). This configuration is usually done via a web-driven software application accessed via a computer.

2. *Wireless network interface card/client*: A wireless card is a piece of hardware plugged into a computer that enables that computer to make a wireless network connection. The card consists of a transmitter, functional circuitry, and a receiver for the wireless signal, all working together to enable communication between the computer, its wireless transmitter/receiver, and its antenna connection. Wireless cards are installed in a computer through various connections, including USB adapters, laptop cardbus (PCMCIA), or desktop peripheral (PCI) cards. As with the WAP, software is loaded onto the user's computer, allowing configuration of the card so that it may operate over the wireless network.

Two of the primary applications for wireless data communications systems are enabling mobile or remote connections to a LAN and establishing wireless communications links between SCADA remote telemetry units (RTUs) and sensors in the field. Wireless car connections are usually used for LAN access from mobile computers. Wireless cards can also be incorporated into RTUs to allow them to communicate with sensing devices located remotely.

20.5 CYBER PROTECTION DEVICES

Critical infrastructure, the energy sector, healthcare, finance, transportation, water, and communications sectors have reported significant cyber incidents in the U.S. Several types of cyber protection devices can be used to combat these incidents.

20.5.1 Anti-Virus and Pest Eradication Software

Antivirus programs are designed to detect, delay, and respond to programs or code specifically designed to harm computers. These programs are known as "malware." Malware can include computer viruses, worms, and Trojan horses (programs that appear benign but have hidden, harmful effects).

Pest eradication tools are designed to detect, delay, and respond to "spyware" (strategies that websites use to track user behavior, such as by sending "cookies" to the user's computer), and hacker tools that track keystrokes (keystroke loggers) or passwords (password crackers).

Viruses and pests can enter a computer system through the Internet, infected floppy discs, or CDs. They can also be placed onto a system by insiders. Some of these programs, such as viruses and worms, move within a computer's drives and files, or between computers if the computers are networked. This malware can deliberately damage files, utilize memory and network capacity, crash application programs, and initiate transmissions of sensitive information from a PC. While the specific mechanisms of these programs differ, they can infect files and even the computer's basic operating system, firmware/hardware.

An anti-virus program's most essential features are its ability to identify potential malware, alert a user before infection occurs, and respond to a virus already resident on a system. Most of these programs provide a log so the user can see what viruses have been detected and where they were detected. After detecting a virus, the anti-virus software may delete it automatically or prompt the user to delete it. Some programs will also fix files or programs damaged by the virus.

Various sources of information are available to inform the general public and computer system operators about new viruses being detected. Since anti-virus programs use signatures (or snippets of code or data) to detect the presence of a virus, periodic updates are required to identify new threats. Many anti-virus software providers offer free upgrades that can detect and respond to the latest viruses.

20.5.2 Firewalls

A firewall is an electronic barrier that keeps computer hackers, intruders, or insiders from accessing specific data files and information on a financial services sector's computer network or other electronic/computer systems. Firewalls operate by evaluating and filtering information from a public network (such as the Internet) into the utility's computer or other electronic system. This evaluation can include identifying the source or destination addresses and ports and allowing or denying access based on this identification.

Firewalls use two methods to limit access to the utility's computers or other electronic systems from the public network:

- The firewall may deny all traffic unless it meets specific criteria
- The firewall may allow all traffic through unless it meets specific criteria.

A simple example of the first method is to screen requests to ensure that they come from an acceptable (i.e., previously identified) domain name and Internet protocol address. Firewalls may also use more complex rules that analyze the application data to determine if traffic should be allowed. For example, the firewall may require user authentication (i.e., password use) to access the system. How a firewall determines what traffic to let through depends on which network layer it operates at and how it is configured. Some pros and cons of various methods to control traffic flowing in and out of the network are provided in Table 20.12.

Firewalls may be a piece of hardware, a software program, or an appliance card that contains both.

Advanced features that can be incorporated into firewalls allow for tracking attempts to log on to the LAN system. For example, a report of successful and unsuccessful long-in attempts

TABLE 20.12
Pros and Cons of Various Firewall Methods for Controlling Network Access

Method	Description	Pros	Cons
Packet filtering	Incoming and outgoing packets (small chunks of data) are analyzed against a set of filters. Packets that make it through the filters are sent to the requesting system, and all others are discarded. There are two types of packet filtering: unstatic (the most common) and dynamic.	Static filtering is relatively inexpensive, and little maintenance is required. It is well-suited for closed environments where access to or from multiple addresses is not allowed.	Leaves permanent open holes in the network; allows direct connection to internal hosts by external sources; offers no user authentication, method can be unstatically manageable in large networks.
Proxy service	Information from the Internet is retrieved by the firewall and then sent to the requesting system and vice versa. In this way, the firewall can limit the information made known to the requesting system, making vulnerabilities less apparent.	Only allows temporary open holes in the network perimeter. Can be used for all types of internal protocol services.	Allows direct connections to internal hosts by external clients; offers no user authentication
Stateful pattern recognition	This method examines and compares the contents of certain key parts of an information packet against a database of acceptable information. Information traveling from inside the firewall to the outside is monitored for specific user-defining characteristics, then incoming information is compared to these characteristics. If the comparison yields a reasonable match, the information is allowed through. If not, the information is discarded.	Provides a limited time window to allow pockets of information to be sent; does not allow any direct connections between internal and external hosts; supports user-defining level authentication.	Slower than packet filtering; does not support all types of connections.

Source: USEPA (2005).

may be generated for the computer specialist to analyze. For systems with mobile users, firewalls allow remote access into the private network through secure log-on procedures and authentication certificates. Most firewalls have a graphical user interface for managing the firewall.

In addition, new Ethernet firewall cards that fit in the slot of an individual computer bundle additional layers of defense (like encryption and permit/deny) for personal computer transmissions to the network interface function. These new cards have a slightly higher cost than traditional network interface cards.

20.5.3 Network Intrusion Hardware/Software

Network intrusion detection and prevention systems are software—and hardware-based programs designed to detect unauthorized attacks on a computer network system.

While other applications, such as firewalls and anti-virus software, share similar objectives with network intrusion systems, network intrusion systems provide a deeper layer of protection beyond the capabilities of these other systems because they evaluate patterns of computer activity rather than specific files.

It is worth noting that attacks may come from either outside or within the system (i.e., from an insider), and that network IDSs may be more applicable for detecting patterns of suspicious activity from inside a facility (i.e., accessing sensitive data) than are other information technology solutions.

Network IDSs employ a variety of mechanisms to evaluate potential threats. The types of search and detection mechanisms depend upon the level of sophistication of the system. Some of the available detection methods include:

- *Protocol analysis*: Protocol analysis is the process of capturing, decoding, and interpreting electronic traffic. The protocol analysis method of network intrusion detection involves analyzing data captured during transactions between two or more systems or devices and evaluating this data to identify unusual activity and potential problems. Once a problem is isolated and recorded, problems or potential threats can be linked to pieces of hardware or software. Sophisticated protocol analysis will also provide statistics and trend information on the captured traffic.
- *Traffic anomaly detection*: Traffic anomaly detection identifies potential threatening activity by comparing incoming traffic to "normal" traffic patterns and identifying deviations. It does this by comparing user characteristics against thresholds and triggers defined by the network administrator. This method is designed to detect attacks that span several connections, rather than a single session.
- *Network honeypot*: This method establishes non-existent services to identify potential hackers. A network honeypot impersonates services that don't exist by sending fake information to people scanning the network. It identifies the attacker when they attempt to connect to the service. Legitimate traffic has no reason to access these resources because they don't exist; therefore, any attempt to access them constitutes an attack.
- *Anti-IDS evasion techniques*: These methods are designed to evade attackers' IDS scanning. They include IP defragmentation, TCP streams reassembly, and deobfuscation.

While these detection systems are automated, they can only indicate patterns of activity, and a computer administrator or other experienced individual must interpret activities to determine whether or not they are potentially harmful. Monitoring the logs generated by these systems can be time-consuming, and there may be a learning curve to determine a baseline of "normal" traffic patterns from which to distinguish potential suspicious activity.

20.6 ROADMAP FOR ENERGY STORAGE CYBER SECURITY

To effectively defend Energy Storage Sector equipment and control networks, it is essential to **identify** hardware and software assets and determine possible vulnerabilities and risks to those system components. Steps to identify possible vulnerabilities and risks include evaluating potential high-impact cyber events. Understanding which attacks are benign or catastrophic helps engineers prioritize which defensive solutions should be implemented first and which assets are most critical. To determine credible threats, modeling can help identify high-value assets and potential vulnerabilities. To assess the status of cyber operations in energy storage, "red teaming" (penetration testing) by internal or external organizations can identify weaknesses in cybersecurity posture in the design phase (Onunkwo et al., 2018).

The next step in the cyber protection roadmap is **protection**, which is accomplished by creating new protection technology to define energy storage systems. Network segmentation in control networks, using firewalls, VPNs, proxies, or other networking technologies, minimizes traffic between enclaves and isolates attacks (Johnson et al., 2019; Suh-Lee, 2019). Moving target defense secures control networks against cyberattacks by rotating network addresses, network parameters, application libraries, or other cryptographic tools without affecting system performance noticeably. Many computing products include tamperproof Trusted Platform Modules (TPMs) or similar integrated circuits designed to secure private keys and function alongside the central processor for cryptographic operations (Chavez et al., 2018). Another measure to protect computing is using two processors: one trusted and one untrusted. This makes the two processors mutually dependent in such a way that any tampering is detected. Encryption of data-at-rest and data-in-flight ensures the confidentiality and integrity of the information (Obert et al., 2019). Physical security is also needed because it is necessary to secure cyberspace operations in the Energy Storage Sector. One protection method being used more often is deception and obfuscation, which disrupts reconnaissance and attack attempts. How is this accomplished? Methods include generating false network traffic to disguise legitimate traffic or creating intentionally complex programs. Moreover, device decoys or networks of decoys (honeypots and honeynets) are routinely inserted into the network to confuse attackers and capture their actions before attacks on physical systems. Finally, research must continue on multi-factor **authentication** mechanisms, one-time-use tokens (one-time passwords, OTP), and other technologies that prevent brute force password attacks.

Once an adversary has penetrated a device or network, quick **detection** of their presence is necessary to mitigate damage. Detection is all about situational awareness or intrusion detection. *Situational awareness* comprises logging software or a security incident and event management system; bump-in-the-wire devices for operational technology (OT), encryption, and logging; open-source software for collecting, analyzing, visualizing, and storing network data; and products that ensure telemetry and end-device data integrity.

Intrusion detection detects adversarial actions on the Energy Storage Sector control networks, and appropriate countermeasures must be implemented. Fortunately, there is a wide range of technologies that can be used for IDSs, but they can generally be divided into three approaches (NIST, 2019):

- *Signature-based*: The IDS monitors data for specific patterns indicative of known malware signatures previously observed. Note that the signatures can be in the form of a specific string match, a match on binary data, or a sequence of events occurring within the data.
- *Anomaly-based*: Focus on recognizing abnormal patterns in data compared to a baseline.
- *Policy-based*: Leverage a logical security policy and an execution trace validation algorithm to identify legal and illegal information flows between the objects of a system.

In the Response Mode, the energy storage sector must have appropriate countermeasures designed to minimize the duration and impact of a cyberattack. Specific capabilities must be developed to effectively respond to an energy storage sector cyber event. These capabilities start with cybersecurity personnel having a dedicated, recognized information sharing platform to be shared among Energy Storage Sector stakeholders to learn of adversary actions. Following a cyberattack, it is necessary to have the ability to dissect the events using cyber-forensics. Energy Storage Sector managers and operators must have established an adaptive response mechanism to withstand a cyberattack and quickly recover to a known operable state. What this means is reverting to default, lower-risk operating modes. The energy storage sector must ensure resilient designs, and the cyber-resilience plan must provide the system with the ability to maintain critical operations in the presence of adversary actions. Dynamic assessment technologies and real-time analytics must be performed on data streams to understand the tactics and approach of any adversary. The final action in response mode is to identify those responsible for cyberattacks and begin criminal proceedings.

After taking a response action to the adversary, the system should be returned to regular operation. Ideally, this recovery process should be quick, coordinated, and preplanned. Energy Storage Sector networks and devices should be designed with the capability to ride through cybersecurity attacks by removing inherent weaknesses and building in attack responsiveness. Even after successful cyberattacks, the Energy Storage Sector should be able to recover from the event quickly. Restoration can be maintained using trusted "gold master" firmware, software, and virtual machines to enable the repair and recovery following a cyber-incident.

20.7 THE BOTTOM LINE

This chapter presents an overview of the current state and future trends of the Energy Storage Sector's physical security and cybersecurity, including fundamental security concepts, standards, state-of-the-art physical security and cybersecurity technology, and ongoing R&D efforts to make energy storage more secure.

Including security as a fundamental component in the energy storage sector culture is critical. Experience in the energy storage sector shows that ignoring security during the new product development cycle may lead to costly and ineffective security solutions when added in later stages of product development. The union of change across the energy storage sector is driving the need for massive innovation in how we build out and manage the energy storage sector of the future.

NOTE

1 It is important to point out that even though the following USEPA security asset and device recommendations were first made for water/wastewater critical infrastructure, they are applicable to all other critical infrastructure sectors, including the Energy sector, where practicable.

REFERENCES

Chavez, A.R., Hamlet, J., and Stout, W. (2018). *Artificial Diversity and Defense Security (ADDSec) Final Report*. Albuquerque, NM: Sandia National Lab.

Garcia, M.L. (2001). *The Design and Evaluation of Physical Protection Systems*. Oxford, UK: Butterworth-Heinemann.

IBWA (2004). *Bottled Water Safety and Security*. Alexandria, VA: International Bottled Water Association.

Johnson, J., Quiroz, J., Concepcion, R., Wilches-Bernal, F., and Reno, M.J. (2019, Jan). Power System Effects and Mitigation Recommendations for DER Cyberattacks. *IET Cyber-Phys Syst Theory Appl.*

NAERC. (2002). *Security Guidelines for the Electricity Sector.* NIST Special Report 180047. Washington, DC: North American Electric Reliability Council.

NIST. (2019). *Situational Awareness for Electric Utilities.* Washington, DC: National Institute of Standards and Technology: Department of Commerce.

Obert, J., et al. (2019). *Recommendations for Trust and Encryption in DER Interoperability Standards.* Albuquerque, NM: Sandia Technical Report.

Onunkwo, I., et al. (2018). *Cybersecurity Assessments on Emulated DET Communication Networks.* Albuquerque, NM: Sandia Technical Report.

Schneier, B. (2000). *Secrets & Lies.* New York: Wiley.

Suh-Lee, C. (2019). *EPRI Security Architecture for the Distributed Energy Resources Integration Network: Risk-based Approach for Network Design.* Palo Alto, CA: Electric Power Research Institute.

USEPA. (2005). *Water and Wastewater Security Product Guide.* Accessed 4/14/25 @ http://cfpub.epa.gov. safewater/watersecurity/guide

Glossary[1]

Note: Besides the terms used within the text, economic terms and definitions related to energy production, distribution, arbitrage, and storage are included in this glossary.

A

Absorber The material readily absorbs photons to generate charge carriers (free electrons or holes) in a photovoltaic device.

Acid Hydrolysis A chemical process in which acid converts cellulose or starch to sugar.

Adequacy The ability of a bulk power system to always supply the aggregate electrical demand and energy requirements of the end-use customers, considering scheduled and expected outages of system components.

Advanced Distribution Management System An advanced distribution management system (ADMS) supports the adoption of distributed energy resources (DERs), such as distributed generation and storage technology, energy efficiency, demand response, demand-side management programs, electric vehicles, and others. ADMSs also assist in maintaining reliability and resilience across the distribution grid. Moreover, ADMSs add levels of communication, intelligence, and visibility to the distribution grid so that the distribution utility can better understand real-time conditions across its distribution service territory. ADMSs provide utilities with several specific functions, such as automated fault location, isolation, service restoration, conservation voltage reduction, and volt/VAR optimization. Installing an ADMS is not merely about better integrating DER; instead, an ADMS will change how a utility operates and where it envisions itself and its customers in the future. As customers continue to adopt technology and DER continues to grow, having the information about the grid that can be gathered from ADMS investments will help the utility meet customer demands while maintaining reliability, resilience, and flexibility. Functionally, an ADMS integrates several utility systems, such as outage management and geographical information systems, advanced metering infrastructure (AMI), and customer information systems, into one enterprise-wide system.

Advanced Metering Infrastructure (AMI) Advanced metering infrastructure (AMI) is an electricity and other parameter metering system that records the customer's electricity (or other parameters) hourly or more frequently and transmits measurements over a communication network to a central collection point daily or more frequently. Advanced meters can measure consumption in 15-minute to 1-hour increments. The meters are connected to a communications network, which then transmits the consumption information to the utility's back office for billing. This differs from the historical metering mode, which usually occurs once a month and includes a physical reading of the meter or collecting the information through a local radio network. With the installation of AMI, implementing electric rate designs like time of use (TOU), critical peak pricing (CPP), and real-time pricing (RTP) becomes possible at lower costs than in the past. An integral part of an AMI system is a communication network, which allows the meter to communicate with the utility, send information like consumption, and receive messages like prices or demand response signals. This two-way flow of information means that the utility can provide customers with usage, price, and cost information over the month rather than only at the end of the month.

Adverse Reliability Impact The impact of an event that results in bulk electric system instability or cascading.

Albedo The ratio of light reflected by a surface to the light falling on it.

Alcohol A general class of hydrocarbons that contains a hydroxyl group (OH). There are many types of alcohol (butanol, ethanol, methanol).

Alcohol Fuels Alcohol can be blended with gasoline for use as a transportation fuel. It may be produced from a wide variety of organic feedstocks. The standard alcohol fuels are methanol and ethanol. Methanol may be produced from coal, natural gas, wood, and organic waste. Ethanol is commonly made from agricultural plants, primarily corn, which contains sugar.

Alkaline Fuel Cell (AFC) In a hydrogen/oxygen fuel cell type, the electrolyte is concentrated potassium hydroxide (KOH), and the hydroxide ions (OH) are transported from the cathode to the anode.

Alternating Current (AC) An electric current that reverses its direction at regularly recurring intervals, usually 50 or 60 times/second.

Alternator A device that turns the rotation of a shaft into alternating current (AC).

Ambient The natural condition of the environment at any given time.

Ampere (Amp) A unit of electrical current can be considered the rate of water flowing through a pipe (liters per minute).

Ampere-Hour Amperes times hour measures energy production over time and battery capacity.

Amphorous Silicon An alloy of silica and hydrogen, with a disordered, noncrystalline internal atomic arrangement, that can be deposited in thin layers (a few micrometers in thickness) by several deposition methods to produce thin-film photovoltaic cells on glass, metal, or plastic substrates.

Anaerobic Digestion A biochemical process by which organic matter is decomposed by bacteria in the absence of oxygen, producing methane and other byproducts.

Ancillary Services The conversion of a series of transactions to an equivalent annuity.

Anemometer A wind speed measurement device sends data to the controller and surveys the wind site.

Angle of Attack In wind turbine operation, the airflow angle is relative to the blade.

Angle of Incidence The angle a sun ray makes with a line perpendicular to the surface. For example, a surface that directly faces the sun has a solar incidence of zero. Still, if the surface is parallel to the sun (for instance, sunrise striking a horizontal rooftop), the incidence is 90 degrees.

Anion A negatively charged ion is an ion that is attracted to the anode.

Annual Removals The net volume of growing stock trees removed from the inventory during a specified year by harvesting, cultural operations such as timber land improvement, or land clearing.

Annualized Growth Rate Calculated as follows: (xn/x1)1/n, where x is the value under consideration and n is the number of periods.

Annuity A series of equal annual payments occurring over some time.

Anode The electrode at which oxidation (a loss of electrons) takes place. The anode is the negative terminal for fuel and other galvanic cells; the anode is the positive terminal for electrolytic cells.

Aquifer Water-bearing stratum of permeable sand, rock, or gravel.

Arbitrage The purchase of a commodity or derivative in one market and the sale of the same or a similar commodity or derivative in another market to exploit price differentials.

Área Control Error The instantaneous difference between a balancing authority's actual and scheduled interchange considers the effects of frequency bias, correction for meter error, and automatic time error correction (ATEC) if operating in ATEC mode.

Asexual Reproduction The naturally occurring ability of some plant species to reproduce asexually through seeds, meaning the embryos develop without a male gamete. This ensures the seeds will produce plants identical to the mother plant.

Asset Valuation Methods Various methods, such as reproduction and replacement costs, determine an asset's value. Related terms include market value and earnings.

Reproduction cost is the estimated cost of duplicating an existing facility in its current form and function, usually at current prices. The valuation method requires that costs be based on reproducing facilities using identical replacements; other facilities that perform the same function cannot be used. Note that precise costs can be challenging to calculate because some facilities may be custom-made or impossible to duplicate.

Replacement cost is the estimated cost, usually at current prices, of duplicating an existing facility in function only. This valuation method allows for replacing facilities with others that may vary considerably in form from existing facilities, while still duplicating the existing facility's functions. Using this method, the new facilities may be redesigned to take advantage of new technology or increase efficiency.

Market value is the value established in the market by exchanges between willing sellers and willing buyers. A reasonably sure market value can be determined when several similar sales occur. When a market value cannot be easily determined due to a lack of transactions, other methods, such as reproduction or replacement costs, may be used to estimate the property's value for sale.

Earnings value, also called the income or revenue method of estimating value, estimates the value of property as the present value of future net earnings expected to result from its ownership.

Original and *historical costs* are also sometimes used to estimate the value of an asset.

Automatic Generation Control A process designed and used to adjust a balancing authority area's demand and resources to help maintain the area's control error within the bounds required by applicable NERC reliability standards.

Availability Factor A percentage represents the number of hours a generating unit is available to produce power (regardless of the amount of power) in each period, compared to the number of hours.

Average Rates Average electric or natural gas rates customers pay over a period are usually calculated for a specific group of customers or a particular geographical or service area.

Avoided Cost The cost that an electric utility would incur to produce or otherwise procure electric power is not incurred because the utility purchases this power from a qualifying facility.

Azimuth Angle The angle between the authentic south and the point on the horizon is directly below the sun.

B

Bagasse Sugar mills often burn the fibrous material remaining after the juice is extracted from sugarcane as an energy source.

Baseload Plants Electricity-generating units are operated to meet the system's constant or minimum load. The energy cost from such units is usually the lowest available to the system.

Binary Cycle Binary geothermal systems heat a secondary fluid, which transfers the heat and drives the power turbine using the extracted hot water or steam.

Biobased Product The term 'biobased product,' as defined by the Farm Security and Rural Investment Act (FSRIA), means a product determined by the U.S. Secretary of Agriculture to be a commercial or industrial product (other than food or feed) that is composed, in whole or in significant part, of biological products or renewable domestic agriculture materials (including plant, animal, and marine materials) or forestry materials.

Biochemical Conversion Fermentation or anaerobic digestion produces fuels and chemicals from organic sources.

Biodiesel Fuel is derived from vegetable oils or animal fats. It is produced when vegetable oil or animal fat chemically reacts with alcohol.

Bioenergy Renewable energy is produced from organic matter, which may be used directly as fuel or processed into liquids and gases.

Biofuels Liquid fuels and blending components produced from biomass (plant) feedstocks are used primarily for transportation.

Biogas A combustible gas derived from decomposing biological waste. Biogas typically consists of 50–60% methane.

Biomass Any organic nonfossil material of biological origin constitutes a renewable energy source. It is produced from organic matter available on a renewable or recurring basis, including crops and trees, wood and wood residues, plants (including aquatic plants), grass, animal manure, municipal residues, and other residue materials. Biomass is generally produced sustainably from water and carbon dioxide by photosynthesis. There are three main categories of biomass: primary, secondary, and tertiary.

Biomass Gas (Biogas) A medium Btu gas containing methane and carbon dioxide results from microorganisms' action on organic materials, such as a landfill.

Biomaterials Products derived from organic (as opposed to petroleum-based) products.

Bio-oil Intermediate fuel derived from fast pyrolysis.

Biopower The use of biomass feedstock to produce electric power to heat through direct combustion of the feedstock, gasification, and then combustion of the resultant gas or other thermal conversion processes. Fuel, turbines, fuel cells, or other equipment generate power.

Biorefinery A facility that processes and converts biomass into value-added products. These products can range from biomaterials to fuel, such as ethanol, or important feedstocks for producing chemicals and other materials. Biorefineries can be based on several processing platforms using mechanical, thermal, chemical, and biochemical processes.

Blackbody It is an ideal substance that absorbs all radiation falling on it, reflecting nothing.

Black Liquor (Pulping Liquor) The alkaline spent liquor was removed from the digesters by chemically pulling wood. After evaporation, the liquor is burned as fuel in a recovery furnace that permits the recovery of certain basic chemicals.

Bone Dry Having zero-percent moisture content.

Borehole Breakouts Stress in the rock surrounding the borehole causes the borehole to fail. The breakout is generally located symmetrically in the wellbore perpendicular to the direction of the most significant horizontal stress on a vertical wellbore.

Brine A geothermal solution containing appreciable amounts of sodium chloride or other salts.

British Thermal Unit (Btu) This is an essential measure of thermal (heat) energy. A Btu is the energy required to increase the temperature of 1 pound of water by 1°F at normal atmospheric pressure. 1 Btu = 1055 joules.

Bulk Density Weight per unit of volume is usually specified in pounds per cubic foot.

C

Cap Rocks Rocks of low permeability overlie a geothermal reservoir.

Capacity Factor The ratio of the electrical energy produced by a generating unit for the period considered to the electrical energy that could have been made at continuous full-power operation during the same period.

Capacity, Gross The full-load continuous rating of a generator, prime mover, or other electric equipment under specified conditions designated by the manufacturer. It is usually indicated on a nameplate attached to the equipment.

Capitol Cost The cost of field development, plant construction, and the equipment required for electricity generation.

Carbon Dioxide (CO_2) A product of combustion, the most common greenhouse gas.

Carbon Monoxide (CO) A colorless, odorless gas produced by incomplete combustion. Carbon monoxide is poisonous when inhaled.

Carbon Sequestration The absorption and storage of carbon dioxide from the atmosphere by naturally occurring plants.

Carnot Cycle An ideal heat engine (conceived by Sadi Carnot) is one in which the sequence of operations forming the working cycle consists of isothermal expansion, adiabatic expansion, isothermal compression, and adiabatic compression back to its initial state.

Cascading Heat A process that uses a stream of geothermal hot water or steam to perform successive tasks requiring low temperatures.

Casing The pipe is placed in a wellbore as a structural interface between the wellbore and the surrounding formation. It typically extends from the top of the well and is cemented to maintain the diameter of the wellbore and provide stability.

Cast Silicon Crystalline silicon is obtained by pouring pure molten silicon into a vertical mold and adjusting the temperature gradient along the mold volume during cooling to get slow, vertically advancing crystallization of the silicon. The polycrystalline ingot thus formed is composed of large, relatively parallel, interlocking crystals. The cast ingots are sawn into wafers for further fabrication into photovoltaic cells. Cast-silicon wafers and ribbon-silicon sheets fabricated into cells are usually called polycrystalline photovoltaic cells.

Cathode The electrode at which reduction (a gain of electrons) occurs. The cathode is the positive terminal for fuel and other galvanic cells; the cathode is the negative terminal for electrolytic cells (where electrolysis occurs).

Cation A positively charged ion.

Cellulose The main carbohydrate in living plants. Cellulose forms the skeletal structure of the plant cell wall.

Chips Woody material is cut into short, thin wafers. Chips are used as raw material for pulping, fiberboard, or biomass fuel.

Closed-Loop Biomass Crops grown sustainably to optimize their value for bioenergy and bioproduct uses include annual crops such as maize and wheat, and perennial crops such as trees, shrubs, and grasses such as switchgrass.

Coarse Materials Wood residues suitable for chipping, such as slabs, edgings, and trimmings.

Co-Firing The practice of introducing biomass into the boilers of coal-fired power plants.

Combined Cycle An electric generating technology in which electricity is produced from otherwise lost waste heat existing from one or more gas (combustion) turbines. The exiting heat is routed to a conventional boiler or a heat recovery steam generator for utilization by a steam turbine in the production of electricity. Such designs increase the efficiency of the electric generating unit.

Combined Heat and Power (CHP) Plant A plant is designed to produce heat and electricity from a single heating source. Note: This term is being used instead of the term "cogenerator" used by EIA in the past. CHP better describes the facilities because some plants did not produce heat and power sequentially and, as a result, did not meet the legal definition of cogeneration specified in the Public Utility Regulatory Policies Act (PURPA).

Commercial Sector An energy-consuming sector comprises service-providing facilities and equipment for businesses, federal, state, local governments, and other private and public organizations, such as religious, social, or fraternal groups. The commercial sector includes institutional living quarters. It also includes sewage treatment facilities. Typical uses of energy associated with this sector include space heating, water heating, air conditioning, lighting, refrigeration, cooking, and running various other equipment. ***Note***: This sector includes generators that produce electricity and/or sound thermal output primarily to support the activities of the above-mentioned commercial establishments.

Commercial Species Tree species are suitable for industrial wood products.

Compressed Natural Gas (CNG) Mixtures of hydrocarbon gases and vapors consist principally of methane in gaseous form that has been compressed.

Concentrator A reflective or refractive device that focuses incident insolation onto an area smaller than the reflective or refractive surface, resulting in increased insolation at the point of focus.

Condensate The condensation of steam forms water.

Condenser Equipment that condenses turbine exhaust steam into condensate.

Conservation Reserve Program (CRP) CRP provides farm owners or operators with an annual per-acre rental payment and half the cost of establishing permanent land cover in exchange for retiring environmentally sensitive cropland from production for 10–15 years. In 1996, Congress reauthorized CRP for an additional round of contracts, limiting enrollment to 36.4 million acres at any time. The 2002 Farm Act increased the enrollment limit to 39 million acres. Producers can offer land for competitive bidding based on an Environmental Benefits Index (EBI) during periodic signups or automatically enroll more limited acreages in processes such as riparian buffers, field windbreaks, and grass strips. CRP is funded through the Commodity Credit Corporation (CCC).

Cooling Tower A structure in which heat is removed from hot condensate.

Concentrates A reflective or refractive device that focuses incident insolation onto an area smaller than the reflective or refractive surface, resulting in increased insolation at the point of focus.

Condenser Equipment that condenses turbine exhaust steam into condensate.

Congestion A condition occurs when insufficient transfer capacity is available to implement all of the preferred schedules for electricity transmission simultaneously.

Congestion Costs The charge was assessed and redistributed due to electricity network constraints.

Conservation A reduction in energy consumption corresponds to a decrease in service demand. Service demand can include building-sector end uses such as lighting, refrigeration, heating, industrial processes, or vehicle transportation. Unlike energy efficiency, which is typically a technological measure, conservation is better associated with behavior. Examples of conservation include adjusting the thermostat to reduce the output of a heating unit, using occupancy sensors that turn off lights and appliances, and carpooling.

Conservation Reserve Program (CRP) CRP provides farm owners or operators with an annual per-acre rental payment and half the cost of establishing permanent land cover in exchange for retiring environmentally sensitive cropland from production for 10–15 years. In 1996, Congress reauthorized CRP for an additional round of contracts, limiting enrollment to 36.4 million acres at any time. The 2002 Farm Act increased the enrollment limit to 39 million acres. Producers can offer land for competitive bidding based on an Environmental Benefits Index (EBI) during periodic signups or automatically enroll more limited acreages in processes such as riparian buffers, field windbreaks, and grass strips. CRP is funded through the Commodity Credit Corporation (CCC).

Constraints Constraints or system requirements are a subset of outcomes that are real-world operation requirements (or their modeling approximations) that bound the valuation process.

Contingency The unexpected failure or outage of a system component, such as a generator, transmission line, circuit breaker, switch, or other electrical element.

Single contingency: the loss of a single system element in any operating condition or anticipated mode of operation. *Most severe single contingency*: that single contingency that results in the most adverse system performance in any operating condition or anticipated mode of operation. *Multiple contingency outages*: the loss of two or more system elements caused by unrelated events or by a single low-probability event occurring within a time interval too short (less than 10 minutes) to permit system adjustment in response to losses.

Contingency Reserve The capacity that may be deployed by the balancing authority to respond to a contingency (e.g., outage) and other contingency requirements (such as energy emergency alerts) to balance system generation and demand and return area control error within the specified range. Contingency reserves are typically deployed within 10 minutes following an outage. Usually, at least 50% of the contingency reserve must be spinning reserve, which automatically responds to frequency deviations.

Control Area An area composed of an electric system or systems bound by interconnection metering and telemetry can control generation to maintain its interchange schedule with other control areas and contribute to frequency regulation of the interconnection.

Control Performance Standard The reliability standard that sets the limits of a balancing authority's area control error over a specified period.

Conventional hydroelectric (hydropower) plant A plant where all the power is produced from natural streamflow as regulated by available storage.

Core A cylinder of rock was recovered from the well by a special coring drill bit.

Crop Failure It consists mainly of the acreage on which crops failed because of weather, insects, and diseases, but includes some land not harvested due to lack of labor, lower market prices, or other factors. The acreage planted to cover and improve crops not intended for harvest is excluded from crop failure and is considered idle.

Cropland Total cropland includes five components: cropland harvested, crop failure, cultivated summer fallow, cropland used only for pasture, and idle cropland.

Cropland Harvested Includes row and closely sown crops; hay and silage crops; tree fruits, small fruits, berries, and tree nuts; vegetables and melons; and miscellaneous other minor crops. In recent years, farmers have double-cropped about 4% of the acreage.

Cropland Pasture Land is used for long-term crop rotation. However, some cropland pastures are marginal for crop uses and may remain in the pasture indefinitely. This category also includes land used for pasture before crops reached maturity and some land used for pasture that could have been cropped without additional improvement.

Cropland Used For Cops Cropland used for crops includes cropland harvested, crop failure, and cultivated summer fallow.

Crust Earth's outer layer of rock. Also called the lithosphere.

Cryogenic Liquefaction The process through which gases such as nitrogen, hydrogen, helium, and natural gas are liquefied under pressure at very low temperatures.

Cull Tree A live tree, 5.0 inches in diameter at breast height (dbh) or larger than that, is non-merchantable for saw logs now or prospectively because of rot, roughness, or species.

Curtailment A reduction in the scheduled capacity or energy delivery of an interchange transaction.

Cut-in Speed The speed at which a shaft must turn to generate electricity and send it over a wire.

D

Daylighting Direct, diffuse, or reflected sunlight provides supplemental lighting for building interiors.

dbh The diameter is measured at approximately breast height from the ground.

Debt Ratio The ratio of debt money to total capitalization.

Decision Analysis Evaluation of decision options and estimation of the value of additional information or testing, using the decision-maker's time and risk preferences.

Decision Tree A decision support tool that uses a tree-like graph or model of decisions, choices, options, or actions, and their possible outcomes.

Demand The rate at which electric energy is delivered to or by a system or part of a system (at a given instant or average over any definite interval of time), or the rate at which the customer is using energy.

Demand Response Changes in electric usage by end-use customers from their standard consumption patterns in response to changes in the price of electricity over time or to incentive payments designed to induce lower electricity use at times of high wholesale market prices, or when system reliability is jeopardized.

Demand Response Programs Incentive-based programs that encourage electric power customers to temporarily reduce their demand for power at certain times in exchange for a reduction in their electricity bills. Some demand response programs allow electric power system operators to directly reduce load, while in others, customers retain control. Customer-controlled reductions in demand may involve curtailing load, operating on-site generation, or shifting electricity use to another period. Demand response programs are one type of demand-side management, which also covers broad, less immediate programs such as promoting energy-efficient equipment in residential and commercial sectors.

Demand-Side Management The term for all activities or programs undertaken by an entity (e.g., utility and customers) to influence their demand (e.g., the amount or timing of electricity they use).

Densification A mechanical process to compress biomass (usually wood waste) into pellets, briquettes, cubes, or densified logs.

Depletion Factor Annual percentage of the depletion of the thermal resource.

Depreciation The accounting mechanism for reducing the value of a capitalized item. The precise definition and the reduction schedule may vary widely, depending on the use and type of asset. It is frequently associated with capital cost deductions for income tax purposes.

Depreciation Period The time required for the original capital investment to be fully recovered.

Depreciation, Accelerated Any depreciation schedule that reduces a sum of money more rapidly than would be done with straight-line depreciation.

Depreciation, Book A component of the carrying charge is the revenue required to repay the original investment. In the utility industry, it is usually calculated on a straight-line basis.

Digester Gas Biogas is produced using a digester, an airtight vessel or enclosure in which bacteria decompose biomass in water to produce biogas.

Diode A solid-state device that acts as a one-way valve for electricity.

Direct Control Load Management Demand-side management is under the system operator's control. Direct control load management may control the electric supply to individual appliances or equipment on customer premises. Direct-control management, as defined here, does not include interruptible demand.

Direct Current An electric current that flows in a constant direction. The magnitude of the current does not vary or has a slight variation.

Direct Methanol Fuel Cell (DMFC) A type of fuel cell in which the fuel is methanol (CH3OH) in gaseous or liquid form. The methanol is oxidized directly at the anode instead of first being reformed to produce hydrogen. The electrolyte is typically a polymer electrolyte membrane (PEM).

Direct Use Use geothermal heat without converting it to electricity, such as for space heating and cooling, food preparation, and industrial processes.

Discount Rate The rate used for computing present values reflects that the value of the cash flow depends on the time at which the flow occurs.

Discounted Payback Period The payback period is computed in a way that accounts for the time value of money.

Distributed Generation (Distributed Energy Resources) This refers to electricity provided by small, modular power generators (typically ranging in capacity from a few kilowatts to 50 megawatts) located at or near customer demand.

Distributed Energy Resources Distributed energy resources include distributed generation and storage technologies, energy efficiency, demand response, demand-side management programs, electric vehicles, and other distributed resources.

District Heating A type of direct use is when a utility system supplies multiple users with hot water or streams from a central plant or well field.

Disturbance An unplanned event that produces an abnormal system condition, such as high or low frequency, abnormal voltage, or oscillations.

Dollar Year The year in which the constant dollar results of an analysis are reported.

Downwind Turbine A turbine that does not face into the wind and whose direction is controlled directly by the wind.

Drag Bit Drilling bit that drills by scraping or shearing the rock with fixed hard surfaces, "cutters."

Drilling Boring into the Earth to access geothermal resources, usually with oil and gas drilling equipment modified to meet geothermal requirements.

Dry Steam Very hot steam that doesn't occur with liquid.

Dump Load A device that allows excess energy to be safely disposed of.

E

E-10 A mixture of 10% ethanol and 90% gasoline, based on volume.

E-85 Based on volume, a mixture of 85% ethanol and 15% gasoline.

Earnings That portion of revenue remains after all charges, including interest.

Economic Dispatch The allocation of demand to individual generating units online affects the most economical electricity production.

Effective Load-Carrying Capability The amount of additional load the power system can supply with a particular generation resource of interest, with no net change in reliability.

Efficiency The ratio of the sound energy output of a machine or other energy-converting plant to the energy input.

Electrical Energy The generation of electric power by a device over a period of time is expressed in kWh, MWh, GWh, or TWh.

Electric Power Sector An energy-consuming sector consisting of electricity only and combined heat and power (CHP) plants whose primary business is to sell electrical energy or electricity and heat to the public—i.e., North American Industry Classification System 22 plants.

Electric Power System A combination of generation, transmission, and distribution components.

Electric Rates The rates paid by end-use customers for electricity service. In addition to the energy charge (based on the customer's energy consumption in kWh), electricity rates may include a capacity or damage charge (based on the customer's peak demand in KW) and a service charge. Common types of electric rates are the following:

Flat electric rate: A flat rate charges customers per unit of consumption (kWh), at the same rate for all units. This rate structure (in combination with a monthly customer charge) is commonly used in rates for residential electric customers. It is the most common form of residential rate design used nationwide today.

Block electric rate: An increasing, inverted, or inclining block rate (IBR) structure is designed to charge customers a higher per unit rate as their usage increases over certain "blocks" within a billing cycle. For example, a three-tier INR would identify three blocks of usage: block one could be 0–150 kWh, block two could be 150–250 kWh, and block three could be all usage over 250 kWh. Each block has a price for all electricity used within it, with the prices increasing as a customer moves through the blocks over a billing period.

Time-variant electric rate: Time-variant rates (TVRs) are designed to recognize differences in a utility's service cost and marginal costs at different times (e.g., hour, day, or season). Generally, a TVR charges customers a higher price during peak hours and a lower price during off-peak hours. Unlike flat rates, customers must be aware of usage throughout the day and the month to respond to the price signals in a TVR design. A customer may increase savings under a TVR compared with a flat rate if that customer uses energy in response to the time-variant price signal, such as shifting usage to lower-cost periods or conservation.

Time-of-use electric rate: A specific kind of time-variant rate, a TOU rate charges customers different prices according to a pre-determined schedule of peak and off-peak hours and rates. For many utilities, TOU rates have been a voluntary option for residential customers for decades, but generally, few customers participate. Many commercial and industrial (C&I) electric customers already receive service under TOU rate designs.

Real-time pricing rate: In a real-time (RTP) plan, the customer is charged for generation at the prices set by the wholesale market (for deregulated utilities or vertically integrated utilities participating in an organized wholesale market) or at the short-run marginal generation costs (for vertically integrated utilities not participating in an organized wholesale market) by the hour. With advanced metering infrastructure, it is possible to implement real-time pricing for residential and smaller C&I customers. RTP is available to residential customers in the Illinois service territories for Commonwealth Edison (ConEd) and Ameren. The real-time rates for these programs are based on the day-ahead hourly wholesale price for the given utility zones.

Critical peak pricing rate: A utility may implement a critical peak pricing (CPP) rate during expected shortages or anticipated high-usage days to mimic peak price increases. The utility will announce, usually the day before, the hours that the CPP rate will be in effect. The CPP rate reflects the higher electricity generation price during those CPP hours or scarcity during the event hours. Generally, the CPP rate is set significantly higher than the non-CPP rate to incentivize customers to reduce consumption. A CPP can be included with a TOU rate or paired with a demand response (DR) program. A CPP event is usually limited to certain peak hours over a year.

Three-part rate/demand charges: Because the utility system is built to serve peak loads, the costs of providing electricity at peak hours are higher than during on-peak hours. Part of this indicates the higher costs associated with maintaining adequate infrastructure and generation necessary to serve customers during peak demand times. To address this situation, a rate structure option is the three-part rate, which adds a demand charge to the existing fixed charge and volumetric rate. This rate recognizes three of the major contributors to a utility's costs. To the extent that each rate component properly reflects its associated costs, the price signal to customers should be improved over the use of flat or block rates. Such rates are commonplace for C&I customers. The demand charge component usually reflects the cost of providing electricity during the peak hour of the month. To identify costs associated with peak hours, a "demand charge" is the way for a utility to send a peak pricing signal over a certain time period (such as a month).

Electric Utility A corporation, person, agency, authority, or other legal entity or instrumentality aligned with distribution facilities for the delivery of electric energy primarily by the public. Investor-owned electric utilities, municipal and State utilities, Federal electric utilities, and rural electric cooperatives are included. A few tariff-based entities that are corporately aligned with companies that own distribution facilities are also included.

Embedded Cost The total current cost of owning, operating, and maintaining an existing electric power system.

Emergency Any abnormal system condition that requires immediate manual or automatic action to prevent loss of firm load, equipment damage, or tripping of system elements that could adversely affect the reliability of the electric system.

Emissions Anthropogenic releases of gases into the atmosphere. In the context of global climate change, they consist of relatively essential greenhouse gases (e.g., the release of carbon dioxide during fuel combustion).

Endothermic A chemical reaction that absorbs or requires energy (usually in the form of heat).

Energy The ability to do work.

Energy Arbitrage In general, storing energy when electricity prices are low and generating when prices are high typically refer to the mode of pumped storage hydropower (PSH) operation in electricity markets, when they pump during low electricity prices and generate during hours with high electricity prices. Some other energy storage technologies can also perform energy arbitrage.

Energy Charge That portion of the charge for electric service is based upon the electric energy (kWh) consumed or billed.

Energy Crops Crops are grown specifically for their fuel value. They include food crops such as corn and sugarcane and nonfood crops such as poplar trees and switchgrass. Currently, low-energy crops are under development: short-rotation woody crops, which are fast-growing hardwood trees harvested in five to eight years, and herbaceous energy crops, such as perennial grasses, which are harvested annually after taking two to three years to reach full productivity.

Energy Efficiency A ratio of service provided to energy input.

Energy Efficiency Programs Programs that reduce the energy used by specific end-use devices and systems, typically without affecting the services provided.

Energy Intensity A ratio of energy consumption to another metric (typically national gross domestic product in the case of a country's energy intensity). Sector-specific intensities may refer to energy consumption per household, per unit of commercial floor space, per dollar value of industrial shipment, or another metric indicative of a sector. Improvements in energy intensity include energy efficiency, conservation, and structural factors unrelated to technology or behavior.

Enhanced Geothermal Systems (EGS) Engineered reservoirs that can extract economic amounts of that from geothermal resources. Rock fracturing, water injection, and water circulation technologies sweep heat from the unproductive areas of existing geothermal fields or new fields lacking sufficient production capacity.

Enthalpy A substance's thermodynamic property is defined as the sum of its internal energy plus the pressure of the substance times its volume, divided by the mechanical equivalent of heat. The total heat content of air is the sum of the enthalpies of dry air and water vapor per unit weight of dry air, measured in Btu per pound (or calories per kilogram).

Environmental Externality Health and environmental impacts on society are generally not internalized in the market price of a good or service.

Environmental Impact Statement A document was created from a study of the expected environmental effects of a new development or installation.

Equity That portion of a company's capitalization results from the sale of common and preferred stock and retained earnings.

Equity Ratio The ratio of equity money to total capitalization. It is also equal to one minus the debt ratio.

Escalation, Apparent The total annual rate of increase in cost. The apparent escalation rate includes the effects of inflation and real escalation.

Escalation, Real The annual rate of expenditure increase is due to factors such as resource depletion, increased demand, and improvements in design or manufacturing (negative rate). The real escalation rate does not include inflation.

Ethanol (also known as Ethyl Alcohol or Grain Alcohol, CH3-CH2OH) A clear, colorless, flammable, oxygenated hydrocarbon with a boiling point of 173.5°F in an anhydrous state. However, it readily forms a binary azeotrope with water, with a boiling point of 172.67°F at a composition of 95.57% by weight ethanol. It is used in the United States as a gasoline octane enhancer and oxygenate (maximum 10% concentration). Ethanol can be used in higher concentrations (E85) in vehicles designed for its use. Ethanol is typically produced chemically from ethylene or biologically from the fermentation of various sugars from carbohydrates found in crops and cellulosic residues from crops or wood. The lower heating value, equal to 76,000 Btu/gallon, is assumed for estimates in this text.

Evacuated Tube In a solar thermal collector, an absorber tube is contained in an evacuated glass cylinder through which collector fluids flow.

Exothermic A chemical reaction that gives off heat.

Expected Value The mean or average value of a variable.

Expense A cost of goods and services that are typically used or consumed in one year or less (e.g., fuel, operation, and maintenance).

Externality A cost or benefit is not accounted for in the price of goods and services. Often, "externality" refers to the cost of pollution and other environmental impacts.

F

Fast Pyrolysis Thermal biomass conversion by rapid heating to 450–600°C without oxygen.

Fault A fracture in rock exhibiting relative movement between the adjoining surfaces.

Federal Energy Regulatory Commission Federal agency with jurisdiction over interstate electricity sales, wholesale electric rates, hydroelectric licensing, natural gas pricing, oil pipeline rates, and gas pipeline certification. FERC is an independent agency within the Department of Energy and is the successor to the Federal Power Commission.

Feller-buncher A self-propelled machine that cuts trees with giant shears near ground level and then stacks the trees into piles to await skidding.

Feedstock A product used as the basis for the manufacture of another product.

Fenestration The whole-building design approach will determine what fenestration products should be used—windows, doors, and skylights. You want to select products that accommodate your building's climate, including insulation, daylighting, heating and cooling, and natural ventilation.

Fermentation Conversion of carbon-containing compounds by microorganisms for producing fuels and chemicals such as alcohol, acids, or energy-rich gases.

Fiber Products Products derived from fibers of herbaceous wood plant material. Examples include pulp, composition board products, and wood chips for export.

Financial Transmission Right A contract entitles the holder to receive or pay compensation for transmission charges that arise when grid congestions cause price differences due to the redispatch of generators.

Fine Materials Planer shavings and sawdust are not suitable for chipping.

Firm Power Power or power-producing capacity is intended to be available at all times during the period covered by a guaranteed commitment to delivery, even in adverse conditions.

Fischer-Tropsch Fuels Liquid hydrocarbon fuels are produced by combining carbon monoxide and hydrogen. This process converts coal, natural gas, and low-value refinery products into a high-value diesel substitute.

First Contingency Reliability Criteria The requirement is that an electric system be planned and operated to safely withstand the loss of the most prominent single system element (i.e., power plant or transmission line).

Fixed Costs Costs or expenses that do not depend on the production output or operation level are incurred even without production or operation. For example, for generating units, fixed costs are mainly the costs of capacity, while variable costs are primarily the costs of operation.

Fixed Charge Rate The factor by which the present value of capital investment is multiplied to obtain the annual cost attributable to capital investment.

Flash Steam Steam is produced when the pressure on a geothermal liquid is reduced. It's also called flashing.

Flat Plate Pumped A medium-temperature solar thermal collector typically consists of a metal frame, glazing, absorbers (usually metal), and insulation. It uses a liquid as the heat-transfer medium and is predominantly used in water heating applications.

Flexibility An electric system's ability to respond to further changes may stress it in the short term and require it to adapt in the long term. Increased variability resulting from the growing share of variable renewable generation, such as wind and solar power, is increasing the need for flexibility in grid planning and operations.

Flexibility Reserve A new type of reserve is being introduced in some electricity markets, mainly to compensate for the variability and uncertainty of variable renewable generation (e.g., wind and solar) and to correct control area exchanges (reduce energy imbalances).

Flexible-fuel Vehicle A vehicle with a single fuel tank designed to run on varying blends of unleaded gasoline with either ethanol or methanol.

Flow Battery An electrochemical energy storage device that utilizes tanks of rechargeable electrolytes to refresh the energy-producing reaction. Since its capacity is limited only by the size of its electrolyte tanks, it is helpful for large-scale backup systems to supplement other intermittent generation forms.

Flow Through Accounting An accounting practice used by regulated utilities in which deferred income taxes are passed on immediately to ratepayers through a decrease in rates or to stakeholders through an increase in earnings (return of equity) is the opposite of normalization accounting.

Forced Outage The condition in which the equipment is unavailable for service due to unanticipated failure or the removal of equipment from service for emergency reasons.

Forest Land Land at least 10% stocked by forest trees of any size, including land that formerly had such tree cover and that will be naturally or artificially regenerated. Forest land includes transition zones, such as areas between heavily forested and nonforested lands that are at least 10% stocked with forest trees and forest areas adjacent to urban and built-up lands. Also included are pinyon-juniper and chaparral areas in the West and the forested regions. The minimum area for the classification of forest land is 1 acre. Roadside, streamside, and shelterbelt strips of trees have a crown width of at least 120 feet to qualify as forest land. Unimproved roads, trails, streams, and clearings in the forest areas are classified as forests if they are less than 120 feet wide.

Fracture Natural or induced break in the rock.

Fracturing Treatments Fracturing treatments are performed by pumping fluid into the subsurface at pressures above the fracture pressure of the reservoir formation to create a highly conductive flow path between the reservoir and the wellbore.

Framework A defined, systematic approach to accounting for and comparing costs and benefits.

Frequency Bias A value, usually given as MW/0.1 Hz, associated with a balancing authority area relates the difference between scheduled and actual frequency to the amount of generation required to correct it.

Frequency Control Also known as frequency regulation, frequency control includes maintaining system frequency within the specified range by continuously regulating system generation and loads. Typically, a three-stage frequency control procedure (primary, secondary, and tertiary control) is applied:

Primary frequency control is the automatic and immediate response of turbine governors and some loads to frequency changes. It assists in stabilizing the system frequency immediately following a disturbance. Primary control, also called frequency response, occurs within the first few seconds following a change in system frequency.

Secondary frequency control: Balancing services are deployed within minutes. Secondary frequency control is accomplished using automatic generation control and the manual actions taken by the system operator to provide additional adjustments. Secondary control maintains the minute-to-minute balance throughout the day and is used to restore frequency to its scheduled value following a disturbance.

Tertiary frequency control: Actions taken to provide relief for the secondary frequency control resources available to handle current and future contingencies. Reserve deployment and reserve restoration following a disturbance are everyday tertiary control actions.

Frequency Regulation Frequency regulation, also known as frequency control, maintains system frequency within the specified range. It typically refers to both the frequency response of turbine governors and automatic generation control. It is provided by online generation units with frequency-responsive governors and by generation and demand resources that can respond rapidly to AGC requests for up and down movements to counterbalance minute-to-minute fluctuations in system load and correct for unintended fluctuations in generator outputs.

Frequency Response The ability of a system or elements of the system to react or respond to a change in system function.

Fuel Cell One or more cells can generate an electrical current by converting the chemical energy of fuel directly into electrical energy. Fuel cells differ from conventional electrical cells in that the active materials, such as fuel and oxygen, are not contained within the cell but are supplied from outside.

Fuel Cell Poisoning A fuel cell's efficiency is lowered due to impurities in the fuel binding to the catalyst.

Fuel Cell Stack Individual fuel cells are connected in a series. Fuel cells are stacked to increase voltage.

Fuel Treatment Evaluator (FTE) A strategic assessment tool capable of aiding the identification, evaluation, and prioritization of fuel treatment opportunities.

Fuelwood Wood and wood products, possibly coppices, scrubs, branches, etc., are bought, gathered, and used by direct combustion.

Full Sun The amount of power density in sunlight is received at the Earth's surface at noon on a clear day (about 1,000 Watts/square meter).

Fumarole A vent or hole in the Earth's surface, usually in a volcanic region, from which steam, gaseous vapors, or hot gases issue.

G

Gallon Fuel oil is measured using a volumetric measure equal to 4 quarts (231 cubic inches). One gallon equals 3.785 liters, and one barrel equals 42 gallons.

Gasification A chemical or heat process converts solid fuel to a gaseous form.

Gasohol A motor vehicle fuel that is a blend of 90% unleaded gasoline and 10% ethanol (by volume).

Gearbox Increases the rpm of the low-speed shaft, transferring its energy to the high-speed shaft to provide enough speed to generate electricity.

Generation (Electricity) The process of producing electric energy from other forms of energy; the amount of electric energy produced is expressed in watt-hours (Wh).

Geothermal Of or relating to the Earth's interior heat.

Geothermal Energy Extraction of the Earth's interior heat from hot water or rocks makes it available to man. At electric power plants, hot water or steam extracted from geothermal reservoirs in the Earth's crust is supplied to steam turbines that drive generators to produce electricity.

Geothermal Gradient The rate of temperature increase on Earth is a function of depth. Temperature increases an average of 1°F for every 75 feet in descent.

Geothermal Heat Pumps Devices that take advantage of the relatively constant temperature of the Earth's interior use it as a source and sink of heat for heating and cooling. When cooling, heat is extracted from space and dissipated into the Earth; when heating, heat is extracted from the Earth and pumped into space.

Geothermal Resources When the Earth's natural heat is collected and transported to the surface, it can be used for beneficial purposes.

Geothermal Plant A plant in which a turbine is driven either by hot water or by natural steam, which derives its energy from heat found in rocks or fluids at various depths beneath the Earth's surface. The fluids are extracted by drilling and/or pumping.

Geyser A spring that shoots jets of hot water and steam into the air.

Glazing Transparent or translucent material (glass or plastic) used to admit light and/or to reduce heat loss; used for building windows, skylights, or greenhouses, or for covering the aperture of a solar collector.

Global Positioning System (GPS) A navigation system using satellite signals to fix the location of a radio receiver on or above the Earth's surface.

Governor The electronic, digital, or mechanical device that implements the primary frequency response of generating units or other system elements.

Grassland Pasture and Range All open land is used primarily for pasture and grazing, including shrub and brush land types of pasture; grazing land with sagebrush and scattered mesquite; and all team and native grasses, legumes, and other forage used for pasture or grazing. Because of the diversity in vegetative composition, grassland pasture and range are not always clearly distinguishable from different types of pasture and range. At one extreme, permanent grassland may merge with cropland pasture, or grassland may often be found in transitional areas with forested grazing land.

Gravimetry The use of precisely measured gravitational force to determine mass differences that can be correlated to subsurface geology.

Green Pricing/Marketing In the case of renewable electricity, green pricing represents a market solution to the various problems associated with regulatory valuation of the non-market benefits of renewables. Green pricing programs allow electricity customers to express their willingness to pay for renewable energy development through direct payments on their monthly utility bills.

Greenhouse Gas Gases that trap the heat of the sun in the Earth's atmosphere, producing the greenhouse effect. The two major greenhouse gases are water vapor and carbon dioxide. Other greenhouse gases include methane, ozone, chlorofluorocarbons, and nitrous oxide.

Grid The layout of an electrical distribution system.

Grid Services The combination and operationalization of performance characteristics to perform a specific action, such as providing spinning reserve or load following. The commonly recognized and services have evolved through time as new challenges have faced the grid. In a market context, performance characteristics are monetized through the procurement of select services via market products. However, not all services (e.g., inertia) currently have market products and so remain unmonetized.

Gross Generation The total amount of electric energy produced by the generating units at a generation station or stations is measured at the generator terminals.

Growing Stock A classification of timber inventory that includes live trees of commercial species meeting specified standards of quality or vigor. Cull trees are excluded. When associated with volume, it contains only trees 5.0 inches in dbh and larger.

H

HDR Hot, dry rock. Subsurface geologic formations of abnormally high heat content that contain little or no water.

Hardwoods Usually broad-leaved and deciduous trees.

Heat Exchanger A device for transferring thermal energy from one fluid to another.

Heat Flow Heat moves from the earth to the surface, where it is dissipated by radiation into the atmosphere, surface water, and space.

Heat Pump A year-round heating and air conditioning system employing a refrigeration cycle. In a refrigeration cycle, a refrigerant is compressed (as a liquid) and expanded (as a vapor) to absorb and reject heat. The heat pump transfers heat to a space to be heated during the winter period and, by reversing the operation, extracts (absorbs) heat from the same space to be cooled during the summer. The refrigerant within the heat pump in the heating mode absorbs the heat to be supplied to the space to be heated from an outside medium (air, ground, or groundwater) and, in the cooling mode, absorbs heat from the space to be cooled to be rejected to the outside medium.

Heat Pump (Air Source) An air-source heat pump is the most common type of heat pump. The heat pump absorbs heat from the outside air and transfers the heat to the space to be heated in the heating mode. In cooling mode, the heat pump absorbs heat from the space to be cooled and rejects the heat to the outside air. In the heating mode, when the outside air approaches 32°F or less, air-source heat pumps lose efficiency and generally require a backup (resistance) heating system.

Heat Pump (Geothermal) A heat pump in which the refrigeration exchanges heat (in a heat exchanger) with a fluid circulating through an earth connection medium (ground or groundwater). The fluid is contained in a variety of loop (pipe) configurations depending on the temperature of the ground and the ground area available. Loops may be installed horizontally or vertically in the ground or submerged in water.

Heat Pump (efficiency) The efficiency of a heat pump, that is, the electrical energy to operate it, is directly related to the temperatures between which it operates. Geothermal heat pumps are more efficient than conventional heat pumps for air conditioners that use outdoor air since the ground or groundwater is a few feet below the Earth's surface and remains relatively constant throughout the year. It is from the atmosphere where the air temperature is much colder, and in summer, waste heat is transferred to the relatively calm ground rather than to the hotter air. Geothermal heat pumps are generally more expensive to install than outside air heat pumps. However, depending on the location, geothermal heat pumps can reduce energy consumption (operating cost) and emissions by more than 20% compared to high-efficiency outside air heat pumps. Geothermal heat pumps also use the waste heat from air conditioning to provide free hot water heating in the summer.

Heat Rate The amount of input energy (e.g., usually expressed in kJ or Btu) required to produce one kWh of electric energy.

Heating Value The maximum amount of energy that is available from burning a substance.

Herbaceous Non-woody type of vegetation, usually lacking permanent strong stems, such as grasses, cereals, and canola (rapeseed).

High-Speed Shaft Transmits force from the gearbox to the generator.

High-Temperature Collector A solar thermal collector is designed to operate at 180°F or higher.

Hub The center part of the rotor assembly connects the blades to the low-speed shaft.

Hurdle Rate The minimum acceptable rate of return on a project.

Hydraulic Stimulation A stimulation technique was performed using fluid.

Hydrothermal About hot water.

Hydrothermal Reservoir An aquifer, or subsurface water, has sufficient heat, permeability, and water to be exploited without stimulation or enhancement.

I

Idle Cropland Land in cover, soil improvement crops, and cropland on which no crops were planted. Some croplands are idle each year for various physical and economic reasons. Acreage diverted from corps to soil-conserving uses (if not eligible for and utilized as cropland pasture) under federal farm programs is included in this component. Cropland enrolled in the Federal Conservation Reserve Program (CRP) is included in the idle cropland.

Imbalance Energy There is a discrepancy between the amount of energy a seller contracted to deliver and the actual amount of energy delivered.

Impacts Metrics measure the changes in outcomes.

Inadvertent Interchange The difference between the control area's net actual and scheduled interchange.

Incident Light Light that shines onto the face of a solar cell or module.

Incremental Cost The change in total costs results when output is increased or decreased by a block or specific increment of units, not just one unit. If the output is increased or decreased by just one unit (a single kW or kWh), the resulting costs are referred to as *marginal costs.*

Independent Power Producer A corporation, person, agency, authority, or other legal entity or instrumentality that owns or operates facilities for the generation of electricity for use primarily by the public and is not an electric utility.

Independent System Operator An independent, federally regulated entity established to coordinate regional transmission in a non-discriminatory manner and ensure the safety and reliability of the electric system.

Induced Seismicity Induced seismicity refers to minor earthquakes and tremors typically caused by human activity altering the stresses and strains on the Earth's crust. Most induced seismicity is of an extremely low magnitude, and in many cases, human activity is merely the trigger for an earthquake that would have occurred naturally.

Industrial Wood All commercial roundwood products except fuelwood.

Inertia The property of a mass that resists speed changes.

Inertia Response The inertial resistance of the rotating mass of the turbine generator resists instantaneous speed changes.

Inflation The rise in price levels caused by an increase in available currency and credit without a proportionate increase in available goods and services of equal quality. Inflation does not include real escalation. Inflation is usually expressed in terms of an annual percentage change.

Injection The process of returning spent geothermal fluids to the subsurface. Sometimes referred to as reinjection.

Integrated Resource Planning (IRP) IRP analyzes utility growth and operation to ensure that energy needs are met through the optimum mix of supply and demand-side resources. It is also called least-cost planning.

Interchange Energy transfers that cross balancing authority boundaries.

Interconnected Power System A network of subsystems of generators, transmission lines, transformers, switching stations, and substations.

Interconnection A geographic area in which the operation of bulk power system components is synchronized.

Interferometric Synthetic Aperture Radar (InSAR) A remote sensing technique that uses radar satellite images to determine the movement of the Earth's surface.

Internal Collector Storage A solar thermal collector is one in which the storage medium absorbs incident solar radiation.

Internal Rate of Return The discount rate required to equate the net present value of a cash flow stream to zero.

Interruptible Load or Demand Demand that the end-use customer make available to its load-serving entity via contract or agreement for curtailment.

Irradiance The direct, diffuse, and reflected solar radiation that strikes the surface.

Investment The expenditure for which returns are expected to extend beyond one year.

Investment Useful Lifetime The estimated useful life of a capital investment.

Investment Tax Credit An immediate reduction in income taxes equal to a percentage of the installed cost of a new investment.

Investment Year The year a capital or equipment investment is entirely constructed or installed and placed into service.

Investor-Owned Utility A privately owned electric utility whose stock is publicly traded. It is rate-regulated and authorized to achieve an allowed rate of return.

J

Joule This is the basic energy unit for the metric system, or in a later, more comprehensive formulation, the International System of Units (SI). It is ultimately defined in terms of the meter, kilogram, and second.

K

Kaplan Turbine A type of turbine that has two blades whose pitch is adjustable. The turbine may have gates to control the fluid flow angle into the blades.

Kilowatt (kW) One thousand watts of electricity.

Kilowatt-hour (kWh) One thousand watt-hours.

L

Landfill Gas Gas is generated by the decomposition of organic material at landfill disposal sites. Landfill gas is approximately 50% methane.

Langley (L) Unit of solar irradiance. One gram calorie per square centimeter. 1 L = 85.93 kWh/m^2.

Least-Cost Planning A process of analyzing the growth and operation of utilities to ensure that energy needs over a specified future period are met through the optimal (least-cost) mix of supply-side and demand-side resources, while satisfying all reliability criteria and other constraints.

Leeward Away from the direction of the wind. Opposite of windward.

Levelization Conversion of a series of transactions to an equivalent value per output unit.

Levelized Cost of Energy The cost per unit of energy, if held constant through the analysis period, would provide the same net present value as the system's net present value.

Life Cycle Analysis The analysis focused on a product's environmental impact throughout its life cycle, from resource extraction to post-consumer waste disposal. It is a comprehensive approach to examine a product or package's ecological impacts.

Life Cycle Cost The present value over the analysis period of all system resultant costs.

Lignin Structural constituent of wood and (to a lesser extent) other plant tissues, which encrusts the cell walls and cements the cells together.

Line Shaft Pump A fluid pump with a pumping mechanism in the wellbore is driven by a shaft connected to a motor on the surface.

Liner A casing string that does not extend to the top of the wellbore but is anchored or suspended from inside the bottom of the previous casing string.

Liquid Collector A medium-temperature solar thermal collector, predominantly employed in water heating, uses pumped liquid as the heat-transfer medium.

Lithology The study and description of rocks in terms of their color, texture, and mineral composition.

Live Cull A classification that includes live cull trees. When associated with volume, the net volume in live cull trees is 5.0 inches in dbh and larger.

Load At any specified point in an electric power system, the simultaneous demand of all customers is required. Stated differently, an end-use device or customer that receives power from the electric system is required.

Load Balancing One of the challenges of wind energy production, which produces energy on a less predictable schedule than other methods, is keeping the amount of electricity produced (the supply) equal to consumption (the demand).

Load Duration Curve A chart showing electric demand in decreasing magnitude plotted against total duration of occurrence over a specific period of time (usually a year).

Load Factor The ratio of the actual energy consumed during a designated period to the energy that would have been consumed if the peak load existed throughout the specified period (i.e., the ratio between the actual and maximum possible consumption in the period). The term describes a characteristic of an individual or aggregated load rather than generation.

Load Following Increase or decrease in generating unit power output to follow longer-term (hourly) changes in electricity demand.

Load Leveling Shifting the load from peak to off-peak periods results in a flatter load profile for the system load.

Load Profile A curve depicting the aggregated system load of all electricity consumers, typically over 24 hours.

Load-Serving Entity Secures energy and transmission service (and related interconnected operations services) to serve its end-use customers' electrical demand and energy requirements.

Local Solar Time A system of astronomical time in which the sun crosses the true north-south meridian at noon and which differs from local time according to longitude, time zone, and equation of time.

Location Marginal Price The market-clearing price for electricity at the location where the energy is delivered or received.

Logging Residues The unused portions of growing and non-growing stock trees were cut or killed by logging and left in the woods.

Long-Term Transmission Planning Horizon The transmission planning period covers years six through ten or beyond, when required, to accommodate any known long-term projects that may take several years to complete.

Loss-of-Load Expectation The expected number of days per year for available generating capacity is insufficient to serve the daily peak or hourly demand (load).

Loss-of-Load Probability The proportion (probability) of days per year, hours per year, or even per season for which available generating capacity is insufficient to serve the daily peak or hourly demand (i.e., the proportion of time that the available generation is expected to be unable to meet the system load).

Lost Circulation Zones in a well that imbibe drilling fluid from the wellbore reduce fluid flow returning to the surface. This loss causes drilled rock particles to build up in the well and can cause problems with cementing the casing.

Low-Speed Shaft Connects the rotor to the gearbox.

Low-Temperature Collectors Metallic or nonmetallic solar thermal collectors generally operate at temperatures below 110°F and use pumped liquid or air as the heat-transfer medium. They usually contain no glazing or insulation and are often made of plastic or rubber, although some are metal.

M

Magma Molten rock within the Earth, from which igneous rock is formed by cooling.

Magnetic Survey Measurements of the Earth's magnetic field are mapped and used to determine subsurface geology.

Magneto-Telluri An electromagnetic method of determining structures below the Earth's surface using electrical currents and the magnetic field.

Mantle The Earth's inner layer of molten rock lies beneath its crust and above its liquid iron and nickel core.

Marginal Cost The change in total costs resulted from increasing or decreasing output by one unit. In the electric power industry, the marginal cost is the change in total costs resulting from producing one additional kW or kWh of electricity.

Marginal Electric Generating Unit In organized wholesale markets, the price of the marginal source of electricity (e.g., generating unit providing the next increment or decrement of energy) usually sets the price for all generation.

Market Clearing Price The price where supply meets demand for day-ahead or hour-ahead markets.

Market-Based Pricing Prices of electric power or other forms of energy determined in an open market system of supply and demand in prices are set solely by agreement as to what buyers will pay and sellers will accept. Such prices could recover less or more than full costs, depending upon what the buyers and sellers see as their relevant opportunities and risks.

Matrix Treatment Treatment performance is below the reservoir fracture pressure. Matrix treatments are generally designed to restore the reservoir's natural permeability following damage to the near-wellbore area. They typically use hydrochloric or hydrofluoric acids to remove mineral material that reduces flow into the well.

Medium-Temperature Collectors Solar thermal collectors are designed to operate in the temperature range of 140–180°F but can also operate at a temperature as low as 110°F. The Prices are solely by agreement as to what buyers will pay and sellers will accept.

Megawatt (WM) One million watts of electricity.

Merchant Generator A generating plant built with no energy sales contracts in place.

Methane Methane is a colorless, flammable, odorless hydrocarbon gas (CH4) that is a major component of natural gas. It is also an essential source of hydrogen in various industrial processes. Methane is a greenhouse gas.

Methanol Methanol, also known as methyl alcohol or wood alcohol, has the chemical formula CH3OH. It is usually produced by chemical conversion at high temperatures and pressures. Although it is generally made from natural gas, methanol can also be produced from gasified biomass.

Metrics Factors that provide an indication of the extent to which an outcome is achieved. Metrics can be quantitative or qualitative but should provide a reasonably objective means of assessing the outcomes and allow comparisons to be made.

Microgrid Microgrids are localized grids that can disconnect from the traditional grid to operate independently. Microgrids can strengthen grid resilience and help mitigate grid disturbances because of their ability to continue operating while the main electric grid is down, thereby functioning as a grid resource for faster system response and recovery.

Micro-Seismicity Small movements of the Earth cause fracturing and movement of rocks. Such seismic activity does not release sufficient energy for the events to be recognized except with sensitive instrumentation.

Mini-Frac A minor fracturing was performed before the primary hydraulic fracturing treatment to acquire stress data and test pre-simulation permeability.

Monetization Presenting a benefit or cost in terms of monetary value, i.e., in terms of dollars.

MSW (Municipal Solid Waste) Residential solid waste and some non-hazardous commercial, institutional, and industrial wastes.

MTBE Methyl tertiary butyl ether is a fuel oxygenate produced by reacting methanol with isobutylene.

Multiplier Effect The multiplier effect is sometimes called the ripple effect because a single expenditure in an economy can have repercussions throughout the entire economy. The multiplier measures how much additional economic activity is generated from an initial expenditure.

N

Nacelle (or Cowling) It contains and protects the gearbox and generator, sometimes large enough for an engineer or technician to stand in while performing maintenance.

Near-Term Transmission Planning Horizon The transmission planning period that covers years one through five.

Net Metering An arrangement permits a facility (using a meter that reads inflows and outflows of electricity) to sell any excess power it generates over its load requirement back to the electrical grid to offset consumption.

Net Present Value The value in the base year (usually the present) of all cash flows associated with a project.

Net Summer Capacity The maximum output, commonly expressed in megawatts (MW), that generating equipment can supply to system load, as demonstrated by a multi-hour test, during summer peak demand. This output reflects a reduction in capacity due to electricity use for station services or auxiliaries.

Nocturnal Cooling The effect of cooling is the radiation of heat from a building to the night sky.

Nominal Dollars The values are stated in nominal or current dollars, accounting for inflation.

Non-Coincidental Peak Load The sum of two or more peak loads on individual systems that do not occur in the same time interval. Meaningful only when considering loads within a limited period of time, such as a day, week, month, or heating or cooling season, and usually for not more than one year.

Non-Energy Impacts Costs or benefits beyond those relating directly to energy, capacity, or ancillary services.

Non-Spinning Reserve The portion of operating reserve that is not connected to the system but is capable of serving the demand within a specified time (typically within 10 minutes), or interruptible load that can be removed from the system in a specified time.

Nonutility Power Producer A corporation, person, agency, authority, or other legal entity or instrumentality that owns or operates facilities for electric generation and is not an electric utility. Nonutility power producers consist of qualifying cogenerators, qualifying small power producers, and other nonutility generators, including independent power producers. Nonutility power producers are without a designated franchised service area and do not file forms listed in the Code of Federal Regulations, Title 18, Part 141.

Nonfirm Power Power or power-producing capacity supplied or available under a commitment having limited or no assured availability.

Nonforested Land Land that has never supported forests and lands formerly forested where the use of time management is precluded by development for other uses. (Note: Includes areas used for crops, improved pasture, residential areas, city parks, improved roads of any width and adjoining clearings, powerline clearings of any width, and 1- to 4.5-acre areas of water classified by the Bureau of the Census as land. If intermingled in forest areas, unimproved roads and nonforested strips must be more than 120 feet wide, and clearings, etc., must be more than 1 acre in area to qualify as nonforested land.)

Nonindustrial Private An ownership class of private lands where the owner does not operate wood-using processing plants.

Nonutility Generation Electric generation by nonutility power is produced to supply electric power for industrial, commercial, and military operations or sales to electric utilities.

Nonutility Power Producer A corporation, person, agency, authority, or other legal entity or instrumentality that owns electric generating capacity and is not an electrical utility. Nonutility power produces include qualifying cogenerators, qualifying small power producers, and other nonutility generators without a designated, franchised service area that do not file forms listed in the Code of Federal Regulations, Title 18, Part 141.

Normalization Accounting An accounting practice used by regulated utilities in which deferred income taxes are accumulated in a reserve account and effectively used to purchase new investments. The rate base is reduced by the accumulated reserve. Normalization accounting is the opposite of flow-through accounting.

North American Electric Reliability Corporation The successor to the North American Electric Reliability Council, a nonprofit corporation formed in 2006 to develop and maintain mandatory reliability standards for the bulk electric system with the fundamental goal of maintaining and improving the reliability of that system.

O

Occupied Space The space within a building or structure that people usually occupy may be conditioned (heated, cooled, and/or ventilated).

Open Access FERC Order No. 888 requires public utilities to provide non-discriminatory transmission service over their transmission facilities to third parties to move bulk power from one point to another on a non-discriminatory basis for a cost-based fee. Order 890 expanded open access to cover the methodology for calculating available transmission transfer capability, improvements that opened a coordinated transmission planning processes, standardization of energy and generation imbalance charges, and other reforms regarding the designation and un-designation of transmission network resources.

Operating Reserve The capability above firm system demands to provide for regulation, load forecasting error, equipment forced and scheduled outages, and local area protection. It consists of spinning reserve and non-spinning reserve.

Opportunity Cost The rate of return on the best alternative investment available.

Other Biomass This category of biomass energy includes agricultural byproducts/crops (agricultural byproducts, straw); other biomass gas (digester waste alcohol); other biomass liquids (fish oil liquid acetonitrile, waste, tall oil, waste alcohol); and other biomass solids (medical waste, solid byproducts, sludge waste, and tires).

Other Forest Land Forest land other than timberland and reserved forest land. It includes available forest land, which cannot produce 20 cubic feet per acre of industrial wood annually under natural conditions because of adverse site conditions such as sterile soils, dry climate, poor drainage, high elevation, steepness, or rockiness.

Other Sources Sources of roundwood products that are not growing stock. These include available dead, rough, and rotten trees, trees of non-commercial species, trees less than 5.0 inches dbh, tops, and roundwood harvested from nonforest land (for example, fence rows).

Outage The period during which a generating unit, transmission line, or other facility is out of service. A forced or unplanned outage refers to the emergency shutdown of a generating unit, transmission line, or other facility. A scheduled or planned outage is the shutdown for inspection or maintenance, in accordance with an advance schedule.

Outcomes The actual or modeled end state of grid operations, as quantified by metrics.

Overnight Construction Cost The value of total plant investment if construction had occurred overnight and all expenditures were made instantaneously.

P

Paper Pellets Paper is compressed and bound into uniform-diameter pellets to be burned in a heating stove.

Parabolic Dish A high-temperature (above 180°F) solar thermal concentrator, generally bow-shaped, with two-axis tracking.

Passive Solar A system in which solar energy alone is used for the transfer of thermal energy. Pumps, blowers, or thither heat-transfer devices that use energy other than solar are not used.

Payback Period The time required for net revenues is associated with an investment to repay the cost of the investment. Can be calculated as simply a payback period or discounted payback period.

Peak Demand The maximum load during a specified period of time.

Peak Watt A manufacturer's unit indicates the amount of power a photovoltaic cell or module will produce at standard test conditions (typically 1,000 Watts/square meter and 25°C).

Peaking Capacity Generating equipment normally operated only during the hours of highest daily, weekly, or seasonal loads.

Peaking Plants Electricity-generating plants that are operated to meet the peak or maximum load on the system. The energy cost from such plants is generally higher than that of baseload plants.

Peat Peat consists of partially decomposed plant debris. It is considered an early stage in the development of coal. Peat is distinguished from lignite by free cellulose and a high moisture content (+70%). The heat content of air-dried peat (about 50% moisture) is about 9 million Btu per ton. Most U.S. peat is used as a soil conditioner.

Performance Characteristics The physical and operational attributes of a technology or system. Simple characteristics would include emissions rates, ramp rates, and storage capabilities. More complex characteristics might include transient responses. In a valuation context, performance characteristics must be modeled with varying levels of granularity depending on the metrics to be quantified.

Permeability The ability of a rock to transmit fluid through its pores or fractures when subjected to a difference in pressure is typically measured in darcies or millidarcies.

Photovoltaic (PV) Cell An electronic device consisting of layers of semiconductor materials fabricated to form a junction (adjacent layers of materials with different electronic characteristics) and electrical contacts capable of converting incident light directly into electricity (direct current).

Photovoltaic (PV) Module An integrated assembly of interconnected photovoltaic cells is designed to deliver a selected level of working voltage and current at its output terminals. It is packaged to protect against environmental degradation and is suited for incorporation in photovoltaic power systems.

Plate Tectonics A theory of global-scale dynamics involving the movement of many rigid plates of the Earth's crust. Tectonic activity is evident along the margins of the plates where buckling, grinding, faulting, and vulcanism occur as the plates are propelled by the forces of deep-seated mantle convection current. Geothermal resources are often associated with tectonic activity since they allow groundwater to encounter deep subsurface heat sources.

Pole Timber Trees Live trees at least 5.0 inches in dbh But smaller than saw timber trees.

Porosity The ratio of the aggregate volume of pore spaces in rock or soil to its total volume is usually stated as a percentage.

Power Purchase Agreement Guarantees a market for power produced by an independent power producer and the price at which it is sold to purchasers. Such an agreement imposes legal obligations on both parties to perform precisely accepted tasks in a predetermined manner.

Present Value The value in the base year (usually the present) of a cash flow is adjusted for the time value differences in those cash flows between the time of the actual flow and the base year.

Present Value Dollars The future amount of money that has been discounted to reflect its present value, as if it existed today. For projects with multiple years of investments and benefits, the costs and benefits in each year of the future are typically presented in present value terms using a constant discount rate per year.

Primary Frequency Response Generating units provide the immediate proportional increase or decrease in actual power output, and the system load provides the natural real power dampening response in response to frequency deviations. This response is in the direction that stabilizes frequency.

Process Heating The direct process end use is when energy is used to raise the temperature of substances involved in the manufacturing process.

Proppant Sized particles are mixed with fracturing fluid to open fractures after hydraulic stimulation.

Public Utility Regulatory Policies Act of 1978 (PURPA) The National Energy Act, PURPA, contains measures designed to encourage energy conservation, more efficient use of resources, and equitable rates. Among these were suggested retail rate reforms, new incentives for cogenerators to produce electricity, and the use of renewable resources.

Pulpwood Roundwood, whole tree chips, or wood residues produce wood pulp.

Pumped Storage Hydroelectric Plant A plant that usually generates electric energy during peak load periods by using water previously pumped into an elevated storage reservoir during off-peak periods when excess generating capacity is available to do so. When additional generating capacity is needed, the water can be released from the reservoir through a conduit to turbine generators located in a power plant at a lower level.

Pyrolysis The thermal decomposition of biomass at high temperatures (greater than 400°F, or 200°C) in the absence of air. The product of pyrolysis is a mixture of solids (char), liquids (oxygenated oils), and gases (methane, carbon monoxide, and carbon dioxide) with proportions determined by operating temperature, pressure, oxygen content, and other conditions.

Q

Quadrillion Btu (Quad) Equivalent to 10 to the 15th power Btu.

Qualifying Facility (QF) A cogeneration or small power production facility that meets specific ownership, operating, and efficiency criteria established by the Federal Energy Regulatory Commission (FERC) under the Public Utility Regulatory Policies Act of 1978 (PURPA).

Quantification Presenting a benefit or cost numerically, regardless of the unit used to quantify it (e.g., tons, MWh, and job-years).

R

Rankine Cycle The thermodynamic cycle is an ideal standard for comparing the performance of heat engines, steam power plants, steam turbines, and heat pump systems that use condensate vapor as the working fluid; efficiency is measured as work done divided by sensible heat supplied.

Real Dollars Real or constant dollars adjusted to remove the effects of inflation.

Real Power The portion of electricity that supplies energy to the load.

Real-Time Market An electricity market that settles—determines the price—for 1-hour periods or less during the day of delivery.

Real-Time Pricing Electricity pricing is instantaneous and based on the cost of the electricity available for use when the customer demands it.

Regional Transmission Organization A voluntary organization of electric transmission owners, transmission users, and other entities approved by regulators to efficiently coordinate electric transmission planning (and expansion), operation, and use on a regional (and interregional) basis. The RTO must operate transmission facilities on a non-discriminatory basis.

Recovery Factor The fraction of total resources that can be extracted for productive uses.

Regulating Margin The amount of spinning reserve required in non-emergency conditions by each control area to bring the area control error to zero at least once every 10 minutes and to hold the average difference over each 10 minutes to less than that control area's allowable limit for average deviation, as defined by the NERC control performance criteria.

Regulating Reserve An amount of reserve responsive to automatic generation control sufficient to provide a normal regulating margin.

Regulation Service The process whereby one balancing authority contacts to provide a corrective response to all or a portion of the area control error of another balancing authority.

Reliability The ability of an electric power system to meet the electricity needs of end-use customers, even when unexpected equipment failure or other conditions reduce the amount of available power supply.

Reliability Must Run A unit that must run for operational or reliability reasons, regardless of economic considerations. Also called a reliability agreement.

Renewable Energy Energy is produced using resources that regenerate quickly or are inexhaustible. Wind energy is considered inexhaustible because while it may blow intermittently, it will never stop.

Renewable Energy Resources Energy resources that are naturally replenishing but flow-limited. They are virtually inexhaustible in duration but limited in the energy available per unit of time. Renewable energy resources include biomass, hydro, geothermal, solar, wind, ocean thermal, wave action, and tidal action.

Renewable Portfolio Standard (RPS) A mandate requiring renewable energy provides a certain percentage of total energy generation or consumption.

Reserve Margin (Operating) The unused available capability of an electric power system (at peak load for a utility system) as a percentage of total capability.

Reserve Margin (Planning) Planning reserve margins measure the generation capacity available to meet expected reserve margins. They have been an industry standard used by planners for decades as a relative indication of adequacy.

Reservoir A natural underground container of liquids, such as water or steam (or, in the petroleum contest, oil, or gas).

Residues Bark and woody materials are produced in primary wood-using mills, while roundwood products are converted into other goods. Examples are slabs, edgings, trimmings, sawdust, shavings, veneer cores and clippings, and pulp screenings. Includes bark and wood residues (coarse and fine materials) but excludes logging residues.

Resilience The ability of an energy power source to resist, absorb, or withstand the impact of changes in conditions that can affect its operation, adapt in response to the change, and recover and restore system functionality rapidly.

Resistivity Survey Resistivity measures a material's ability to resist or inhibit the flow of an electrical current, measured in ohm-meters. The voltage between two electrodes is measured while an electrical current is generated between two other electrodes. Resistivity surveys can be used to delineate the boundaries of geothermal fields.

Resource Base All the given material in the Earth's crust, whether its existence is known or unknown, and regardless of cost considerations.

Resource Planner The entity that develops a long-term (generally one year and beyond) plan for the resource adequacy of specific loads (customer demand and energy requirements) within a planning authority area.

Ribbon Silicon Single-crystal silicon is derived by fabricating processes that produce sheets or ribbons. These processes include edge-defined film-fed growth, dendritic web growth, and ribbon-to-ribbon growth.

Roller Cone Bit A drill bit that drills by crushing the rock with studded rotating cones attached to the bit.

Rotten Tree A live tree of commercial species that does not contain a saw log now or prospectively primarily because of rot (that is, when rot accounts for more than 50% of the total cull volume).

Rough Tree (a) A live tree of commercial species that does not contain a saw log now or prospectively primarily because of roughness (that is, when sound cull, due to such factors as poor form, splits, or cracks, accounts for more than 50% of the total cull volume) or (b) a live tree of non-commercial species.

Roundwood Wood is cut precisely for use as fuel.

Roundwood Products Logs and other round timber are generated from harvesting trees for industrial or consumer use.

S

Salinity A measure of the quantity or concentration of dissolved salts in water.

Saplings Live trees 1.0 inches through 4.9 inches in dbh.

Secondary Wood Processing Mills A mill that uses primary wood products to manufacture finished wood products, such as cabinets, moldings, and furniture.

Seismic About the nature of, or caused by, an earthquake or Earth vibration, natural or human-made.

Seismicity The phenomenon of Earth movements. Also, the frequency, distribution, and intensity of earthquakes.

Seismometer An electrical device measures the magnitude and direction of seismic events on the surface and within wellbores.

Self-Potential Self-potential in geothermal systems measures currents induced in the subsurface because of the flow of fluids.

Silicon A semiconductor material made from silica, purified for photovoltaic applications.

Skylight A window is located on a structure's roof and provides natural daylight, warmth, and ventilation to interior building spaces.

Slim Hole Drill holes with a nominal inside diameter of less than 6 inches.

Slotted Liner Liner that has slots or holes in it to let fluid pass between the wellbore and surrounding rock.

Sludge A dense, slushy, liquid-to-semi-fluid product that accumulates because of an industrial or technological process designed to purify a substance. Industrial sludges are produced from the processing of energy-related raw materials, chemical products, water, mined ores, sewerage, and other natural and human-made products. Sludges can also form from natural processes, such as the runoff produced by rainfall, which accumulates on the bottom of bogs, streams, lakes, and tidelands.

Solar Energy The sun's radiant energy can be converted into other forms of energy, such as heat or electricity.

Solar Heat Gain Coefficient (SHGC) Measures how well a product blocks heat caused by sunlight. The SHGC is expressed as a number between 0 and 1. The lower the SHGC, the less solar heat it transmits.

Solar Thermal Collector A device designed to receive solar radiation and convert it into thermal energy. Typically, a solar thermal collector includes a frame, glazing, an absorber, and the appropriate insulation. The heat collected by the solar thermal collector may be used immediately or stored for later use.

Solar Thermal Collector, Special An evacuated tube collector or a concentrating (focusing) collector. Special collectors operate at temperatures ranging from low concentration (for pool heating) to several hundred degrees Fahrenheit (high concentration for air conditioning and specialized industrial processes).

Spark Spread A measurement of the difference between the price that a generator can obtain from selling one MWh of electricity and the cost of the natural gas needed to generate the MWh of electricity. Spark spread measures the potential profit for generating electricity on a particular day.

Spinning Reserve A reserve of generation capacity, where generators are kept only, but at idle, in anticipation of an unexpected increase in demand or decrease in supply.

Spot Market The natural gas market involves short-term contracts, typically lasting a month or less, starting shortly, often the next day or within a few days. Spot markets are usually organized for day-ahead and real-time electricity, run by an independent system operator or regional transmission organization.

Stability The ability of an electric system to maintain a state of equilibrium during normal and abnormal conditions or disturbances.

Start-Up Speed The speed at which a rotor begins to rotate.

Stress The forces act on the rock. In the subsurface, the most significant force or stress is generally vertical, caused by the weight of the overlying rock.

Structural Discontinuity A discontinuity of the rock fabric can be a fracture, fault, intrusion, or differing adjacent rock type.

Submersible Sump The pump, including the pumping mechanism and a driving electric motor, is suspended together at depth in the well.

Subsidence A sinking of an area of the Earth's crust due to fluid withdrawal and pressure decline.

Sunk Cost Any cost incurred by a prior decision that cannot be affected by the current course of action.

Supervisory Control and Data Acquisition (SCADA) A remote control and telemetry system monitors and controls the transmission system.

Sustainability The ability of an electric system to provide electric services to customers with minimal impacts on natural resources, human health, or safety.

Syngas A synthesis gas is produced through the gasification of biomass. Syngas is similar to natural gas and can be cleaned and conditioned to form a feedstock for methanol production.

System One organization controls integrated electrical facilities, including generation, transmission, and distribution.

System Characteristics The current configuration of the power system. These characteristics have physical (i.e., current transmission topology and generators), regulatory (i.e., market structure), and policy (i.e., incentives) dimensions.

System Load Total aggregated demand of all electricity consumers in an electric system at a given time (e.g., instantaneous load, and within a particular hour).

System Power Value A forecast of the value in an electric system for the next incremental power generation unit is typically expressed in dollars per MWh for energy and dollars per kWh for capacity.

System Requirements System requirements or constraints are a subset of outcomes that are real-world operational requirements (or their modeling approximations) that bound the valuation process.

T

TDS Total dissolved solids. Used to describe the amount of solid materials in water.

Tall Oil The oily mixture of rosin acids, fatty acids, and other materials is obtained by acid treatment of the alkaline liquors from the digesting (pulping) of pine wood.

Taxable Income The portion of revenue remaining after all deductions permitted under the Internal Revenue Service or a state revenue code has been taken.

Taxonomy of Grid Services and Technologies A classification scheme for grid-related technologies and services that provides a common language for the discussion of valuation.

Thermal Drawdown Decline in formation temperature due to geothermal production.

Thermal Gradient The temperature increase rate is a function of depth into the Earth's crust.

Thermosiphon System A solar collector system for water heating circulates the collection fluid through a storage loop solely by the temperature and density difference between the hot and cold fluids.

Thin-Film Silicon A technology in which amorphous or polycrystalline material is used to make photovoltaic (PV) cells.

Tiltmeter The device can measure minimal changes in its horizontal rotation. The "tilt" measured by an array of tiltmeters placed over a stimulation allows delineation of inflation and fracturing caused by the stimulation.

Tipping Fee A fee for the disposal of waste.

Total Plant Investment Total plant costs as modified by escalation and interest during construction.

Tower Steel structures support the turbine assembly. Higher towers allow for longer glades and the capture of faster-moving air at higher altitudes.

Tracer A chemical is injected into the flow stream of a production or injection well to determine fluid path and velocity.

Transformer Used to increase or decrease AC voltage or current.

Transmission An interconnected group of lines and associated equipment for the movement or transfer of electric energy between points of supply and points at which it is transformed for delivery to customers or is delivered to other electric systems.

Transmission Constraint A limitation on one or more transmission elements that may be reached during regular or contingency system operation.

Transmission Deferral Deferral of transmission system investments or upgrades.

Transmission Line Structures and conductors that carry bulk supplies of electrical energy from power-generating units.

Transmission Loading Relief A NERC procedure that allows reliability coordinators to curtail transactions (among other activities to mitigate potential or actual operational security) limit violations while respecting transmission service reservation priorities.

Transmission System (Electric) An interconnected group of electric transmission lines and associated equipment moves or transfers electric energy in bulk between points of supply and points at which it is transformed for delivery over the distribution system lines to consumers or other electrical systems.

Transportation Sector An energy-consuming sector consists of all vehicles whose primary purpose is transporting people and/or goods from one physical location to another. Included are automobiles, trucks, buses, motorcycles, trains, subways, other rail vehicles, aircraft, ships, barges, and other waterborne vehicles. Vehicles whose primary purpose is not transportation (e.g., construction cranes and bulldozers, farming vehicles, warehouse tractors, and forklifts) are classified in the sector of their primary use.

Turbine A machine for generating rotary mechanical power from the energy of a stream of fluid (such as water, steam, or hot gas). Turbines convert the kinetic energy of fluids to mechanical energy through the principles of impulse and reaction or a mixture of the two.

Two-Settlement System A system in which the price for electricity on any given day is established and settled both day-ahead and in real time. Day-ahead prices are based on forecast energy demand and transmission and generation availability. Real-time prices reflect not only day-ahead anticipated events but also what occurs in real time—for example, generation or transmission failures and differences between forecasted and actual loads.

U

U-Factor Measures the rate of heat loss or how well a product prevents heat from escaping. It includes the thermal properties of the frame as well as the glazing. The R-value indicates the insulating value, the inverse of the U-factor. U-factor ratings generally fall between 0.20 and 1.20. The lower the U-factor, the greater a product's resistance to heat flow and the better its insulation value.

Uniform Capital Recovery Factor As a fraction of the original investment cost, the uniform periodic payment will fully repay a loan, including all interest, over the loan term.

Upwind Turbine A turbine facing the wind requires a wind vane and yaw drive to maintain proper orientation relative to the wind.

Useful Heat Heat is stored above room temperature (in a solar heating system).

Useful Thermal Output Thermal energy is available in any industrial or commercial process or heating or cooling application; that is, total thermal energy made available for processes and applications other than electrical generation.

V

Vapor Dominated A geothermal reservoir system in which subsurface pressures are controlled by vapor rather than by liquid is sometimes referred to as a dry-steam reservoir.

Viewshed The scenic characteristics of an area are referred to as a resource.

Visible Light Transmittance The amount of visible light that passes through the glazing material of a window is expressed as a percentage.

Voltage The measure of electrical potential difference.

Voltage Collapse A power system at a given operating state and subject to a given disturbance undergoes voltage collapse if post-disturbance equilibrium voltages are below acceptable limits. Voltage collapse may be total (blackout) or practical and is associated with voltage and/or angular instability.

Voltage Instability A system state in which an increase in load, disturbance, or system change causes voltage to decay quickly or drift downward; the automatic and manual system controls cannot halt the decay. Voltage decay may take anywhere from a few seconds to tens of minutes. Unabated voltage decay can result in angular instability or voltage collapse.

W

Watt (Electric) The electrical unit of power. The rate of energy transfer is equivalent to 1 ampere of electric current flowing under a pressure of 1 volt at unity power factor.

Watt (Thermal) A unit of power in the metric system, expressed in terms of energy per second, equals the work done at a rate of 1 joule/second.

Watt-hour (Wh) The electrical energy unit of measure equals 1 Watt of power supplied to, or taken from, an electric circuit steadily for 1 hour.

Well Log Logging includes measurements of the well's diameter and various electrical, mass, and nuclear properties of the rock, which can be correlated with the rock's physical properties. The OK log is a chart of the measurements relative to the well's depth.

Wind Energy Wind motion produces energy that can be converted to mechanical energy for driving pumps, mills, and electric power generators. Wind pushes against sails, vanes, or blades radiating from a central rotating shaft.

Wind Power Plant A group of wind turbines interconnected to a standard utility system through transformers, distribution lines, and (usually) one substation. Operation, control, and maintenance functions are often centralized through a network of computer monitoring systems, supplemented by visual inspection. This is a term commonly used in the United States. In Europe, it is called a generating station.

Wind Rose A diagram that indicates the average monthly or annual percentage of time that the wind blows from different directions.

Wind Turbine or Windmill A device for harnessing the wind's kinetic energy and using it to do work or generate electricity.

Wind Vane A wind direction measurement device sends data to the yaw drive.

Windward Into or facing the direction of the wind. Opposite Leeward.

Wood/Wood Waste The category of biomass energy includes black liquor, wood/wood waste liquids (red liquor, sludge wood, spent sulfite liquor), and wood/wood waste solids (peat, paper pellets, railroad ties, utility poles, wood/wood waste).

Wood Energy Wood and wood products used as fuel, including roundwood (cord wood), limb wood, wood chips, bark, sawdust, forest residues, charcoal, pulp waste, and spent pulping liquor.

Wood Pellets Sawdust is compressed into uniform-diameter pellets to be burned in a heating stove.

Y

Yaw The rotation of a horizontal-axis wind turbine around its tower occurs along the vertical axis.
Yaw Drive The motor that keeps an upwind turbine facing into the wind.

Z

Zonal Isolation Various methods selectively partition portions of the wellbore for stimulation, testing, flow restriction, or other purposes.
Zone A zone is an area within the interior space of a building, such as an individual room(s), that can be cooled, heated, or ventilated. It has a thermostat to regulate airflow.

NOTE

1 Definitions retrieved 01/05/2024 @ http://www1.eere.enegy.gov/site_adminstaration/glossary.html from DOE's EERE Glossary of Energy-Related Terms.

Index

Note: Locators in *italics* represent figures and **bold** indicate tables in the text.

F

G

W

Y

Z

For Product Safety Concerns and Information please contact our EU representative GPSR@taylorandfrancis.com Taylor & Francis Verlag GmbH, Kaufingerstraße 24, 80331 München, Germany

For Product Safety Concerns and Information please contact our
EU representative GPSR@taylorandfrancis.com Taylor & Francis
Verlag GmbH, Kaufingerstraße 24, 80331 München, Germany